Global Warming

The Complete Briefing
Third Edition

Global warming and the resulting climate change are among the most serious environmental problems facing the world community. *Global Warming: The Complete Briefing* is the most comprehensive guide available to the subject. A world-renowned expert, Sir John Houghton explores the scientific basis of global warming and the likely impacts of climate change on human society, before addressing the action that could be taken by governments, by industry and by individuals to mitigate the effects. The first two editions received excellent reviews, and this completely updated new edition will prove to be the best briefing the student or interested general reader could wish for.

SIR JOHN T. HOUGHTON CBE, FRS is a former Chairman of the Scientific Assessment Working Group of the Intergovernmental Panel on Climate Change, Chairman of the UK's Royal Commission on Environmental Pollution, Vice President of the World Meteorological Organization, President of the Royal Meteorological Society, and Professor of Atmospheric Physics at Oxford University. He was Chief Executive of the UK Meteorological Office from 1983 to his retirement in 1991. As well as the previous editions of this book, he is author of *The Physics of Atmospheres* (Cambridge University Press, in three editions), and has published numerous research papers and contributed to many influential research documents. Sir John and his wife Sheila live in Wales.

From reviews of previous editions

'It is difficult to imagine how Houghton's exposition of this complex body of information might be substantially improved upon . . . Seldom has such a complex topic been presented with such remarkable simplicity, directness and crystalline clarity . . . Houghton's complete briefing is without doubt the best briefing the concerned citizen could hope to find within the pages of a pocketable book.'

John Perry, *Bulletin of the American Meteorological Society*

'I can recommend (this book) to anyone who wants to get a better perspective on the topic of global warming . . . a very readable and comprehensive guide to the changes that are occurring now, and could occur in the future, as a result of human action . . . brings the global warming debate right up to date . . . Read Houghton's book if you really want to understand both the scientific and political issues involved.'

William Harston, *The Independent*

'. . . precise account of the science, accompanied by figures, graphs, boxes on specific points, and summaries at the end of each chapter, with questions for students . . . ranges beyond the science into the diplomacy, politics, economics and ethics of the problem, which together present a formidable challenge to human understanding and capacity for action.'

Sir Crispin Tickell, *The Times Higher Education Supplement*

'. . . a widely praised book on global warming and its consequences.'

The Economist

'. . . an interesting account of the topic for the general reader.'

Environmental Assessment

'. . . very thorough and presents a balanced, impartial picture.'

Jonathan Shanklin, *Journal of the British Astronomical Association*

'I would thoroughly recommend this book to anyone concerned about global warming. It provides an excellent essentially non-technical guide on scientific and political aspects of the subject. It is an essential briefing for students and science teachers.'

Tony Waters, *The Observatory*

'For the non-technical reader, the best program guide to the political and scientific debate is John Houghton's book *Global Warming: The Complete Briefing*. With this book in hand you are ready to make sense of the debate and reach your own conclusions.'

Alan Hecht, *Climate Change*

'This is a remarkable book . . . It is a model of clear exposition and comprehensible writing . . . Quite apart from its value as a background reader for science teachers and students, it would make a splendid basis for a college general course.'

Andrew Bishop, *Association for Science Education*

'*Global Warming* remains the best single-volume guide to the science of climate change.'

Greg Terrill, *Times Literary Supplement*

'This very readable and informative book is valuable for anyone wanting a broad overview of what we know about climate change, its potential impacts on society and the natural world, and what could be done to mitigate or adapt to global warming. To this end, discussion questions are included at the end of each chapter. The paperback edition is especially good value . . . Houghton's compact book is an accessible, well-researched, and broadly based introduction to the immensely complicated global warming problem.'

Dennis L. Hartmann, Department of Atmospheric Sciences, University of Washington, Seattle, USA

'I have no hesitation in endorsing this important book.'

Wilfrid Bach, *International Journal of Climatology*

'...a useful book for students and laymen to understand some of the complexities of the global warming issue. Questions and essay topics at the end of each chapter provide useful follow-up work and the range of material provided under one cover is impressive. At a student-friendly price, this is a book to buy for yourself and not rely on the library copy.'

Allen Perry, *Holocene*

'In summary I would thoroughly recommend this book to anyone concerned about global warming. It provides an excellent non-technical guide on scientific and political aspects of the subject. It is an essential briefing for students and science teachers.'

Tony Waters, *Weather*

'This book is one of the best I have encountered, that deal with climate change and some of its anthropogenic causes. Well written, well organised, richly illustrated and referenced, it should be required reading for anybody concerned with the fate of our planet.'

Elmar R. Reiter, *Meteorology and Atmospheric Physics*

'Sir John Houghton is one of the few people who can legitimately use the phrase "the complete briefing" as a subtitle for a book on global warming...Sir John has done us all a great favour in presenting such a wealth of material so clearly and accessibly and in drawing attention to the ethical underpinnings of our interpretation of this area of environmental science.'

Progress in Physical Geography

'...this complete briefing on global warming is remarkably factual and inclusive. Houghton's concern about planet Earth and its people blends well with this his hopes for global cooperation in concert with the spirit of the Intergovernmental Panel on Climate Change.'

Choice

'Throughout the book this argument is well developed and explained in a way that the average reader could understand especially because there are many diagrams, tables, graphs and maps which are easy to interpret.'

SATYA

'...this book is the most comprehensive guide available. Ignore it at your peril.'

The Canadian Field-Naturalist

Global Warming

The Complete Briefing

THIRD EDITION

Sir John Houghton

CAMBRIDGE
UNIVERSITY PRESS

PUBLISHED BY THE PRESS SYNDICATE OF THE UNIVERSITY OF CAMBRIDGE
The Pitt Building, Trumpington Street, Cambridge, United Kingdom

CAMBRIDGE UNIVERSITY PRESS
The Edinburgh Building, Cambridge CB2 2RU, UK
40 West 20th Street, New York, NY 10011–4211, USA
477 Williamstown Road, Port Melbourne, VIC 3207, Australia
Ruiz de Alarcón 13, 28014 Madrid, Spain
Dock House, The Waterfront, Cape Town 8001, South Africa

http://www.cambridge.org

First published 1994 by Lion Publishing plc
Second edition published 1997 by Cambridge University Press
Third edition published 2004
Fourth printing 2006

Printed in the United Kingdom at the University Press, Cambridge

Typefaces Times New Roman 10/13 pt. and Stone Sans   *System* LaTeX 2_ε [TB]

A catalogue record for this book is available from the British Library

Library of Congress Cataloguing in Publication data

Houghton, John Theodore.
Global warming: the complete briefing / John T. Houghton – 3rd ed.
  p.   cm.
Includes bibliographical references and index.
ISBN 0 521 81762 5 – ISBN 0 521 52874 7 (paperback)
1. Global warming.   2. Climatic changes.   I. Title.
QC981.8.G56H68 2004
363.738'74–dc22   2003068735

ISBN 0 521 81762 5 hardback
ISBN 0 521 52874 7 paperback

To my grandchildren,
Daniel, Hannah, Esther, Max, Jonathan, Jemima and Sam and
their generation

Contents

Figures

SI unit prefixes

Quantity	Prefix	Symbol
10^{12}	tera	T
10^{9}	giga	G
10^{6}	mega	M
10^{3}	kilo	k
10^{2}	hecto	h
10^{-2}	centi	c
10^{-3}	milli	m
10^{-6}	micro	μ
10^{-9}	nano	n

Chemical symbols

CFCs	chlorofluorocarbons
CH_4	methane
CO	carbon monoxide
CO_2	carbon dioxide
H_2	molecular hydrogen
HCFCs	hydrochlorofluorocarbons
H_2O	water
N_2	molecular nitrogen
N_2O	nitrous oxide
NO	nitric oxide
NO_2	nitrogen dioxide
O_2	molecular oxygen
O_3	ozone
OH	hydroxyl radical
SO_2	sulphur dioxide

Preface to the First Edition

Climate change and global warming are well up on the current political agenda. There are urgent questions everyone is asking: are human activities altering the climate? Is global warming a reality? How big are the changes likely to be? Will there be more serious disasters; will they be more frequent? Can we adapt to climate change or can we change the way we do things so that we can slow down the change or even prevent it occurring?

Because the Earth's climate system is highly complex, and because human behaviour and reaction to change is even more complex, providing answers to these questions is an enormous challenge to the world's scientists. As with many scientific problems only partial answers are available, but our knowledge is evolving rapidly, and the world's scientists have been addressing the problems with much energy and determination.

Three major pollution issues are often put together in people's minds: global warming, ozone depletion (the ozone hole) and acid rain. Although there are links between the science of these three issues (the chemicals which deplete ozone and the particles which are involved in the formation of acid rain also contribute to global warming), they are essentially three distinct problems. Their most important common feature is their large scale. In the case of acid rain the emissions of sulphur dioxide from one nation's territory can seriously affect the forests and the lakes of countries which may be downwind of the pollution. Global warming and ozone depletion are examples of global pollution – pollution in which the activities of one person or one nation can affect all people and all nations. It is only during the last thirty years or so that human activities have been of such a kind or on a sufficiently large scale that their effects can be significant globally. And because the problems are global, all nations have to be involved in their solution.

The key intergovernmental body which has been set up to assess the problem of global warming is the Intergovernmental Panel on Climate Change (IPCC), formed in 1988. At its first meeting in November of that year in Geneva, the Panel's first action was to ask for a scientific report so that, so far as they were known, the scientific facts about global warming could be established. It was imperative that politicians were given a solid scientific base from which to develop the requirements for action.

That first scientific report was published at the end of May 1990. On Monday 17 May I presented a preview of it to the then British Prime Minister, Mrs Margaret Thatcher, and members of her Cabinet at 10, Downing Street in London. I had been led to expect many interruptions and questions during my presentation. But the thirty or so Cabinet members and officials in the historic Cabinet room heard me in silence. They were clearly very interested in the report, and the questions and discussion afterwards demonstrated a large degree of concern for the world's environmental problems.

Since then the interest of many political leaders has been aroused – as has been shown by their attendance at two important world conferences concerned with global warming: the Second World Climate Conference in Geneva in 1990 and the United Nations Conference on Environment and Development (UNCED) in Rio de Janeiro in 1992. The Rio conference with over 25 000 people attending the main sessions and the many side meetings, was the largest conference ever held. Never before had a single conference seen so many of the world's leaders, and for that reason it is often referred to as the Earth Summit.

Much of the continuing assessment of climate change has been focused on the IPCC and its three working groups dealing respectively with science, impacts and response strategies. The IPCC's first report published in 1990 was a key input to the international negotiations which prepared the agenda for the UNCED Conference in Rio de Janeiro; it was that IPCC assessment which provided much of the impetus for the Framework Convention on climate change signed at Rio by over 160 countries. As chairman or co-chairman of the Science Working Group I have been privileged to work closely with hundreds of scientific colleagues in many countries who readily gave of their time and expertise to contribute to the IPCC work.

For this book I have drawn heavily on the 1990 and 1992 reports of all three working groups of IPCC. Further, in putting forward options for action I have followed the logic of the Climate Convention. What I have said I believe to be consistent with the IPCC reports and with the implications of the Climate Convention. However, I must also emphasise that the choice of material and any particular views I put forward are entirely my own and should in no way be construed as the views of the IPCC.

During the preparation of both IPCC reports so far there has been considerable scientific debate about just how much can be said about likely climate change next century. Some researchers initially felt that the uncertainties were such that scientists should refrain from making any estimates or predictions for the future. However, it soon became clear that scientists have a responsibility to communicate the best possible information about the likely magnitude of climate change, along with clear statements of the assumptions made and the level of uncertainty in the estimates. Like weather forecasters, their results will not be entirely accurate, but can provide useful guidance.

Many books have been published on global warming. This book differs from the others because I have attempted to describe the science of global warming, its impacts and what action might be taken in a way which the intelligent non-scientist can understand. Although there are many numbers in the book – I believe the quantification of the problem to be very important – there are no mathematical equations. I have also used the minimum of jargon in the main text. Some technical explanations which would be of interest to the scientifically trained are included in some of the boxes. Others contain further material of specific interest.

I am grateful to many who have helped me with the provision and preparation of particular material for this book and to those who have read and helpfully commented on my drafts. There have been those who have been involved with the IPCC: Bert Bolin, the IPCC Chairman, Gylvan Meira Filho, my co-chairman on the IPCC Science Working Group, Robert Watson, co-chairman of the IPCC Working Group on Impacts and Response Strategies, Bruce Callander, Chris Folland, Neil Harris, Katherine Maskell, John Mitchell, Martin Parry, Peter Rowntree, Catherine Senior and Tom Wigley. Others I wish to thank are Myles Allen, David Carson, Jonathan Gregory, Donald Hay, David Fisk, Kathryn Francis, Michael Jefferson, Geoffrey Lean and John Twidell. The staff at Lion Publishing, Rebecca Winter, Nicholas Rous and Sarah Hall, have been most helpful in preparing the book for publication, especially in ensuring that it is as attractive and readable as possible. Finally, I owe an especial debt to my wife, Sheila, who gave me strong encouragement to write the book in the first place, and who has continued her encouragement and support through the long hours of its production.

Preface to the Second Edition

Since the publication of the first edition nearly three years ago, interest in the issue of Global Warming and concern about it has continued to grow. The Framework Convention on Climate Change (FCCC) agreed at the Earth Summit in 1992 has been ratified and machinery for its implementation is gradually being developed. At the end of 1995, the IPCC produced a further comprehensive report updating the 1990 report. Although the main conclusions have not changed, much has been added to the detail of our knowledge regarding all aspects of the issue, the science, the impacts and the possible response. This revised edition takes into account this further information from the 1995 IPCC reports.

In the first edition I included a chapter, Chapter 8, with the heading 'Why should we be concerned?' which addresses the question of the responsibility of humans for the Earth and for looking after the environment. In it I presented something of the basis for my personal motivation as a Christian for being concerned with environmental problems. Although I believe that it is important that science is presented in the broad context of human values, I realised that the inclusion of such a chapter was something of a departure and wondered how it would be received.

Some have expressed surprise that in the middle of a science book, there should be, unusually, a chapter of this kind which deals with ethical and religious issues. However, it has been pleasing that scientific colleagues and reviewers of the book have referred favourably to the chapter stressing the value and importance of placing environmental science in the context of the reasons for its pursuit. For instance, John Perry, in the *Bulletin of the American Meteorological Society*, writes:

> Many scientists, including avowed agnostics such as myself, will find this forth-right declaration of religious belief and divine purpose a bit startling in an otherwise rigorously scientific volume. However, in a line of argument that I have no difficulty whatever in supporting, Houghton demonstrates that the domains of science and religion are simply complementary ways of looking at truth. The former deals with how the world works and the latter with why. In Houghton's framework, we and the earth are each other's reasons for existence in a divine plan that we must

struggle to understand but must inescapably follow. Thus, Houghton holds that we have no choice but to care for the earth solicitously as its 'gardeners' in a 'partnership with God'. His lucid precis of the complex factual substance of global warming is an authoritative guide to the issue's scientific dimensions; his inspiring synthesis of science, faith and stewardship is an even more illuminating handbook to its moral and ethical dimensions. Together, they constitute a uniquely valuable Baedeker to one of the most important issues of our science and our time.

In revising Chapter 8 for this edition, I have been somewhat more objective and less personal – which I felt was more appropriate for student readers from a wide range of disciplines, for whom the edition is particularly suited. As a didactic aid I have also included a number of problems and questions for discussion at the end of all the chapters.

Some of my colleagues sometimes comment on how formidable is the task of sewardship of the Earth feeling that it is perhaps beyond the capability of the human race to tackle it adequately. I feel optimistic about it, however, for three main reasons. Firstly, I have seen how the world's scientists, coming from very different countries, cultures and backgrounds, have worked closely and responsibly in the IPCC to provide a consensus presentation of the science of global warming. Secondly, the technologies required to provide for greater efficiency in the use of fossil fuels and for their replacement with renewable sources of energy are available and, when developed on thc necessary scale, also affordable. Thirdly, my belief in God's commitment to the material world coupled with his offer of partnership in caring for it, makes stewardship of the Earth an especially exciting and challenging activity.

In the preparation of this revised volume I wish to express again my gratitude to the scientific colleagues with whom I have worked in the ongoing activity of the IPCC and from whom I have learnt much. My thanks are also due to John Twidell and Michael Banner who have commented on particular chapters, and to Catherine Flack, Matt Lloyd and other staff of the Cambridge University Press for their competence, courtesy and assistance in the preparation of the book.

John Houghton
1997

Preface to the Third Edition

Since the Second Edition seven years ago, research and debate on the issue of human-induced climate change have grown at a rapidly increasing pace. Observations of climate during this period have provided further information about the warming Earth and there has been substantial improvement in the models that simulate both past and future climate. Although the main messages regarding the fact of human-induced climate change and its impact have not changed significantly (on the whole they have been strengthened) more detailed understanding has been achieved regarding the basic science (including the uncertainties), the likely impacts and the imperative for action. Hence the need to update this book.

In 2001 the Intergovernmental Panel on Climate Change (IPCC) published its Third Assessment Report–even more thorough and comprehensive than the first two. As co-chair of the scientific assessment working group for all three of the IPCC reports, I have been privileged to be a part of the IPCC process, which has been so effective in informing the scientific community. Then, through that community, information has been spread to decision makers and others regarding what is known about climate change with some degree of certainty and also about the areas where there remains much uncertainty. I have leant heavily on the IPCC 2001 Report in revising this text and wish to express my deep gratitude to those many IPCC colleagues with whom I have worked and from whom I have learnt so much. I have also benefited greatly from my association with the UK Hadley Centre for Climate Prediction and Research, which has become the world's premier centre for climate modelling research.

My especial thanks are due to those who have provided me with particular new material; Peter Cox, Chris Jones, Colin Prentice and Jo House for Chapter 3; Chris Folland and Alan Dickinson for Chapters 4 and 5; Tim Palmer and Jonathan Gregory for Chapter 6; Martin Parry and Rajendra Pachauri for Chapter 7; Stephen Briggs for material regarding Envisat for Chapter 9; Aubrey Meyer for Chapter 10; Mark Akhurst, Andre Romeyn, Robert Kleiburg, Gert Jan Kramer, Chris West, Peter Smith and Chris Llewellyn Smith for Chapter 11; and William Clark for Chapter 12. John Mitchell, Terry Barker and Susan Baylis kindly read and commented on some of the draft chapters. I am also particularly grateful to David Griggs, Geoffrey Jenkins, Philippe Rekacewicz and Paul van der Linden who assisted with the sourcing and preparation of the figures. Finally, I wish to thank Matt Lloyd, Carol Miller, Sarah Price and other staff of Cambridge University Press who have carefully steered the book through its gestation and production.

In January of this year I attended the World Economic Forum in Davos and engaged in discussion and debate regarding global warming and climate change. Nearly everyone there accepted the fact of climate change due to human activities and the need for action to reduce greenhouse gas emissions in order to reduce its impact. However, many participants knew little of the likely impacts of climate change or of the extent of the action required to address it; they just believed that it was one of those problems that would have to be addressed sometime. I hope that this book will assist in making the necessary information more readily available and so help to provide the foundation for the urgent action that is required.

John Houghton

Chapter 1
Global warming and climate change

The phrase 'global warming' has become familiar to many people as one of the important environmental issues of our day. Many opinions have been expressed concerning it, from the doom-laden to the dismissive. This book aims to state the current scientific position on global warming clearly, so that we can make informed decisions on the facts.

Is the climate changing?

In the year 2060 my grandchildren will be approaching seventy; what will their world be like? Indeed, what will it be like during the seventy years or so of their normal life span? Many new things have happened in the last seventy years that could not have been predicted in the 1930s. The pace of change is such that even more novelty can be expected in the next seventy. It is fairly certain that the world will be even more crowded and more connected. Will the increasing scale of human activities affect the environment? In particular, will the world be warmer? How is its climate likely to change?

Before studying future climate changes, what can be said about climate changes in the past? In the more distant past there have been very large changes. The last million years has seen a succession of major ice ages interspersed with warmer periods. The last of these ice ages began to come to an end about 20 000 years ago and we are now in what is called an interglacial period. Chapter 4 will focus on these times far back in the past. But have there been changes in the very much shorter period of living memory – over the past few decades?

Variations in day-to-day weather are occurring all the time; they are very much part of our lives. The climate of a region is its average weather over a period that may be a few months, a season or a few years. Variations in climate are also very familiar to us. We describe summers as wet or dry, winters as mild, cold or stormy. In the British Isles, as in many parts of the world, no season is the same as the last or indeed the same as any previous season, nor will it be repeated in detail next time round. Most of these variations we take for granted; they add a lot of interest to our lives. Those we particularly notice are the extreme situations and the climate disasters (for instance, Figure 1.1 shows the significant climate events and disasters during the year 1998). Most of the worst disasters in the world are, in fact, weather- or climate-related. Our news media are constantly bringing them to our notice as they occur in different parts of the world – tropical cyclones (called hurricanes or typhoons), wind-storms, floods, tornadoes and droughts whose effects occur more slowly, but which are probably the most damaging disasters of all.

The remarkable last decades of the twentieth century

The 1980s and 1990s were unusually warm. Globally speaking, the decades have been the warmest since accurate records began somewhat over a hundred years ago and these unusually warm years are continuing into the twenty-first century. In terms of global average near-surface air temperature, the year 1998 was the warmest in the instrumental record and the nine warmest years in that record have occurred since 1990.

The period has also been remarkable (just how remarkable will be considered later) for the frequency and intensity of extremes of weather and climate. For example, periods of unusually strong winds have been experienced in western Europe. During the early hours of the morning of 16 October 1987, over fifteen million trees were blown down in south-east England and the London area. The storm also hit Northern France, Belgium and The Netherlands with ferocious intensity; it turned out to be the worst storm experienced in the area since 1703. Storm-force winds of similar or even greater intensity but covering a greater area of western Europe have struck since – on four occasions in 1990 and three occasions in December 1999.[1]

But those storms in Europe were mild by comparison with the much more intense and damaging storms other parts of the world have experienced during these years. About eighty hurricanes and typhoons – other names for tropical cyclones – occur around the tropical oceans each year,

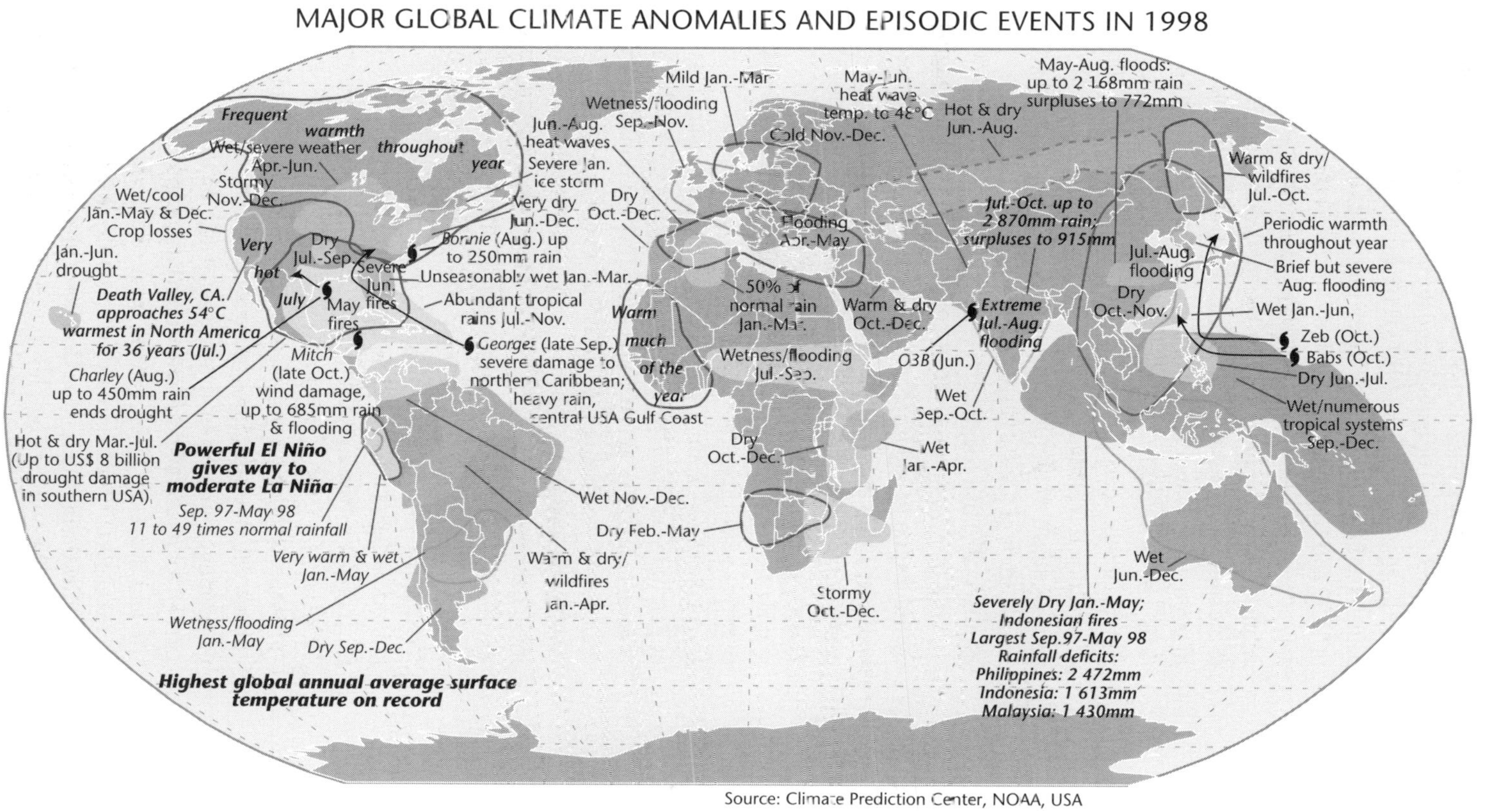

MAJOR GLOBAL CLIMATE ANOMALIES AND EPISODIC EVENTS IN 1998

Figure 1.1 Significant climate anomalies and events during 1998 as recorded by the Climate Prediction Center of the National Oceanic and Atmosphere Administration (NOAA) of the United States.

Source: Climate Prediction Center, NOAA, USA

familiar enough to be given names. Hurricane Gilbert that caused devastation on the island of Jamaica and the coast of Mexico in 1988, Typhoon Mireille that hit Japan in 1991, Hurricane Andrew that caused a great deal of damage in Florida and other regions of the southern United States in 1992 and Hurricane Mitch that caused great devastation in Honduras and other countries of central America in 1998 are notable recent examples. Low-lying areas such as Bangladesh are particularly vulnerable to the storm surges associated with tropical cyclones; the combined effect of intensely low atmospheric pressure, extremely strong winds and high tides causes a surge of water which can reach far inland. In one of the worst such disasters in the twentieth century over 250 000 people were drowned in Bangladesh in 1970. The people of that country experienced another storm of similar proportions in 1999 as did the neighbouring Indian state of Orissa also in 1999, and smaller surges are a regular occurrence in that region.

The increase in storm intensity during recent years has been tracked by the insurance industry, which has been hit hard by recent disasters. Until the mid 1980s, it was widely thought that windstorms or hurricanes with insured losses exceeding one billion (thousand million) US dollars were only possible, if at all, in the United States. But the gales that hit western Europe in October 1987 heralded a series of windstorm disasters which make losses of ten billion dollars seem commonplace. Hurricane Andrew, for instance, left in its wake insured losses estimated at nearly twenty-one billion dollars (1999 prices) with estimated total economic losses of nearly thirty-seven billion dollars. Figure 1.2 shows the costs of weather-related disasters[2] over the past fifty years as calculated by the insurance industry. It shows an increase in economic losses in such events by a factor of over 10 in real terms between the 1950s and the 1990s. Some of this increase can be attributed to the growth in population in particularly vulnerable areas and to other social or economic factors; the world community has undoubtedly become more vulnerable to disasters. However, a significant part of it has also arisen from the increased storminess in the late 1980s and 1990s compared with the 1950s.

Windstorms or hurricanes are by no means the only weather and climate extremes that cause disasters. Floods due to unusually intense or prolonged rainfall or droughts because of long periods of reduced rainfall (or its complete absence) can be even more devastating to human life and property. These events occur frequently in many parts of the world especially in the tropics and sub-tropics. There have been notable examples during the last two decades. Let me mention a few of the floods. In 1988, the highest flood levels ever recorded occurred in Bangladesh, eighty per cent of the entire country was affected; China experienced

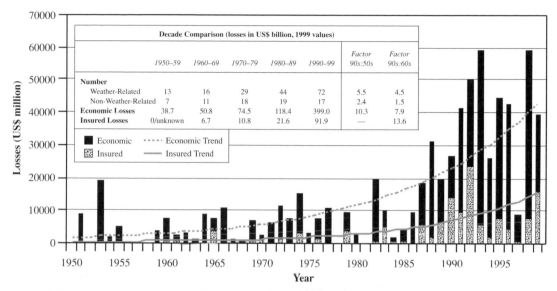

Figure 1.2 The total economic costs and the insured costs of catastrophic weather events for the second half of the twentieth century as recorded by the Munich Re insurance company. Both costs show a rapid upward trend in recent decades. The number of non-weather-related disasters is included for comparison. Tables 7.2 and 7.3 in Chapter 7 provide some regional detail and list some of the recent disasters with the greatest economic and insured losses.

devastating floods affecting many millions of people in 1991, 1994–5 and 1998; in 1993, flood waters rose to levels higher than ever recorded in the region of the Mississippi and Missouri rivers in the United States, flooding an area equivalent in size to one of the Great Lakes; major floods in Venezuela in 1999 led to a large landslide and left 30 000 people dead, and two widespread floods in Mozambique occurred within a year in 2000–1 leaving over half a million homeless. Droughts during these years have been particularly intense and prolonged in areas of Africa, both north and south. It is in Africa especially that they bear on the most vulnerable in the world, who have little resilience to major disasters. Figure 1.3 shows that in the 1980s droughts accounted for more deaths in Africa than all other disasters added together and illustrates the scale of the problem.

El Niño events

Rainfall patterns which lead to floods and droughts especially in tropical and semi-tropical areas are strongly influenced by the surface temperature of the oceans around the world, particularly the pattern of ocean surface temperature in the Pacific off the coast of South America

Figure 1.3 Recorded disasters in Africa, 1980–9, estimated by the Organization for African Unity.

THE EFFECT OF VOLCANIC ERUPTIONS

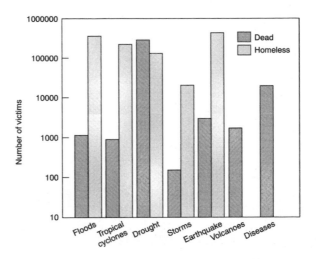

(see Chapter 5 and Figure 5.9). About every three to five years a large area of warmer water appears and persists for a year or more. Because they usually occur around Christmas these are known as El Niño ('the boy child') events.[3] They have been well known for centuries to the countries along the coast of South America because of their devastating effect on the fishing industry; the warm top waters of the ocean prevent the nutrients from lower, colder levels required by the fish from reaching the surface.

A particularly intense El Niño, the second most intense in the twentieth century, occurred in 1982–3; the anomalous highs in ocean surface temperature compared to the average reached 7 °C. Droughts and floods somewhere in almost all the continents were associated with that El Niño (Figure 1.4). Like many events associated with weather and climate, El Niños often differ very much in their detailed character; that has been particularly the case with the El Niño events of the 1990s. For instance, the El Niño event that began in 1990 and reached maturity early in 1992, apart from some weakening in mid 1992, continued to be dominated by the warm phase until 1995. The exceptional floods in the central United States and in the Andes, and droughts in Australia and Africa are probably linked with this unusually protracted El Niño. This, the longest El Niño of the twentieth century, was followed in 1997–8 by the century's most intense El Niño that brought exceptional floods to China and to the Indian sub-continent and drought to Indonesia – that in turn brought extensive forest fires creating an exceptional blanket of thick smog that was experienced over a thousand miles away (Figure 1.1).

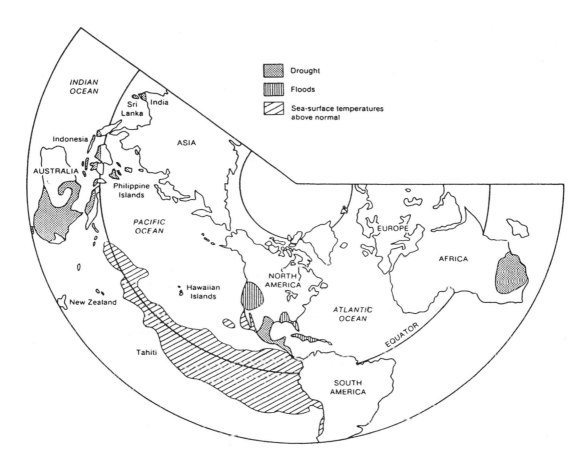

Figure 1.4 Regions where droughts and floods occurred associated with the 1982–3 El Niño.

Studies with computer models of the kind described later in Chapter 5 provide a scientific basis for links between the El Niño and these extreme weather events; they also give some confidence that useful forecasts of such disasters will in due course be possible. A scientific question that is being urgently addressed is the possible link between the character and intensity of El Niño events and global warming due to human-induced climate change.

The effect of volcanic eruptions on temperature extremes

Volcanoes inject enormous quantities of dust and gases into the upper atmosphere. Large amounts of sulphur dioxide are included, which through photochemical reactions using the Sun's energy are transformed to sulphuric acid and sulphate particles. Typically these particles remain in the stratosphere (the region of atmosphere above about 10 km in altitude) for several years before they fall into the lower atmosphere and are quickly washed out by rainfall. During this period they disperse around the whole globe and cut out some of the radiation from the Sun, thus tending to cool the lower atmosphere.

One of the largest volcanic eruptions in the twentieth century was that from Mount Pinatubo in the Philippines on 12 June 1991 which injected about twenty million tonnes of sulphur dioxide into the stratosphere together with enormous amounts of dust. This stratospheric dust caused spectacular sunsets around the world for many months following the eruption. The amount of radiation from the Sun reaching the lower atmosphere fell by about two per cent. Global average temperatures lower by about a quarter of a degree Celsius were experienced for the following two years. There is also evidence that some of the unusual weather patterns of 1991 and 1992, for instance unusually cold winters in the Middle East and mild winters in western Europe, were linked with effects of the volcanic dust.

Vulnerable to change

Over the centuries different human communities have adapted to their particular climate; any large change to the average climate tends to bring stress of one kind or another. It is particularly the extreme climate events and climate disasters which emphasise the importance of climate to our lives and which demonstrate to countries around the world their vulnerability to climate change – a vulnerability which is enhanced by rapidly increasing demands on resources.

But the question must be asked: how remarkable are these events? Do they point to a changing climate due to human activities? Do they provide evidence for global warming because of the increased carbon dioxide and other greenhouse gases being emitted into the atmosphere by burning fossil fuels?

Here a note of caution must be sounded. The range of normal natural climate variation is large. Climate extremes are nothing new. Climate records are continually being broken. In fact, a month without a broken record somewhere would itself be something of a record! Changes in climate that indicate a genuine long-term trend can only be identified after many years.

However, we know for sure that, because of human activities especially the burning of fossil fuels, carbon dioxide in the atmosphere has been increasing over the past two hundred years and more substantially over the past fifty years. To identify climate change related to this carbon dioxide increase, we need to look for trends in global warming over similar lengths of time. They are long compared with both the memories of a generation and the period for which accurate and detailed records exist. Although, therefore, it can be ascertained that there was more storminess, for instance, in the region of the north Atlantic during the 1980s and 1990s than in the previous three decades, it is difficult to know just how exceptional those decades were compared with other periods in previous centuries. There is even more difficulty in tracking detailed climate trends in many other parts of the world, owing to the lack of adequate records; further, trends in the frequency of rare events are not easy to detect.

The generally cold period worldwide during the 1960s and early 1970s caused speculation that the world was heading for an ice age. A British television programme about climate change called 'The ice age cometh' was prepared in the early 1970s and widely screened – but the cold trend soon came to an end. We must not be misled by our relatively short memories.

What is important is continually to make careful comparisons between practical observations of the climate and its changes and what scientific knowledge leads us to expect. During the last few years, as the occurrence of extreme events has made the public much more aware of environmental issues,[4] scientists in their turn have become more sure about just what human activities are doing to the climate. Later chapters will look in detail at the science of global warming and at the climate changes that we can expect, as well as investigating how these changes fit in with the recent climate record. Here, however, is a brief outline of our current scientific understanding.

The problem of global warming

Human activities of all kinds whether in industry, in the field (e.g. deforestation) or concerned with transport or the home are resulting in emissions of increasing quantities of gases, in particular the gas carbon dioxide, into the atmosphere. Every year these emissions currently add to the carbon already present in atmospheric carbon dioxide a further seven thousand million tonnes, much of which is likely to remain there for a period of a hundred years or more. Because carbon dioxide is a good absorber of heat radiation coming from the Earth's surface, increased carbon dioxide acts like a blanket over the surface, keeping it warmer than it would otherwise be. With the increased temperature the amount of water vapour in the atmosphere also increases, providing more blanketing and causing it to be even warmer.

Being kept warmer may sound appealing to those of us who live in cool climates. However, an increase in global temperature will lead to global climate change. If the change were small and occurred slowly enough we would almost certainly be able to adapt to it. However, with rapid expansion taking place in the world's industry the change is unlikely to be either small or slow. The estimate I present in later chapters is that, in the absence of efforts to curb the rise in the emissions of carbon dioxide, the global average temperature will rise by about a third of a degree Celsius every ten years – or about three degrees in a century.

This may not sound very much, especially when it is compared with normal temperature variations from day to night or between one day and the next. But it is not the temperature at one place but the temperature averaged over the whole globe. The predicted rate of change of three degrees a century is probably faster than the global average temperature has changed at any time over the past ten thousand years. And as there is a difference in global average temperature of only about five or six degrees between the coldest part of an ice age and the warm periods in between ice ages (see Figure 4.4), we can see that a few degrees in this global average can represent a big change in climate. It is to this change and especially to the very rapid rate of change that many ecosystems and human communities (especially those in developing countries) will find it difficult to adapt.

Not all the climate changes will in the end be adverse. While some parts of the world experience more frequent or more severe droughts, floods or significant sea level rise, in other places crop yields may increase due to the fertilising effect of carbon dioxide. Other places, perhaps for instance in the sub-arctic, may become more habitable. Even there, though, the likely rate of change will cause problems: large damage to buildings will occur in regions of melting permafrost, and trees in sub-arctic forests like trees elsewhere will need time to adapt to new climatic regimes.

Scientists are confident about the fact of global warming and climate change due to human activities. However, substantial uncertainty remains about just how large the warming will be and what will be the patterns of change in different parts of the world. Although some indications can be given, scientists cannot yet say in precise detail which regions will be most affected. Intensive research is needed to improve the confidence in scientific predictions.

Adaptation and mitigation

An integrated view of anthropogenic climate change is presented in Figure 1.5 where a complete cycle of cause and effect is shown. Begin in the lower right-hand corner where economic activity, both large and small scale, whether in developed or developing countries, results in emissions of greenhouse gases (of which carbon dioxide is the most

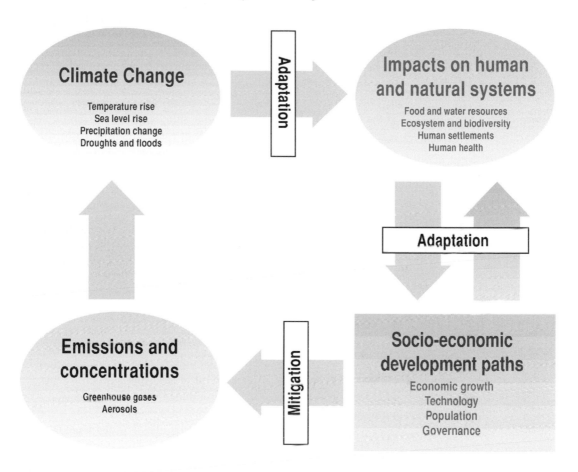

Figure 1.5 Climate change – an integrating framework; see text for explanation.

important) and aerosols. Moving in a clockwise direction around the diagram, these emissions lead to changes in atmospheric concentrations of important constituents that alter the energy input and output of the climate system and hence cause changes in the climate. These climate changes impact both humans and natural ecosystems altering patterns of resource availability and affecting human livelihood and health. These impacts in their turn affect human development in all its aspects. An anticlockwise arrow represents other effects of development on human communities and natural systems, for instance changes in land use that lead to deforestation and loss of biodiversity.

Figure 1.5 also shows how both causes and effects can be changed through *adaptation* and *mitigation*. In general adaptation is aimed at reducing the effects and mitigation is aimed at reducing the causes of climate change, in particular the emissions of the gases that give rise to it.

Uncertainty and response

Predictions of the future climate are surrounded with considerable uncertainty that arises from our imperfect knowledge both of the science of climate change and of the future scale of the human activities that are its cause. Politicians and others making decisions are therefore faced with the need to weigh all aspects of uncertainty against the desirability and the cost of the various actions that can be taken in response to the threat of climate change. Some mitigating action can be taken easily at relatively little cost (or even at a net saving of cost), for instance the development of programmes to conserve and save energy, and many schemes for reducing deforestation and encouraging the planting of trees. Other actions such as a large shift to energy sources that are free from significant carbon dioxide emissions (for example, renewable sources – biomass, hydro, wind, or solar energy) both in the developed and the developing countries of the world will take some time. Because however of the long timescales that are involved in the development of new energy infrastructure and in the response of the climate to emissions of gases like carbon dioxide, there is an urgency to begin these actions now. As we shall argue later (Chapter 9), to 'wait and see' is an irresponsible response.

In the following chapters I shall first explain the science of global warming, the evidence for it and the current state of the art regarding climate prediction. I shall then go on to say what is known about the likely impacts of climate change on human life – on water and food supplies for instance. The questions of why we should be concerned for the environment and what action should be taken in the face of scientific uncertainty are followed by consideration of the technical possibilities for large reductions in the emissions of carbon dioxide and how these might affect our energy sources and usage, including means of transport.

Finally I will address the issue of the 'global village'. So far as the environment is concerned, national boundaries are becoming less and less important; pollution in one country can now affect the whole world. Further, it is increasingly realised that problems of the environment are linked to other global problems such as population growth, poverty, the overuse of resources and global security. All these pose global challenges that must be met by global solutions.

Questions

1　Look through recent copies of newspapers and magazines for articles which mention climate change, global warming or the greenhouse effect. How many of the statements made are accurate?

2   Make up a simple questionnaire about climate change, global warming and the greenhouse effect to find out how much people know about these subjects, their relevance and importance. Analyse results from responses to the questionnaire in terms of the background of the respondents. Suggest ways in which people could be better informed.

Notes for Chapter 1

1   See Table 8.3 in Vellinga, P. V., Mills, E., Bowers, L., Berz, G., Huq, S., Kozak, L., Paultikof, J., Schanzenbacker, B., Shida, S., Soler, G., Benson, C., Bidan, P., Bruce, J., Huyck, P., Lemcke, G., Peara, A., Radevsky, R., van Schoubroeck, C., Dlugolecki, A. 2001. Insurance and other financial services. In McCarthy, J. J., Canziani, O., Leary, N. A., Dokken, D. J., White, K. S. (eds.) 2001. *Climate Change 2001: Impacts, Adaptation and Vulnerability. Contribution of Working Group II to the Third Assessment Report of the Intergovernmental Panel on Climate Change*. Cambridge: Cambridge University Press, Chapter 8.
2   Including windstorms, hurricanes or typhoons, floods, tornadoes, hailstorms, blizzards but not including droughts because their impact is not immediate and occurs over an extended period.
3   A description of the variety of El Niño events and their impacts on different communities worldwide over centuries of human history can be found in a recent paperback by Ross Couiper-Johnston, *El Niño: the Weather Phenomena that Changed the World*. 2000. London: Hodder and Stoughton.
4   A gripping account of some of the changes over the last decades can be found in a recent book by Mark Lynas, *High Tides: News from a Warming World*. 2004. London: Flamingo.

Chapter 2
The greenhouse effect

The basic principle of global warming can be understood by considering the radiation energy from the Sun that warms the Earth's surface and the thermal radiation from the Earth and the atmosphere that is radiated out to space. On average these two radiation streams must balance. If the balance is disturbed (for instance by an increase in atmospheric carbon dioxide) it can be restored by an increase in the Earth's surface temperature.

How the Earth keeps warm

To explain the processes that warm the Earth and its atmosphere, I will begin with a very simplified Earth. Suppose we could, all of a sudden, remove from the atmosphere all the clouds, the water vapour, the carbon dioxide and all the other minor gases and the dust, leaving an atmosphere of nitrogen and oxygen only. Everything else remains the same. What, under these conditions, would happen to the atmospheric temperature?

The calculation is an easy one, involving a relatively simple radiation balance. Radiant energy from the Sun falls on a surface of one square metre in area outside the atmosphere and directly facing the Sun at a rate of about 1370 watts – about the power radiated by a reasonably sized domestic electric fire. However, few parts of the Earth's surface face the Sun directly and in any case for half the time they are pointing away from the Sun at night, so that the average energy falling on one square metre of a level surface outside the atmosphere is only one-quarter of this[1] or about 343 watts. As this radiation passes through the atmosphere a small

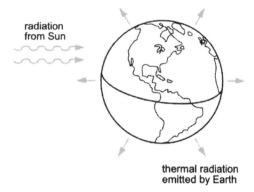

radiation
from Sun

thermal radiation
emitted by Earth

Figure 2.1 The radiation balance of planet Earth. The net incoming solar radiation is balanced by outgoing thermal radiation from the Earth.

amount, about six per cent, is scattered back to space by atmospheric molecules. About ten per cent on average is reflected back to space from the land and ocean surface. The remaining eighty-four per cent, or about 288 watts per square metre on average, remains actually to heat the surface – the power used by three good-sized incandescent electric light bulbs.

To balance this incoming energy, the Earth itself must radiate on average the same amount of energy back to space (Figure 2.1) in the form of thermal radiation. All objects emit this kind of radiation; if they are hot enough we can see the radiation they emit. The Sun at a temperature of about 6000 °C looks white; an electric fire at 800 °C looks red. Cooler objects emit radiation that cannot be seen by our eyes and which lies at wavelengths beyond the red end of the spectrum – infrared radiation (sometimes called long-wave radiation to distinguish it from the short-wave radiation from the Sun). On a clear, starry winter's night we are very aware of the cooling effect of this kind of radiation being emitted by the Earth's surface into space – it often leads to the formation of frost.

The amount of thermal radiation emitted by the Earth's surface depends on its temperature – the warmer it is, the more radiation is emitted. The amount of radiation also depends on how absorbing the surface is; the greater the absorption, the more the radiation. Most of the surfaces on the Earth, including ice and snow, would appear 'black' if we could see them at infrared wavelengths; that means that they absorb nearly all the thermal radiation which falls on them instead of reflecting it. It can be calculated[2] that, to balance the energy coming in, the average temperature of the Earth's surface must be –6 °C to radiate the right amount.[3] This is much colder than is actually the case. In fact, an average of temperatures measured near the surface all over the Earth – over the oceans as well as over the land – averaging, too, over the whole year, comes to about 15 °C. Some factor not yet taken into account is needed to explain this discrepancy.

Table 2.1 *The composition of the atmosphere, the main constituents (nitrogen and oxygen) and the greenhouse gases as in 2001*

Gas	Mixing ratio or mole fraction[a] expressed as fraction* or parts per million (ppm)
Nitrogen (N_2)	0.78*
Oxygen (O_2)	0.21*
Water vapour (H_2O)	Variable (0–0.02*)
Carbon dioxide (CO_2)	370
Methane (CH_4)	1.8
Nitrous oxide (N_2O)	0.3
Chlorofluorocarbons	0.001
Ozone (O_3)	Variable (0–1000)

[a] For definition see Glossary.

The greenhouse effect

The gases nitrogen and oxygen that make up the bulk of the atmosphere (Table 2.1 gives details of the atmosphere's composition) neither absorb nor emit thermal radiation. It is the water vapour, carbon dioxide and some other minor gases present in the atmosphere in much smaller quantities (Table 2.1) that absorb some of the thermal radiation leaving the surface, acting as a partial blanket for this radiation and causing the difference of 21 °C or so between the actual average surface temperature on the Earth of about 15 °C and the figure of −6 °C which applies when the atmosphere contains nitrogen and oxygen only.[4] This blanketing is known as the *natural greenhouse effect* and the gases are known as greenhouse gases. It is called 'natural' because all the atmospheric gases (apart from the chlorofluorocarbons – CFCs) were there long before human beings came on the scene. Later on I will mention the *enhanced greenhouse effect*: the added effect caused by the gases present in the atmosphere due to human activities such as the burning of fossil fuels and deforestation.

The basic science of the greenhouse effect has been known since early in the nineteenth century (see box) when the similarity between the radiative properties of the Earth's atmosphere and of the glass in a greenhouse (Figure 2.2) was first pointed out – hence the name 'greenhouse effect'. In a greenhouse, visible radiation from the Sun passes almost

Pioneers of the science of the greenhouse effect[5]

The warming effect of the greenhouse gases in the atmosphere was first recognised in 1827 by the French scientist Jean-Baptiste Fourier, best known for his contributions to mathematics. He also pointed out the similarity between what happens in the atmosphere and in the glass of a greenhouse, which led to the name 'greenhouse effect'. The next step was taken by a British scientist, John Tyndall, who, around 1860, measured the absorption of infrared radiation by carbon dioxide and water vapour; he also suggested that a cause of the Ice Ages might be a decrease in the greenhouse effect of carbon dioxide. It was a Swedish chemist, Svante Arrhenius, in 1896, who calculated the effect of an increasing concentration of greenhouse gases; he estimated that doubling the concentration of carbon dioxide would increase the global average temperature by 5 °C to 6 °C, an estimate not too far from our present understanding.[6] Nearly fifty years later, around 1940, G. S. Callendar, working in England, was the first to calculate the warming due to the increasing carbon dioxide from the burning of fossil fuels.

 The first expression of concern about the climate change which might be brought about by increasing greenhouse gases was in 1957, when Roger Revelle and Hans Suess of the Scripps Institute of Oceanography in California published a paper which pointed out that in the build-up of carbon dioxide in the atmosphere, human beings are carrying out a large-scale geophysical experiment. In the same year, routine measurements of carbon dioxide were started from the observatory on Mauna Kea in Hawaii. The rapidly increasing use of fossil fuels since then, together with growing interest in the environment, has led to the topic of global warming moving up the political agenda through the 1980s, and eventually to the Climate Convention signed in 1992 – of which more in later chapters.

unimpeded through the glass and is absorbed by the plants and the soil inside. The thermal radiation that is emitted by the plants and soil is, however, absorbed by the glass that re-emits some of it back into the greenhouse. The glass thus acts as a 'radiation blanket' helping to keep the greenhouse warm.

 However, the transfer of radiation is only one of the ways heat is moved around in a greenhouse. A more important means of heat transfer is due to convection, in which less dense warm air moves upwards and more dense cold air moves downwards. A familiar example of this process is the use of convective electric heaters in the home, which heat a room by stimulating convection in it. The situation in the greenhouse

Figure 2.2 A greenhouse has a similar effect to the atmosphere on the incoming solar radiation and the emitted thermal radiation.

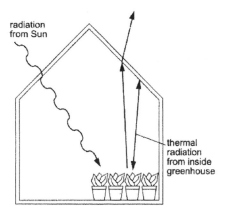

is therefore more complicated than would be the case if radiation were the only process of heat transfer.

Mixing and convection are also present in the atmosphere, although on a much larger scale, and in order to achieve a proper understanding of the greenhouse effect, convective heat transfer processes in the atmosphere must be taken into account as well as radiative ones.

Within the atmosphere itself (at least in the lowest three-quarters or so of the atmosphere up to a height of about 10 km which is called the troposphere) convection is, in fact, the dominant process for transferring heat. It acts as follows. The surface of the Earth is warmed by the sunlight it absorbs. Air close to the surface is heated and rises because of its lower density. As the air rises it expands and cools – just as the air cools as it comes out of the valve of a tyre. As some air masses rise, other air masses descend, so the air is continually turning over as different movements balance each other out – a situation of convective equilibrium. Temperature in the troposphere falls with height at a rate determined by these convective processes; the fall with height (called the lapse-rate) turns out on average to be about 6 °C per kilometre of height (Figure 2.3).

A picture of the transfer of radiation in the atmosphere may be obtained by looking at the thermal radiation emitted by the Earth and its atmosphere as observed from instruments on satellites orbiting the Earth (Figure 2.4). At some wavelengths in the infrared the atmosphere – in the absence of clouds – is largely transparent, just as it is in the visible part of the spectrum. If our eyes were sensitive at these wavelengths we would be able to peer through the atmosphere to the Sun, stars and Moon above, just as we can in the visible spectrum. At these wavelengths all the radiation originating from the Earth's surface leaves the atmosphere.

At other wavelengths radiation from the surface is strongly absorbed by some of the gases present in the atmosphere, in particular by water vapour and carbon dioxide.

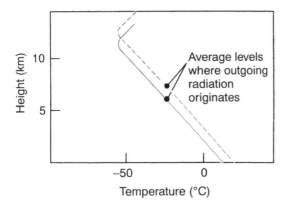

Figure 2.3 The distribution of temperature in a convective atmosphere (full line). The broken line shows how the temperature increases when the amount of carbon dioxide present in the atmosphere is increased (in the diagram the difference between the lines is exaggerated – for instance, for doubled carbon dioxide in the absence of other effects the increase in temperature is about 1.2 °C). Also shown for the two cases are the average levels from which thermal radiation leaving the atmosphere originates (about 6 km for the unperturbed atmosphere).

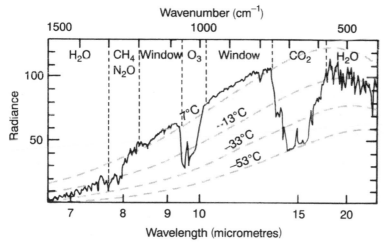

Figure 2.4 Thermal radiation in the infrared region (the visible part of the spectrum is between about 0.4 and 0.7 μm) emitted from the Earth's surface and atmosphere as observed over the Mediterranean Sea from a satellite instrument orbiting above the atmosphere, showing parts of the spectrum where different gases contribute to the radiation. Between the wavelengths of about 8 and 14 μm, apart from the ozone band, the atmosphere, in the absence of clouds, is substantially transparent; this is part of the spectrum called a 'window' region. Superimposed on the spectrum are curves of radiation from a black body at 7 °C, −13 °C, −33 °C and −53 °C. The units of radiance are watts per square metre per steradian per wavenumber.

Figure 2.5 The blanketing effect of greenhouse gases.

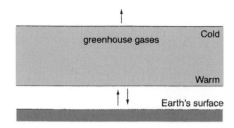

Objects that are good absorbers of radiation are also good emitters of it. A black surface is both a good absorber and a good emitter, while a highly reflecting surface absorbs rather little and emits rather little too (which is why highly reflecting foil is used to cover the surface of a vacuum flask and why it is placed above the insulation in the lofts of houses).

Absorbing gases in the atmosphere absorb some of the radiation emitted by the Earth's surface and in turn emit radiation out to space. The amount of thermal radiation they emit is dependent on their temperature.

Radiation is emitted out to space by these gases from levels somewhere near the top of the atmosphere – typically from between 5 and 10 km high (see Figure 2.3). Here, because of the convection processes mentioned earlier, the temperature is much colder – 30 to 50 °C or so colder – than at the surface. Because the gases are cold, they emit correspondingly less radiation. What these gases have to do, therefore, is absorb some of the radiation emitted by the Earth's surface but then to emit much less radiation out to space. They, therefore, act as a radiation blanket over the surface (note that the outer surface of a blanket is colder than inside the blanket) and help to keep it warmer than it would otherwise be[7] (Figure 2.5).

There needs to be a balance between the radiation coming in and the radiation leaving the top of the atmosphere – as there was in the very simple model with which this chapter started. Figure 2.6 shows the various components of the radiation entering and leaving the top of the atmosphere for the real atmosphere situation. On average, 240 watts per square metre of solar radiation are absorbed by the atmosphere and the surface; this is less than the 288 watts mentioned at the beginning of the chapter, because now the effect of clouds is being taken into account. Clouds reflect some of the incident radiation from the Sun back out to space. However, they also absorb and emit thermal radiation and have a blanketing effect similar to that of the greenhouse gases. These two effects work in opposite senses: one (the reflection of solar radiation) tends to cool the Earth's surface and the other (the absorption of thermal radiation) tends to warm it. Careful consideration of these two effects shows that on average the net effect of clouds on the total budget of radiation results in a slight cooling of the Earth's surface.[8]

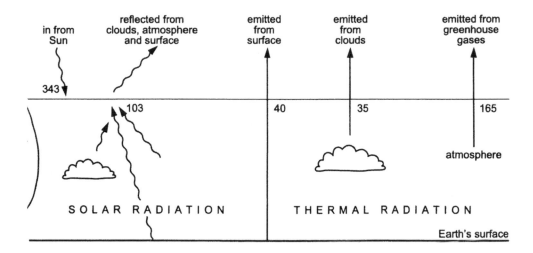

The numbers in Figure 2.6 demonstrate the required balance – 240 watts per square metre on average coming in and 240 watts per square metre on average going out. The temperature of the surface and hence of the atmosphere above adjusts itself to ensure that this balance is maintained. It is interesting to note that the greenhouse effect can only operate if there are colder temperatures in the higher atmosphere. Without the structure of decreasing temperature with height, therefore, there would be no greenhouse effect on the Earth.

Figure 2.6 Components of the radiation (in watts per square metre) which on average enter and leave the Earth's atmosphere and make up the radiation budget for the atmosphere.

Mars and Venus

Similar greenhouse effects also occur on our nearest planetary neighbours, Mars and Venus. Mars is smaller than the Earth and possesses, by Earth's standards, a very thin atmosphere. A barometer on the surface of Mars would record an atmospheric pressure less than one per cent of that on the Earth. Its atmosphere, which consists almost entirely of carbon dioxide, contributes a small but significant greenhouse effect.

The planet Venus, which can often be seen fairly close to the Sun in the morning or evening sky, has a very different atmosphere to Mars. Venus is about the same size as the Earth. A barometer for use on Venus would need to survive very hostile conditions and would need to be able to measure a pressure about one hundred times as great as that on the Earth. Within the Venus atmosphere, which consists very largely of carbon dioxide, deep clouds consisting of droplets of almost pure sulphuric acid completely cover the planet and prevent most of the sunlight from reaching the surface. Some Russian space probes that have landed there have recorded what would be dusk-like conditions on the Earth – only one or two per cent of the sunlight present above the clouds penetrates that far. One might suppose, because of the small amount of solar energy available to keep the surface warm, that it would be rather cool; on the contrary, measurements from the same Russian space probes find a temperature there of about 525 °C – a dull red heat, in fact.

The reason for this very high temperature is the greenhouse effect. Because of the very thick absorbing atmosphere of carbon dioxide, little of the thermal radiation from the surface can get out. The atmosphere acts as such an effective radiation blanket that, although there is not much solar energy to warm the surface, the greenhouse effect amounts to nearly 500 °C.

The 'runaway' greenhouse effect

What occurs on Venus is an example of what has been called the 'runaway' greenhouse effect. It can be explained by imagining the early history of the Venus atmosphere, which was formed by the release of gases from the interior of the planet. To start with it would contain a lot of water vapour, a powerful greenhouse gas (Figure 2.7). The greenhouse effect of the water vapour would cause the temperature at the surface to rise. The increased temperature would lead to more evaporation of water from the surface, giving more atmospheric water vapour, a larger greenhouse effect and therefore a further increased surface temperature. The process would continue until either the atmosphere became saturated with water vapour or all the available water had evaporated.

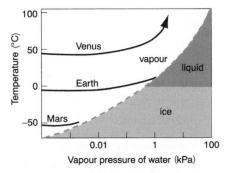

Figure 2.7 Illustrating the evolution of the atmospheres of the Earth, Mars and Venus. In this diagram, the surface temperatures of the three planets are plotted against the vapour pressure of water in their atmospheres as they evolved. Also on the diagram (dashed) are the phase lines for water, dividing the diagram into regions where vapour, liquid water or ice are in equilibrium. For Mars and the Earth the greenhouse effect is halted when water vapour is in equilibrium with ice or liquid water. For Venus no such halting occurs and the diagram illustrates the 'runaway' greenhouse effect.

A runaway sequence something like this seems to have occurred on Venus. Why, we may ask, has it not happened on the Earth, a planet of about the same size as Venus and, so far as is known, of a similar initial chemical composition? The reason is that Venus is closer to the Sun than the Earth; the amount of solar energy per square metre falling on Venus is about twice that falling on the Earth. The surface of Venus, when there was no atmosphere, would have started off at a temperature of just over 50 °C (Figure 2.7). Throughout the sequence described above for Venus, water on the surface would have been continuously boiling. Because of the high temperature, the atmosphere would never have become saturated with water vapour. The Earth, however, would have started at a colder temperature; at each stage of the sequence it would have arrived at an equilibrium between the surface and an atmosphere saturated with water vapour. There is no possibility of such runaway greenhouse conditions occurring on the Earth.

The enhanced greenhouse effect

After our excursion to Mars and Venus, let us return to Earth! The natural greenhouse effect is due to the gases water vapour and carbon dioxide present in the atmosphere in their natural abundances as now on Earth. The amount of water vapour in our atmosphere depends mostly on the temperature of the surface of the oceans; most of it originates through evaporation from the ocean surface and is not influenced directly by human activity. Carbon dioxide is different. Its amount has changed substantially – by about thirty per cent so far – since the Industrial Revolution, due to human industry and also because of the removal of forests (see Chapter 3). Future projections are that, in the absence of controlling factors, the rate of increase in atmospheric carbon dioxide will accelerate and that its atmospheric concentration will double from its pre-industrial value within the next hundred years (Figure 6.2).

This increased amount of carbon dioxide is leading to global warming of the Earth's surface because of its enhanced greenhouse effect. Let us imagine, for instance, that the amount of carbon dioxide in the atmosphere suddenly doubled, everything else remaining the same (Figure 2.8). What would happen to the numbers in the radiation budget presented earlier (Figure 2.6)? The solar radiation budget would not be affected. The greater amount of carbon dioxide in the atmosphere means that the thermal radiation emitted from it will originate on average from a higher and colder level than before (Figure 2.3). The thermal radiation budget will therefore be reduced, the amount of reduction being about 4 watts per square metre (a more precise value is 3.7).[9]

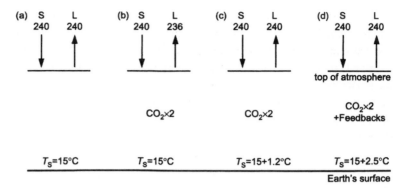

Figure 2.8 Illustrating the enhanced greenhouse gas effect. Under natural conditions (a) the net solar radiation coming in (S = 240 watts per square metre) is balanced by thermal radiation (L) leaving the top of the atmosphere; average surface temperature (T_s) is 15 °C. If the carbon dioxide concentration is suddenly doubled (b), L is decreased by 4 watts per square metre. Balance is restored if nothing else changes (c) apart from the temperature of the surface and lower atmosphere, which rises by 1.2 °C. If feedbacks are also taken into account (d), the average temperature of the surface rises by about 2.5 °C.

This causes a net imbalance in the overall budget of 4 watts per square metre. More energy is coming in than going out. To restore the balance the surface and lower atmosphere will warm up. If nothing changes apart from the temperature – in other words, the clouds, the water vapour, the ice and snow cover and so on are all the same as before – the temperature change turns out to be about 1.2 °C.

In reality, of course, many of these other factors will change, some of them in ways that add to the warming (these are called positive feedbacks), others in ways that might reduce the warming (negative feedbacks). The situation is therefore much more complicated than this simple calculation. These complications will be considered in more detail in Chapter 5. Suffice it to say here that the best estimate at the present time of the increased average temperature of the Earth's surface if carbon dioxide levels were to be doubled is about twice that of the simple calculation: 2.5 °C. As the last chapter explained, for the global average temperature this is a large change. It is this global warming expected to result from the enhanced greenhouse effect that is the cause of current concern.

Having dealt with a doubling of the amount of carbon dioxide, it is interesting to ask what would happen if all the carbon dioxide were removed from the atmosphere. It is sometimes supposed that the outgoing radiation would be changed by 4 watts per square metre in the other direction and that the Earth would then cool by one or two degrees Celsius. In fact, that would happen if the carbon dioxide amount were to be halved. If it were to be removed altogether, the change in outgoing radiation would be around 25 watts per square metre – six times as big – and the temperature change would be similarly increased. The reason for this is that with the amount of carbon dioxide currently present in the atmosphere there is maximum carbon dioxide absorption over much of the region of the spectrum where it absorbs (Figure 2.4), so that a big change in gas concentration leads to a relatively small change in the amount of radiation it absorbs.[10] This is like the situation in a pool of water: when it is clear, a small amount of mud will make it appear muddy, but when it is muddy, adding more mud only makes a small difference.

An obvious question to ask is: has evidence of the enhanced greenhouse effect been seen in the recent climatic record? Chapter 4 will look at the record of temperature on the Earth during the last century or so, during which the Earth has warmed on average by rather more than half a degree Celsius. We shall see in Chapters 4 and 5 that there are good reasons for attributing most of this warming to the enhanced greenhouse effect, although because of the size of natural climate variability the exact amount of that attribution remains subject to some uncertainty.

To summarise the argument so far:

- No one doubts the reality of the natural greenhouse effect, which keeps us over 20 °C warmer than we would otherwise be. The science of it is well understood; it is similar science that applies to the enhanced greenhouse effect.
- Substantial greenhouse effects occur on our nearest planetary neighbours, Mars and Venus. Given the conditions that exist on those planets, the sizes of their greenhouse effects can be calculated, and good agreement has been found with those measurements which are available.
- Study of climates of the past gives some clues about the greenhouse effect, as Chapter 4 will show.

First, however, the greenhouse gases themselves must be considered. How does carbon dioxide get into the atmosphere, and what other gases affect global warming?

Questions

1   Carry out the calculation described in Note 4 (refer also to Note 2) which obtains an equilibrium average temperature of −18 °C for an Earth partially covered with clouds such that thirty per cent of the incoming solar radiation is reflected. If clouds are assumed to cover half the Earth and if the reflectivity of the clouds increases by one per cent what change will this make in the resulting equilibrium average temperature?

2   It is sometimes argued that the greenhouse effect of carbon dioxide is negligible because its absorption band in the infrared is so close to saturation that there is very little additional absorption of radiation emitted from the surface. What are the fallacies in this argument?

3   Use the information in Figure 2.4 to estimate approximately the surface temperature that would result if carbon dioxide were completely removed from the atmosphere. What is required is that the total energy radiated by the Earth plus atmosphere should remain the same, i.e. the area under the radiance curve in Figure 2.4 should be unaltered. On this basis construct a new curve with the carbon dioxide band absent.[11]

4   Using information from books or articles on climatology or meteorology describe why the presence of water vapour in the atmosphere is of such importance in determining the atmosphere's circulation.

5   Estimates of regional warming due to increased greenhouse gases are generally larger over land areas than over ocean areas. What might be the reasons for this?

6   (For students with a background in physics) What is meant by Local Thermodynamic Equilibrium (LTE),[12] a basic assumption underlying calculations of radiative transfer in the lower atmosphere appropriate to discussions of the greenhouse effect? Under what conditions does LTE apply?

Notes for Chapter 2

1   It is about one-quarter because the area of the Earth's surface is four times the area of the disc, which is the projection of the Earth facing the Sun; see Figure 2.1.

2   The radiation by a black body is the Stefan–Boltzmann constant (5.67×10^{-8} J m^{-2} K^{-4} s^{-1}) multiplied by the fourth power of the body's absolute temperature in Kelvin. The absolute temperature is the temperature in degrees Celsius plus 273 (1 K $=$ 1 °C).

3   These calculations using a simple model of an atmosphere containing nitrogen and oxygen only have been carried out to illustrate the effect of the other gases, especially water vapour and carbon dioxide. It is not, of course, a model that can exist in reality. All the water vapour could not be removed from the atmosphere above a water or ice surface. Further, with an average surface temperature of −6 °C, in a real situation the surface would have much more ice cover. The additional ice would reflect more solar energy out to space leading to a further lowering of the surface temperature.

4   The above calculation is often carried out using a figure of thirty per cent for the average reflectivity of the Earth and atmosphere, rather than the sixteen per cent assumed here; the calculation of surface temperature then gives −18 °C for the average surface temperature rather than the −6 °C found here. The higher figure of thirty per cent for the Earth's average reflectivity is applicable when clouds are also included, in which case the average temperature of −18 °C is not applicable to the Earth's surface but to some appropriate level in the atmosphere. Further, clouds not only reflect

solar radiation but also absorb thermal radiation, and so have a blanketing effect similar to greenhouse gases. For the purposes of illustrating the effect of greenhouse gases, therefore, it is more correct to omit the effect of clouds from this initial calculation.

5 Further details can be found in Mudge, F. B. The development of greenhouse theory of global climate change from Victorian times. 1997. *Weather*, **52**, pp. 13–16.

6 A range of 1.5 to 4.5 °C is quoted in Chapter 6, page 120.

7 The formal theory of the greenhouse effect is presented in Houghton, J. T. 2002. *The Physics of Atmospheres*, third edition, Chapter 2. Cambridge: Cambridge University Press. See also Chapter 14 of that book.

8 More detail of the radiative effects of clouds is given in Chapter 5; see Figures 5.14 and 5.15.

9 More detailed information about the enhanced greenhouse effect can be found in Houghton, J. T. 2002. *The Physics of Atmospheres*, third edition, Chapter 14. Cambridge: Cambridge University Press.

10 The dependence of the absorption on the concentration of gas is approximately logarithmic.

11 For some helpful diagrams and more information about the infrared spectrum of different greenhouse gases, see Harries, J. E. 1996. The greenhouse Earth: a view from space. *Quarterly Journal of the Royal Meteorological Society*, **122**, pp. 799–818.

12 For information about LTE see, for instance, Houghton, J. T. 2002. *The Physics of Atmospheres*, third edition. Cambridge: Cambridge University Press.

Chapter 3
The greenhouse gases

The greenhouse gases are those gases in the atmosphere which, by absorbing thermal radiation emitted by the Earth's surface, have a blanketing effect upon it. The most important of the greenhouse gases is water vapour, but its amount in the atmosphere is not changing directly because of human activities. The important greenhouse gases that are directly influenced by human activities are carbon dioxide, methane, nitrous oxide, the chlorofluorocarbons (CFCs) and ozone. This chapter will describe what is known about the origin of these gases, how their concentration in the atmosphere is changing and how it is controlled. Also considered will be particles in the atmosphere of anthropogenic origin that can act to cool the surface.

Which are the most important greenhouse gases?

Figure 2.4 illustrated the regions of the infrared spectrum where the greenhouse gases absorb. Their importance as greenhouse gases depends both on their concentration in the atmosphere (Table 2.1) and on the strength of their absorption of infrared radiation. Both these quantities differ greatly for various gases.

Carbon dioxide is the most important of the greenhouse gases that are increasing in atmospheric concentration because of human activities. If, for the moment, we ignore the effects of the CFCs and of changes in ozone, which vary considerably over the globe and which are therefore more difficult to quantify, the increase in carbon dioxide (CO_2) has contributed about seventy per cent of the enhanced greenhouse effect to date, methane (CH_4) about twenty-four per cent, and nitrous oxide (N_2O) about six per cent (Figure 3.8).

Radiative forcing

In this chapter we shall use the concept of *radiative forcing* to compare the relative greenhouse effects of different atmospheric constituents. It is necessary therefore at the start to define radiative forcing.

In Chapter 2 we noted that, if the carbon dioxide in the atmosphere were suddenly doubled, everything else remaining the same, a net radiation imbalance near the top of the atmosphere of 3.7 W m^{-2} would result. This radiation imbalance is an example of radiation forcing, which is defined as the change in average net radiation at the top of the troposphere[1] (the lower atmosphere; for definition see Glossary) which occurs because of a change in the concentration of a greenhouse gas or because of some other change in the overall climate system; for instance, a change in the incoming solar radiation would constitute a radiative forcing. As we saw in the discussion in Chapter 2, over time the climate responds to restore the radiative balance between incoming and outgoing radiation A positive radiative forcing tends on average to warm the surface and a negative radiative forcing tends on average to cool the surface.

Carbon dioxide and the carbon cycle

Carbon dioxide provides the dominant means through which carbon is transferred in nature between a number of natural carbon reservoirs – a process known as the carbon cycle. We contribute to this cycle every time we breathe. Using the oxygen we take in from the atmosphere, carbon from our food is burnt and turned into carbon dioxide that we then exhale; in this way we are provided with the energy we need to maintain our life. Animals contribute to atmospheric carbon dioxide in the same way; so do fires, rotting wood and decomposition of organic material in the soil and elsewhere. To offset these processes of respiration whereby carbon is turned into carbon dioxide, there are processes involving photosynthesis in plants and trees which work the opposite way; in the presence of light, they take in carbon dioxide, use the carbon for growth and return the oxygen back to the atmosphere. Both respiration and photosynthesis also occur in the ocean.

Figure 3.1 is a simple diagram of the way carbon cycles between the various reservoirs – the atmosphere, the oceans (including the ocean biota), the soil and the land biota (biota is a word that covers all living things – plants, trees, animals and so on – on land and in the ocean, which

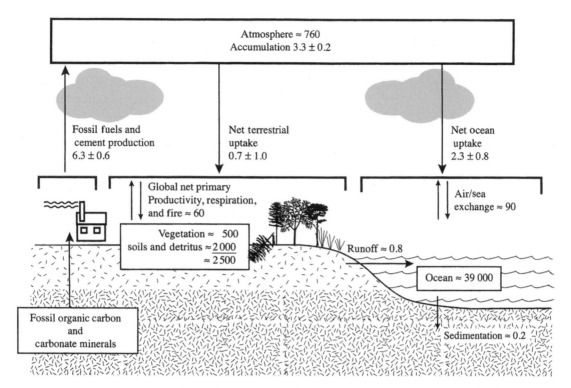

Figure 3.1 The global carbon cycle, showing the carbon stocks in reservoirs (in Gt) and carbon flows (in Gt year^{-1}) relevant to the anthropogenic perturbation as annual averages over the decade from 1989 to 1998. Net ocean uptake of the anthropogenic perturbation equals the net air/sea input plus run-off minus sediment. The units are thousand millions of tonnes or gigatonnes (Gt).

make up a whole known as the biosphere). The diagram shows that the movements of carbon (in the form of carbon dioxide) into and out of the atmosphere are quite large; about one-fifth of the total amount in the atmosphere is cycled in and out each year, part with the land biota and part through physical and chemical processes across the ocean surface. The land and ocean reservoirs are much larger than the amount in the atmosphere; small changes in these larger reservoirs could therefore have a large effect on the atmospheric concentration; the release of just two per cent of the carbon stored in the oceans would double the amount of atmospheric carbon dioxide.

It is important to realise that on the timescales with which we are concerned anthropogenic carbon emitted into the atmosphere as carbon dioxide is not destroyed but redistributed among the various carbon reservoirs. Carbon dioxide is therefore different from other greenhouse gases that are destroyed by chemical action in the atmosphere. The carbon reservoirs exchange carbon between themselves on a wide range of timescales determined by their respective turnover times – which range from less than a year to decades (for exchange with the top layers of the ocean and the land biosphere) to millennia (for exchange with the deep ocean or long-lived soil pools). These timescales are generally much longer than the average time a particular carbon dioxide molecule spends in the atmosphere, which is only about four years. The large range of turnover times means that the time taken for a perturbation in the atmospheric carbon dioxide concentration to relax back to an equilibrium cannot be described by a single time constant. Although a lifetime of about a hundred years is often quoted for atmospheric carbon dioxide so as to provide some guide, use of a single lifetime can be very misleading.

Before human activities became a significant disturbance, and over periods short compared with geological timescales, the exchanges between the reservoirs were remarkably constant. For several thousand years before the beginning of industrialisation around 1750, a steady balance was maintained, such that the mixing ratio (or mole fraction, for definition see Glossary) of carbon dioxide in the atmosphere as measured from ice cores (see Chapter 4) kept within about ten parts per million of a mean value of about 280 parts per million (ppm) – see Figure 3.2(a)

The Industrial Revolution disturbed this balance and since its beginning in about 1700 approximately 600 thousand million tonnes (or gigatonnes, Gt) of carbon have been emitted into the atmosphere from fossil fuel burning. This has resulted in a concentration of carbon dioxide in the atmosphere that has increased by about thirty per cent, from 280 ppm around 1700 to a value of over 370 ppm at the present day (Figure 3.2(b)). Accurate measurements, which have been made since 1959 from an observatory near the summit of Mauna Loa in Hawaii, show that carbon dioxide is currently increasing on average each year by about 1.5 ppm, although there are large variations from year to year (Figure 3.2(c)). This increase spread through the atmosphere adds about 3.3 Gt to the atmospheric carbon reservoir each year.

It is easy to establish how much coal, oil and gas is being burnt worldwide each year. Most of it is to provide energy for human needs: for heating and domestic appliances, for industry and for transport (considered in detail in Chapter 11). The burning of these fossil fuels has increased rapidly since the Industrial Revolution (Figure 3.3 and Table 3.1); currently the annual total is between 6 and 7 Gt of carbon, nearly all of which enters the atmosphere as carbon dioxide. Another contribution to atmospheric carbon dioxide due to human activities comes from land-use change, in particular from the burning and decay of forests balanced in part by aforestation or forest regrowth. This contribution is not easy to quantify but some estimates are given in Figure 3.3 and Table 3.1. For the 1980s (see Table 3.1), annual anthropogenic emissions from

Figure 3.2 Atmospheric carbon dioxide concentration, (a) over the last 10 000 years from the Taylor Dome Antarctic ice core; (b) from Antarctic ice cores for the past millennium; recent atmospheric measurements from the Mauna Loa observatory in Hawaii are also shown; (c) monthly changes in atmospheric carbon dioxide concentration, filtered so as to remove seasonal cycle; vertical arrows denote El Niño events; note the small rate of growth from 1991 to 1994 which may be connected with events such as the Pinatubo volcanic eruption in 1991 or the unusual extended El Niño of 1991–4 (denoted by horizontal line).

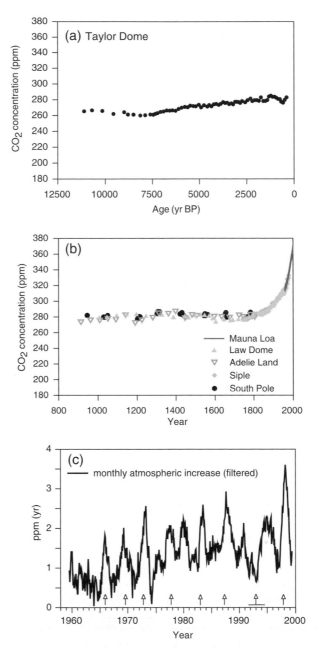

fossil fuel burning, cement manufacture and land-use change amounted to about 7.1 Gt; over three-quarters of these resulted from fossil fuel burning. Since the annual net increase in the atmosphere was about 3.3 Gt, about forty-five per cent of the 7.1 Gt of new carbon remained to increase the atmospheric concentration. The other fifty-five per cent was taken up between the other two reservoirs: the oceans and the land biota.

Table 3.1 *Components of annual average global carbon budget for 1980s and 1990s – in Gt of carbon per year* [a] *(positive values are fluxes to the atmosphere, negative values represent uptake from the atmosphere)*

	1980s	1990s
Emissions (fossil fuel, cement)	5.4 ± 0.3	6.4 ± 0.4
Atmospheric increase	3.3 ± 0.1	3.2 ± 0.1
Ocean–atmosphere flux	-1.9 ± 0.6	-1.7 ± 0.5
Land–atmosphere flux*	-0.2 ± 0.7	-1.4 ± 0.7
partitioned as follows		
Land-use change	1.7 (0.6 to 2.5)	1.4 to 3.0
Residual terrestrial sink	-1.9 (-3.8 to 0.3)	-4.8 to -1.6

[a] The entries in the first four rows are from Table 3.3 in Prentice *et al.* 2001 (see Note 2). Note that the ranges quoted represent sixty-seven per cent certainty. The entries in the 'partitioning of land-atmosphere flux' are from House *et al.* 2003 (see Note 2).

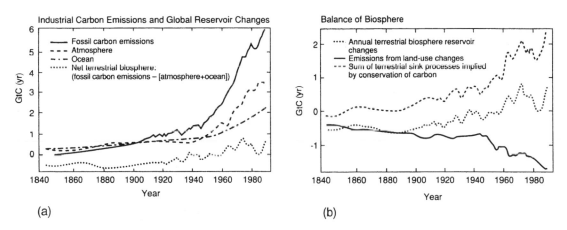

(a)　　　　　　　　　　　　　　　　(b)

Figure 3.3 (a) Fossil carbon emissions (based on statistics of fossil fuel and cement production) and estimates of global reservoir changes: atmosphere (deduced from direct observations and ice core measurements), ocean (calculated with the Geophysical Fluid Dynamics Laboratory (GFDL), University of Princeton, ocean carbon model) and net terrestrial biosphere (calculated as remaining imbalance) from 1840 to 1990. The calculation implies that the terrestrial biosphere was a net source to the atmosphere prior to 1940 (negative values) and has been a net sink since about 1960. (b) Estimates of contributions to the carbon balance of the terrestrial biosphere. The curve showing the terrestrial reservoir changes is taken from (a). Emissions from land-use changes (including tropical deforestation) are plotted negatively because they represent a loss of biospheric carbon. These estimates are subject to large uncertainties (see uncertainty estimates in Table 3.1).

Figure 3.5 shows that these fractions may change substantially in the future.

About ninety-five per cent of fossil fuel burning occurs in the northern hemisphere, so there is more carbon dioxide there than in the southern hemisphere. The difference is currently about two parts per million and, over the years, has grown in parallel with fossil fuel emissions, thus adding further compelling evidence that the atmospheric increase in carbon dioxide levels results from these emissions.

We turn now to what happens in the oceans. We know that carbon dioxide dissolves in water; carbonated drinks make use of that fact. Carbon dioxide is continually being exchanged with the air above the ocean across the whole ocean surface (about 90 Gt per year is so exchanged – Figure 3.1), particularly as waves break. An equilibrium is established between the concentration of carbon dioxide dissolved in the surface waters and the concentration in the air above the surface. The chemical laws governing this equilibrium are such that if the atmospheric concentration changes by ten per cent the concentration in solution in the water changes by only one-tenth of this: one per cent.

This change will occur quite rapidly in the upper waters of the ocean, the top hundred metres or so, so enabling part of the anthropogenic (i.e. human generated) carbon dioxide added to the atmosphere (most of the ocean's share of the fifty-five per cent mentioned above) to be taken up quite rapidly. Absorption in the lower levels in the ocean takes longer; mixing of surface water with water at lower levels takes up to several hundred years or for the deep ocean over a thousand years. This process whereby carbon dioxide is gradually drawn from the atmosphere in the ocean's lower levels is sometimes known as the *solubility pump*.

So the oceans do not provide as immediate a sink for increased atmospheric carbon dioxide as might be suggested by the size of the exchanges with the large ocean reservoir. For short-term changes only the surface layers of water play a large part in the carbon cycle.

Biological activity in the oceans also plays an important role. It may not be immediately apparent, but the oceans are literally teeming with life. Although the total mass of living matter within the oceans is not large, it has a high rate of turnover. Living material in the oceans is produced at some thirty to forty per cent of the rate of production on land. Most of this production is of plant and animal plankton that go through a rapid series of life cycles. As they die and decay some of the carbon they contain is carried downwards into lower levels of the ocean adding to the carbon content of those levels. Some is carried to the very deep water or to the ocean bottom where, so far as the carbon cycle is concerned, it is out of circulation for hundreds or thousands of years.

The *biological pump* in the oceans[3]

In temperate and high latitudes there is a peak each spring in ocean biological activity. During the winter, water rich in nutrients is transferred from deep water to levels near the surface. As sunlight increases in the spring an explosive growth of the plankton population occurs, known as the 'spring bloom'. Pictures of the colour of the ocean taken from satellites orbiting the Earth can demonstrate dramatically where this is happening.

Plankton are small plants (phytoplankton) and animals (zooplankton) that live in the surface waters of the ocean; they range in size between about one-thousandth of a millimetre across and the size of typical insects on land. Herbivorous zooplankton graze on phytoplankton; carnivorous zooplankton eat herbivorous zooplankton. Plant and animal debris from these living systems sinks in the ocean. While sinking, some decomposes and returns to the water as nutrients, some (perhaps about one per cent) reaches the deep ocean or the ocean floor, where it is lost to the carbon cycle for hundreds, thousands or even millions of years. The net effect of the 'biological pump' is to move carbon from the surface waters to lower levels in the ocean. As the amount of carbon in the surface waters is reduced, more carbon dioxide from the atmosphere can be drawn down in order to restore the surface equilibrium. It is thought that the 'biological pump' has remained substantially constant in its operation during the last century unaffected by the increase in carbon dioxide availability.

Evidence of the importance of the 'biological pump' comes from the paleoclimate record from ice cores (see Chapter 4). One of the constituents from the atmosphere trapped in bubbles in the ice is the gas methyl sulphonic acid, which originates from decaying ocean plankton; its concentration is therefore an indicator of plankton activity. As the global temperature began to increase when the last Ice Age receded nearly 20 000 years ago and as the carbon dioxide in the atmosphere began to increase (Figure 4.4), the methyl sulphonic acid concentration decreased. An interesting link is thereby provided between the carbon dioxide in the atmosphere and marine biological activity. During the cold periods of the Ice Ages, enhanced biological activity in the ocean could have been responsible for maintaining the atmospheric carbon dioxide at a lower level of concentration – the 'biological pump' was having an effect.

There is some evidence from the paleo record of the biological activity in the ocean being stimulated by the presence of iron-containing dust blown over the oceans from the land surface. This has led to some proposals in recent years to enhance the 'biological pump' through artificially introducing iron over suitable parts of the ocean. While an interesting idea, it seems from careful studies that even a very large-scale operation would not have a large practical effect.

The question then remains as to why the Ice Ages should be periods of greater marine biological activity than the warm periods in between. A British oceanographer, Professor John Woods, has suggested that the key may lie in what happens in the winter as nutrients are fed into the upper ocean ready for the spring bloom. When there is less atmospheric carbon dioxide, the cooling by radiation from the surface of the ocean increases. Since convection in the upper layers of the ocean is driven by cooling at the surface, the increased cooling results in a greater depth of the mixed layer near the top of the ocean where all the biological activity occurs. This is an example of a positive biological feedback; a greater depth of layer means more plankton growth. Woods calls it the 'plankton multiplier'.[4]

This process, whose contribution to the carbon cycle is known as the *biological pump* (see box), was important in determining the changes of carbon dioxide concentration in both the atmosphere and the ocean during the ice ages (see Chapter 4).

Computer models – which calculate solutions for the mathematical equations describing a given physical situation, in order to predict its behaviour (see Chapter 5) – have been set up to describe in detail the exchanges of carbon between the atmosphere and different parts of the ocean. To test the validity of these models, they have also been applied to the dispersal in the ocean of the carbon isotope ^{14}C that entered the ocean after the nuclear tests of the 1950s; the models simulate this dispersal quite well. From the model results, it is estimated that about 2 Gt (± 0.8 Gt) of the carbon dioxide added to the atmosphere each year ends up in the oceans (Table 3.1 and Figure 3.3). Observations of the relative distribution of the other isotopes of carbon in the atmosphere and in the oceans also confirm this estimate (see Box below).

Further information regarding the broad partitioning of added atmospheric carbon dioxide between the atmosphere, the oceans and the land biota as presented in Table 3.1 comes from comparing the trends in atmospheric carbon dioxide concentration with the trends in very accurate measurements of the atmospheric oxygen/nitrogen ratio.[5] This possibility arises because the relation between the exchanges of carbon dioxide and oxygen with the atmosphere over land is different to that over the ocean. On land, living organisms through photosynthesis take in carbon dioxide from the atmosphere and build up carbohydrates, returning the oxygen to the atmosphere. In the process of respiration they also take in oxygen from the atmosphere and convert it to carbon dioxide. In the ocean, by contrast, carbon dioxide taken from the atmosphere is dissolved, both the carbon and the oxygen in the molecules being removed. How such measurements can be interpreted for the period 1990–4 is shown in Figure 3.4. These data are consistent with budget for the 1990s shown in Table 3.1.

The global land–atmosphere flux in Table 3.1 represents the balance of a net flux due to land-use changes which has generally been positive or a source of carbon to the atmosphere (Figure 3.3) and a residual component that is, by inference, a negative flux or carbon sink. The estimates of land-use changes (Table 3.1) are dominated by deforestation in tropical regions although some uptake of carbon has occurred through forest regrowth in temperate regions of the northern hemisphere and other changes in land management. The main processes that contribute to the residual carbon sink are believed to be the carbon dioxide 'fertilisation' effect (increased carbon dioxide in the atmosphere leads to increased growth in some plants – see box in Chapter 7 on page 166),

What we can learn from carbon isotopes

Isotopes are chemically identical forms of the same element but with different atomic weights. Three isotopes of carbon are important in studies of the carbon cycle: the most abundant isotope ^{12}C which makes up 98.9 per cent of ordinary carbon, ^{13}C present at about 1.1 per cent and the radioactive isotope ^{14}C which is present only in very small quantities. About 10 kg of ^{14}C is produced in the atmosphere each year by the action of particle radiation from the Sun; half of this will decay into nitrogen over a period of 5730 years (the 'half-life' of ^{14}C).

When carbon in carbon dioxide is taken up by plants and other living things, less ^{13}C is taken up in proportion than ^{12}C. Fossil fuel such as coal and oil was originally living matter so also contains less ^{13}C (by about eighteen parts per thousand) than the carbon dioxide in ordinary air in the atmosphere today. Adding carbon to the atmosphere from burning forests, decaying vegetation or fossil fuel will therefore tend to reduce the proportion of ^{13}C.

Because fossil fuel has been stored in the Earth for much longer than 5730 years (the half-life of ^{14}C), it contains no ^{14}C at all. Therefore, carbon from fossil fuel added to the atmosphere reduces the proportion of ^{14}C the atmosphere contains.

By studying the ratio of the different isotopes of carbon in the atmosphere, in the oceans, in gas trapped in ice cores and in tree rings, it is possible to find out where the additional carbon dioxide in the atmosphere has come from and also what amount has been transferred to the ocean. For instance, it has been possible to estimate for different times how much carbon dioxide has entered the atmosphere from the burning or decay of forests and other vegetation and how much from fossil fuels.

Similar isotopic measurements on the carbon in atmospheric methane provide information about how much methane from fossil fuel sources has entered the atmosphere at different times.

the effects of increased use of nitrogen fertilisers and of some changes in climate. The magnitudes of these contributions (Table 3.1 and Figure 3.3) are difficult to estimate directly and are subject to much more uncertainty than their total, which can be inferred from the requirement to balance the overall carbon cycle budget.

A clue to the uptake of carbon by the land biosphere is provided from observations of the atmospheric concentration of carbon dioxide which, each year, show a regular cycle; the seasonal variation, for instance, at the observatory site at Mauna Loa in Hawaii approaches about 10 ppm. Carbon dioxide is removed from the atmosphere during the growing season and is returned as the vegetation dies away in the winter. Since there

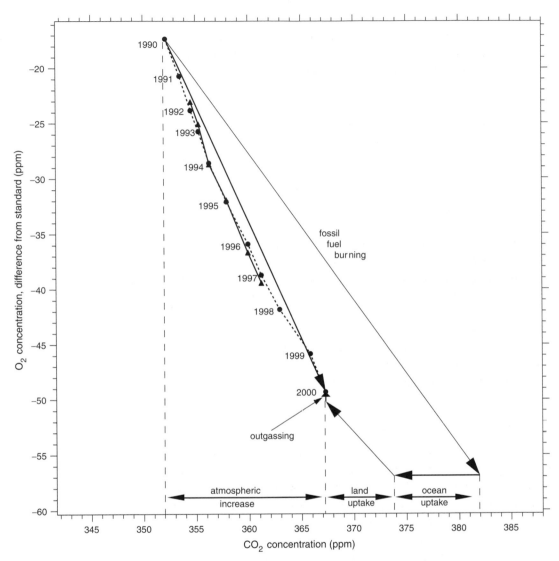

Figure 3.4 Partitioning of fossil fuel carbon dioxide uptake using oxygen measurements. Shown is the relationship between changes in carbon dioxide and oxygen concentrations. Observations are shown by solid circles and triangles. The arrow labelled 'fossil fuel burning' denotes the effect of the combustion of fossil fuels based on the $O_2 : CO_2$ stoichiometric relation of the different fuel types. Uptake by land and ocean is constrained by the stoichiometric ratio associated with these processes, defining the slopes of the respective arrows.

is more vegetation growth in the northern hemisphere than the southern, a minimum in the annual cycle of carbon dioxide in the atmosphere occurs in the northern summer. Estimates from carbon cycle models of the uptake by the land biosphere are constrained by these observations of the difference between the hemispheres.[6]

The carbon dioxide fertilisation effect is an example of a biological feedback process. It is a negative feedback because, as carbon dioxide increases, it tends to increase the take-up of carbon dioxide by plants and therefore reduce the amount in the atmosphere, decreasing the rate of global warming. Positive feedback processes, which would tend to accelerate the rate of global warming, also exist; in fact there are more potentially positive processes than negative ones (see box on page 40). Although scientific knowledge cannot yet put precise figures on them, there are strong indications that some of the positive feedbacks could be large, especially if carbon dioxide were to continue to increase, with its associated global warming, through the twenty-first century into the twenty-second. Carbon dioxide provides the largest single contribution to anthropogenic radiative forcing. Its radiative forcing from pre-industrial times to the present is shown in Figure 3.8. A useful formula for the radiative forcing R from atmospheric carbon dioxide when its atmospheric concentration is C ppm is: $R = 5.3 \ln (C/C_0)$ where C_0 is its pre-industrial concentration of 280 ppm.

Future emissions of carbon dioxide

To obtain information about future climate we need to estimate the future atmospheric concentrations of carbon dioxide, which depend on future anthropogenic emissions. In these estimates, the long time constants associated with the response of atmospheric carbon dioxide to change have important implications. Suppose, for instance, that all emissions into the atmosphere from human activities were suddenly halted. No sudden change would occur in the atmospheric concentration, which would decline only slowly. We could not expect it to approach its pre-industrial value for several hundred years.

But emissions of carbon dioxide are not halting, nor are they slowing; their increase is, in fact, becoming larger each year. The atmospheric concentration of carbon dioxide will therefore also increase more rapidly. Later chapters (especially Chapter 6) will present estimates of climate change during the twenty-first century due to the increase in greenhouse gases. A prerequisite for such estimates is the knowledge of what changes in carbon dioxide emissions there are likely to be. Estimating what will happen in the future is, of course, not easy. Because nearly everything we do has an influence on the emissions of carbon dioxide, it means

Feedbacks in the biosphere

As the greenhouse gases carbon dioxide and methane are added to the atmosphere because of human activities, biological or other feedback processes occurring in the biosphere (such as those that arise from the climate change that has been induced) influence the rate of increase of the atmospheric concentration of these gases. These processes will either tend to add to the anthropogenic increase (positive feedbacks) or to subtract from it (negative feedbacks).

Two feedbacks, one positive (the plankton multiplier in the ocean) and one negative (carbon dioxide fertilisation), have already been mentioned in the text. Three other positive feedbacks are potentially important, although our knowledge is currently insufficient to quantify them precisely.

One is the effect of higher temperatures on respiration, especially through microbes in soils, leading to increased carbon dioxide emissions. Evidence regarding the magnitude of this effect has come from studies of the short-term variations of atmospheric carbon dioxide that have occurred during El Niño events and during the cooler period following the Pinatubo volcanic eruption in 1991. These studies, which covered variations over a few years, indicate a relation such that a change of 5 °C in average temperature leads to a forty per cent change in global average respiration rate[7] – a substantial effect. A question that needs to be resolved is whether this relation still holds over longer-term changes of the order of several decades to a century.

A second positive feedback is the reduction of growth or the die-back especially in forests because of the stress caused by climate change, which may be particularly severe in Amazonia[8] (see box in Chapter 7 on page 173). As with the last effect, this will increase as the amount of climate change becomes larger. A number of carbon cycle models show that, through these two effects, during the second half of the twenty-first century the residual terrestrial sink (Table 3.1) could change sign and become a substantial net source (Figure 3.5).

The third positive feedback is the release of methane, as temperatures increase – from wetlands and from very large reservoirs of methane trapped in sediments in a hydrate form (tied to water molecules when under pressure) – mostly at high latitudes. Methane has been generated from the decomposition of organic matter present in these sediments over many millions of years. Because of the depth of the sediments this latter feedback is unlikely to become operative to a significant extent during the twenty-first century. However, were global warming to continue to increase substantially for more than a hundred years, releases from hydrates could make a large contribution to methane emissions into the atmosphere and act as a large positive feedback on the climate.

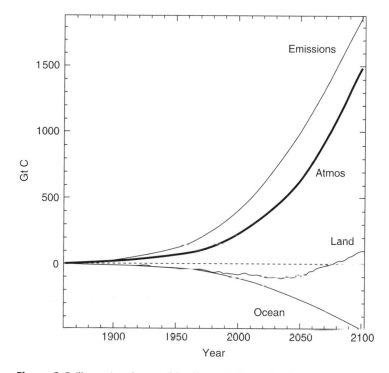

Figure 3.5 Illustrating the possible effects of climate feedbacks on the carbon cycle. Results are shown of the changing budgets of carbon (in gigatonnes of carbon) in the atmosphere, land and ocean in an ocean–atmosphere model coupled to an ocean carbon cycle model (which includes the transfer of carbon dioxide to depth through both the solubility pump and the biological pump) and a dynamic global vegetation model (which includes the exchange of carbon with the soil and with five different types of plant). The model was run with the fossil fuel carbon dioxide emissions from 1860 to the present and then projected to 2100 assuming the IS 92a scenario shown in Figure 6.1. Note that because of climate feedbacks, the terrestrial biosphere changes from being a net sink of carbon to being a net source around the middle of the twenty-first century. Note also as this source becomes stronger, by 2100 the atmospheric carbon content is increasing at about the same rate as the total emissions (i.e. the 'airborne fraction', or the fraction of fossil fuel emissions that remains in the atmosphere, has changed from being about a half in the year 2000 to being about unity in 2100). Note also that an atmospheric carbon content of 1500 Gt more than it was in 1860 is equivalent to a concentration of nearly 1000 ppm.

estimating how human beings will behave and what their activities are likely to be. For instance, assumptions need to be made about population growth, economic growth, energy use, the development of energy sources and the likely influence of pressures to preserve the environment. These assumptions are required for all countries of the world, both developing as well as developed ones. Further, since any assumptions made are unlikely to be fulfilled accurately in practice, it is necessary to make a variety of different assumptions, so that we can get some idea of the range of possibilities. Such possible futures are called *scenarios*.

In Chapter 6 are presented two sets of emission scenarios as developed respectively by the Intergovernmental Panel on Climate Change (IPCC) and the World Energy Council (WEC). These emission scenarios are then turned into future projections of atmospheric carbon dioxide concentrations through the application of a computer model of the carbon cycle that includes descriptions of all the exchanges already mentioned. Further in that chapter, through the application of computer models of the climate (see Chapter 5), projections of the resulting climate change from different scenarios are also presented.

Other greenhouse gases

Methane

Methane is the main component of natural gas. Its common name used to be marsh gas because it can be seen bubbling up from marshy areas where organic material is decomposing. Data from ice cores show that for at least two thousand years before 1800 its concentration in the atmosphere was about 700 ppb. Since then its concentration has more than doubled (Figure 3.6). During the 1980s it was increasing at about 10 ppb year^{-1} but during the 1990s the average rate of increase fell to around 5 ppb year^{-1}.[9] Although the concentration of methane in the atmosphere is much less than that of carbon dioxide (less than 2 ppm compared with about 370 ppm for carbon dioxide), its greenhouse effect is far from negligible. That is because the enhanced greenhouse effect caused by a molecule of methane is about eight times that of a molecule of carbon dioxide.[10]

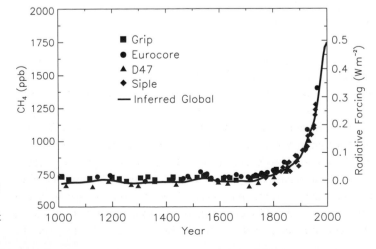

Figure 3.6 Change in methane concentration (mole fraction in ppb) over the last 1000 years determined from ice cores, glacier ice and whole air samples. Radiative forcing since the pre-industrial era due to the methane increase is plotted on the right axis.

Table 3.2 *Estimated sources and sinks of methane in millions of tonnes per year.*[a] *The first column of data shows the best estimate from each source; the second column illustrates the uncertainty in the estimates by giving a range of values*

Source	Best estimate	Uncertainty
Natural		
Wetlands	150	(90–240)
Termites	20	(10–50)
Ocean	15	(5–50)
Other (including hydrates)	15	(10–40)
Human-generated		
Coal mining, natural gas, petroleum industry	100	(75–110)
Rice paddies	60	(30–90)
Enteric fermentation	90	(70–115)
Waste treatment	25	(15–70)
Landfills	40	(30–70)
Biomass burning	40	(20–60)
Sinks		
Atmospheric removal	545	(450–550)
Removal by soils	30	(15–45)
Atmospheric increase	22	(35 40)

[a] From Prather, M., Ehhalt, D. *et al.* 2001. Atmospheric chemistry and greenhouse gases. Chapter 4 in Houghton *et al.*, *Climate Change 2001*. See also Prather, M. *et al.* 1995. Other trace gases and atmospheric chemistry. In *Climate Change 1994*. Cambridge: Cambridge University Press. The figure for atmospheric increase is an average for the 1990s.

The main natural source of methane is from wetlands. A variety of other sources result directly or indirectly from human activities, for instance from leakage from natural gas pipelines and from oil wells, from generation in rice paddy fields, from enteric fermentation (belching) from cattle and other livestock, from the decay of rubbish in landfill sites and from wood and peat burning. Details of the best estimates of the sizes of these sources are shown in Table 3.2. Attached to many of the numbers is a wide range of uncertainty. It is, for instance, difficult to estimate the amount produced in paddy fields averaged on a worldwide basis. The amount varies enormously during the rice growing season and also very widely from region to region. Similar problems arise when trying to estimate the amount produced by animals. Measurements of the proportions of the different isotopes of carbon (see Box on page 37) in atmospheric methane assist considerably in helping to tie down the proportion which comes from fossil fuel sources, such as leakage from mines and from natural gas pipelines.

The main process for the removal of methane from the atmosphere is through chemical destruction. It reacts with hydroxyl (OH) radicals, which are present in the atmosphere because of processes involving sunlight, oxygen, ozone and water vapour. The average lifetime of methane in the atmosphere is determined by the rate of this loss process. At about twelve years,[11] it is much shorter than the lifetime of carbon dioxide.

Although most methane sources cannot be identified very precisely, the largest sources apart from natural wetlands are closely associated with human activities. It is interesting to note that the increase of atmospheric methane (Figure 3.6) follows very closely the growth of human population since the Industrial Revolution. However, even without the introduction of deliberate measures to control human-related sources of methane because of the impact on climate change, it is not likely that this simple relationship with human population will continue. The IPCC Special Report on Emission Scenarios (SRES) presented in Chapter 6 include a wide range of estimates of the growth of human-related methane emissions during the twenty-first century – from approximately doubling over the century to reductions of about twenty-five per cent. In Chapter 10 (page 253) ways are suggested in which methane emissions could be reduced and methane concentrations in the atmosphere stabilised.

Nitrous oxide

Nitrous oxide, used as a common anaesthetic and known as laughing gas, is another minor greenhouse gas. Its concentration in the atmosphere of about 0.3 ppm is rising at about 0.25 per cent per year and is about sixteen per cent greater than in pre-industrial times. It possesses a relatively long atmospheric lifetime of about 115 years. The largest emissions to the atmosphere are associated with natural and agricultural ecosystems; those linked with human activities are probably due to increasing fertiliser use. Biomass burning and the chemical industry (for example, nylon production) also play some part. The sink of nitrous oxide is photodissociation in the stratosphere and reaction with electronically excited oxygen atoms, leading to an atmospheric lifetime of about 120 years.

Chlorofluorocarbons (CFCs) and ozone

The CFCs are man-made chemicals which, because they vaporise just below room temperature and because they are non-toxic and non-flammable, appear to be ideal for use in refrigerators, the manufacture of insulation and aerosol spray cans. Since they are so chemically unreactive, once they are released into the atmosphere they remain for a long time – one or two hundred years – before being destroyed. As their use increased rapidly through the 1980s their concentration in the atmosphere has been building up so that they are now present (adding together all the different CFCs) in about 1 ppb (part per thousand million – or billion – by volume). This may not sound very much, but it is quite enough to cause two serious environmental problems.

The first problem is that they destroy ozone.[12] Ozone (O_3), a molecule consisting of three atoms of oxygen, is an extremely reactive gas present in small quantities in the stratosphere (a region of the atmosphere between about 10 km and 50 km in altitude). Ozone molecules are formed through the action of ultraviolet radiation from the Sun on molecules of oxygen. They are in turn destroyed by a natural process as they absorb solar ultraviolet radiation at slightly longer wavelengths – radiation which would otherwise be harmful to us and to other forms of life at the Earth's surface. The amount of ozone in the stratosphere is determined by the balance between these two processes, one forming ozone and one destroying it. What happens when CFC molecules move into the stratosphere is that some of the chlorine atoms they contain are stripped off, also by the action of ultraviolet sunlight. These chlorine atoms readily react with ozone, reducing it back to oxygen and adding to the rate of destruction of ozone. This occurs in a catalytic cycle – one chlorine atom can destroy many molecules of ozone.

The problem of ozone destruction was brought to world attention in 1985 when Joe Farman, Brian Gardiner and Jonathan Shanklin at the British Antarctic Survey discovered a region of the atmosphere over Antarctica where, during the southern spring, about half the ozone overhead disappeared. The existence of the 'ozone hole' was a great surprise to the scientists; it set off an intensive investigation into its causes. The chemistry and dynamics of its formation turned out to be complex. They have now been unravelled, at least as far as their main features are concerned, leaving no doubt that chlorine atoms introduced into the atmosphere by human activities are largely responsible. Not only is there depletion of ozone in the spring over Antarctica (and to a lesser extent over the Arctic) but also substantial reduction, of the order of five per cent, of the total column of ozone – the amount above one square metre at a given point on the Earth's surface – at mid latitudes in both hemispheres.

Because of these serious consequences of the use of CFCs, international action has been taken. Many governments have signed the Montreal Protocol set up in 1987 which, together with the Amendments agreed in London in 1991 and in Copenhagen in 1992, required that manufacture of CFCs be phased out completely by the year 1996 in industrialised countries and by 2006 in developing countries. Because of this action the concentration of CFCs in the atmosphere is no longer increasing. However, since they possess a long life in the atmosphere, little decrease will be seen for some time and substantial quantities will be present well over a hundred years from now.

So much for the problem of ozone destruction. The other problem with CFCs and ozone, the one which concerns us here, is that they are both greenhouse gases.[13] They possess absorption bands in the region known as the longwave atmospheric window (see Figure 2.4) where few other gases absorb. Because, as we have seen, the CFCs destroy some ozone, the greenhouse effect of the CFCs is partially compensated by the reduced greenhouse effect of atmospheric ozone.

First considering the CFCs on their own, a CFC molecule added to the atmosphere has a greenhouse effect five to ten thousand times greater than an added molecule of carbon dioxide. Thus, despite their very small concentration compared, for instance, with carbon dioxide, they have a significant greenhouse effect. It is estimated that due to the CFCs now present in the atmosphere the radiative forcing in the tropics (at higher latitudes there is a compensating effect due to ozone reduction which is explained below) is about 0.25 W m^{-2} – or about twenty per cent of the radiative forcing due to all greenhouse gases. This forcing will only decrease very slowly next century.

Turning now to ozone, the effect from ozone depletion is complex because the amount by which ozone greenhouse warming is reduced depends critically on the height in the atmosphere at which it is being destroyed. Further, ozone depletion is concentrated at high latitudes while the greenhouse effect of the CFCs is uniformly spread over the globe. In tropical regions there is virtually no ozone depletion so no change in the ozone greenhouse effect. At mid latitudes, very approximately, the greenhouse effects of ozone reduction and of the CFCs compensate for each other. In polar regions the reduction in the greenhouse effect of ozone more than compensates for the greenhouse warming effect of the CFCs.[14]

As CFCs are phased out, they are being replaced to some degree by other halocarbons – hydrochloro-fluorocarbons (HCFCs) and hydro-fluorocarbons (HFCs). In Copenhagen in 1992, the international community decided that HCFCs would also be phased out by the year 2030. While being less destructive to ozone than the CFCs, they are still greenhouse gases. The HFCs contain no chlorine or bromine, so they do not destroy ozone and are not covered by the Montreal Protocol. Because of their shorter lifetime, typically tens rather than hundreds of years, the concentration in the atmosphere of both the HCFCs and the HFCs, and therefore their contribution to global warming for a given rate of emission, will be less than for the CFCs. However, since their rate of manufacture could increase substantially their potential contribution to greenhouse warming is being included alongside other greenhouse gases (see Chapter 10 page 247).

Concern has also extended to some other related compounds which are greenhouse gases, the perfluorocarbons (e.g. CF_4, C_2F_6) and sulphur hexafluoride (SF_6), which are produced in some industrial processes. Because they possess very long atmospheric lifetimes, probably more than 1000 years, all emissions of these gases accumulate in the atmosphere and will continue to influence climate for thousands of years. They are also therefore being included as potentially important greenhouse gases.

Ozone is also present in the lower atmosphere or troposphere, where some of it is transferred downwards from the stratosphere and where some is generated by chemical action, particularly as a result of the action of sunlight on the oxides of nitrogen. It is especially noticeable in polluted atmospheres near the surface; if present in high enough concentration, it can become a health hazard. In the northern hemisphere the limited observations available together with model simulations of the chemical reactions leading to ozone formation suggest that ozone concentrations in the troposphere have doubled since pre-industrial times – an increase which is estimated to have led to a global average radiative forcing of between 0.2 and 0.6 W m^{-2} (Figure 3.8). Ozone is also generated at levels in the upper troposphere as a result of the nitrogen oxides emitted from aircraft exhausts; nitrogen oxides emitted from aircraft are more effective at producing ozone in the upper troposphere than are equivalent emissions at the surface. The radiative forcing in northern mid latitudes from aircraft due to this additional ozone[15] is of similar magnitude to that from the carbon dioxide emitted from the combustion of aviation fuel which is about three per cent of current global fossil fuel consumption.

Gases with an indirect greenhouse effect

I have described all the gases present in the atmosphere which have a direct greenhouse effect. There are also gases which through their chemical action on greenhouse gases, for instance on methane or on lower atmospheric ozone, have an influence on the overall size of greenhouse warming. Carbon monoxide (CO) and the nitrogen oxides (NO and NO_2) emitted, for instance, by motor vehicles are some of these. Carbon monoxide has no direct greenhouse effect of its own but, as a result of chemical reactions, it forms carbon dioxide. These reactions also affect the amount of the hydroxyl radical (OH) which in turn affects the concentration of methane. Substantial research has been carried out on the chemical processes in the atmosphere that lead to these indirect effects on greenhouse gases.[16] It is of course important to take them properly into account, but it is also important to recognise that their combined effect is much less than that of the major contributors to human-generated greenhouse warming, namely carbon dioxide and methane.

Particles in the atmosphere[17]

Small particles suspended in the atmosphere (often known as *aerosol*) affect its energy balance because they both absorb radiation from the Sun and scatter it back to space. We can easily see the effect of this on a bright day in the summer with a light wind when downwind of an industrial area. Although no cloud appears to be present, the sun appears hazy. We call it 'industrial haze'. Under these conditions a significant proportion of the sunlight incident at the top of the atmosphere is being lost as it is scattered back and out of the atmosphere by the millions of small particles (typically between 0.001 and 0.01 mm in diameter) in the haze.

Atmospheric particles come from a variety of sources. They arise partially from natural causes; they are blown off the land surface, especially in desert areas; they result from forest fires and they come from sea spray. From time to time large quantities of particles are injected into the upper atmosphere from volcanoes – the Pinatubo volcano which erupted in 1991 provides a good example (see Chapter 5). Some particles are also formed in the atmosphere itself, for instance sulphate particles from the sulphur-containing gases emitted from volcanoes.

Other particles arise from human activities – from biomass burning (e.g. the burning of forests) and the sulphates and soot resulting from the burning of fossil fuels. The sulphate particles are particularly important. They are formed as a result of chemical action on sulphur dioxide, a gas that is produced in large quantities by power stations and other industries in which coal and oil (both of which contain sulphur in varying quantities) are burnt. Because these particles remain in the atmosphere only for about five days on average, their effect is mainly confined to regions near the sources of the particles, i.e. the major industrial regions of the northern hemisphere (Figure 3.7). Over limited regions of the northern hemisphere the radiative effect of these particles is comparable in size, although opposite in effect, to that of human-generated greenhouse gases up to the present time. Estimates of the direct radiative forcing, averaged over the globe, due to the particles from various human-generated sources are shown in Figure 3.8. It will be seen that there are substantial uncertainties associated with these estimates.

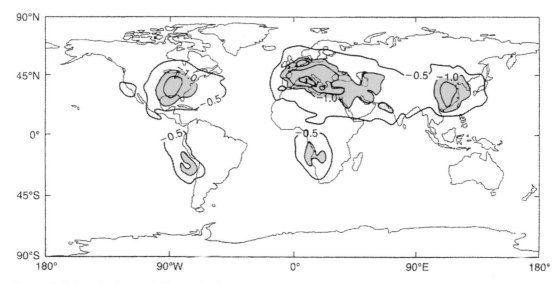

Figure 3.7 Modelled geographic distribution of estimates of the annual mean direct radiative forcing (in watts per square metre) from anthropogenic sulphate aerosols in the troposphere. The radiative forcing, which is negative, is largest in regions close to industrial activity.

So far for aerosol we have been describing *direct* radiative forcing. There is a further way by which particles in the atmosphere could influence the climate; that is through their effect on cloud formation that is described as *indirect* radiative forcing. Two mechanisms of indirect forcing have been proposed. The first is the influence of the number of particles and their size on cloud radiative properties. It arises as follows. If particles are present in large numbers when clouds are forming, the resulting cloud consists of a large number of smaller drops – smaller than would otherwise be the case – this is similar to what happens as polluted fogs form in cities. Such a cloud will be more highly reflecting to sunlight than one consisting of larger particles, thus further increasing the energy loss resulting from the presence of the particles. The second mechanism arises because of the influence of droplet size and number on precipitation efficiency, the lifetime of clouds and hence the geographic extent of cloudiness. There is observational evidence for both of these mechanisms but the processes involved are difficult to model and will vary a great deal with the particular situation. Estimates of their magnitude as shown in Figure 3.8 therefore remain very uncertain. To refine these estimates, more studies need to be made especially by making careful measurements on suitable clouds.

The estimates for the radiative effects of particles as in Figure 3.8 can be compared with the global average radiative forcing to date due

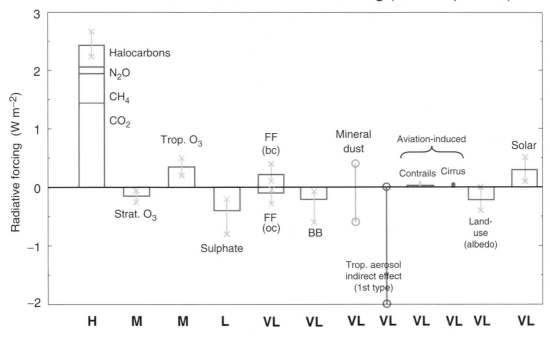

Figure 3.8 Global, annual mean radiative forcings ($W m^{-2}$) due to a number of agents for the period from pre-industrial (1750) to 2000. The height of the rectangular bar denotes a best estimate value while its absence denotes no best estimate is possible because of large uncertainties. The vertical lines with 'x' or 'o' delimiters indicates estimates of the uncertainty ranges. A 'level of scientific understanding (LOSU)' index is accorded to each forcing, with H, M, L and VL denoting high, medium, low and very low levels respectively. This represents a judgement about the reliability of the forcing estimate involving factors such as the assumptions necessary to evaluate the forcing, the degree of knowledge of the mechanisms determining the forcing and the uncertainties surrounding the quantitative estimate of the forcing. The well-mixed greenhouse gases are grouped together into a single rectangular bar with the individual contributions shown. The second and third bars apply to stratospheric and tropospheric ozone. The next bars denote the direct effect of aerosols. FF denotes aerosol from fossil fuel burning and is separated into black carbon (bc) and organic carbon (oc) components. BB denotes aerosols from biomass burning. The sign of the effects due to mineral dust is itself an uncertainty. Only the first indirect aerosol effect is estimated as little quantitative evidence exists regarding the second. All the forcings have distinct spatial and seasonal variations (see Figure 3.7) so that they cannot be added up and viewed a priori as providing offsets in terms of complete global climate impact.

to the increase in greenhouse gases of about 2.6 W m^{-2}. Comparing global average forcings, however, is not the whole story. Because of the large regional variation of particles in the atmosphere (Figure 3.7), any effect they have on the climate can be expected to be substantially different from the effect of increases in greenhouse gases, which is much more uniform over the globe. This will be illustrated in Chapter 5 where we consider the likely pattern of anthropogenic climate change to date. More consideration of it will also be given in Chapter 6 when discussing projections of climate change on regional scales that can depend critically on assumptions about the likely concentrations of atmospheric particles in the future.

An important factor that will influence the future concentrations of sulphate particles is 'acid rain' pollution, caused mainly by sulphur dioxide emissions. This leads to the degradation of forests and fish stocks in lakes especially in regions downwind of major industrial areas. Serious efforts are therefore under way, especially in Europe and North America, to curb these emissions to a substantial degree. Although the amount of sulphur-rich coal being burnt elsewhere in the world, for instance in Asia, is increasing rapidly, the damaging effects of sulphur pollution are such that tight controls on sulphur emissions are being extended to these regions also. For the globe as a whole therefore, sulphur emissions are likely to rise much less rapidly than emissions of carbon dioxide; the IPCC SRES scenarios of future emissions as presented in Chapter 6 anticipate these changes. The climate change resulting from an increase in sulphate particles therefore will become increasingly less by comparison with that from the likely increase of greenhouse gases.

Estimates of radiative forcing

This chapter has summarised current scientific knowledge about the sources and sinks of the main greenhouse gases and the exchanges which occur between the components of the climate system – the atmosphere, the ocean and the land surface – including the close balances that are maintained between the different components and the way in which these balances are being disturbed by human-generated emissions. Different assumptions about future emissions have been used to generate emission scenarios. From these scenarios estimates have been made (for carbon dioxide, for instance, using a computer model of the carbon cycle) of likely increases in greenhouse gas concentrations in the future.

Given information about the possible increases in greenhouse gases, the next step is to calculate the effect of these increases on the amounts of thermal (infrared) radiation absorbed and emitted by the atmosphere.

This is done using information about how the different gases absorb radiation in the infrared part of the spectrum, as mentioned in Chapter 2. The radiative forcing associated with the increases in each of the gases can then be calculated. In Figure 3.8 are brought together estimates of global average radiative forcing for the period from 1750 to 2000 for the different greenhouse gases and for tropospheric aerosols of different origins which we have been considering in this chapter.

It is useful to be able to compare the radiative forcing generated by the different greenhouse gases. Because of their different life-times, the future profile of radiative forcing due to releases of greenhouse gases varies very much from gas to gas. An index called the global warming potential (GWP) has been defined for greenhouse gases that takes the ratio of the time-integrated radiative forcing from the instantaneous release of 1 kg of a given gas to that from the release of 1 kg of carbon dioxide. A time horizon has also to be specified for the period over which the integration is carried out. The GWPs of the six greenhouse gases included in the Kyoto Protocol are listed in Table 10.2. Applying the GWPs to the emissions from a mixture of greenhouse gases enables the mixture to be considered in terms of an equivalent amount of carbon dioxide. However, because the GWPs for different time horizons are very different, GWPs are of limited application and must be used with care.

When considering radiative forcing of climate, the question is bound to be asked as to whether variations have occurred or are likely to occur in the amount of energy from the Sun that is incident on the Earth and that could, therefore, affect the climate. We shall see, for instance, in the next chapter that ice ages in the past have been triggered by variations in the geometry of the Earth's orbit. It is considered possible that the Sun's output itself could vary by small amounts over time (see box in Chapter 6 page 138). Figure 3.8 (see also Figure 6.12) indicates the range of estimates of solar variability that may have occurred since 1850 showing that its influence is much less than that of the increase in greenhouse gases.

Also included in Figure 3.8 are the possible effects of aviation on radiative forcing and effects due to land-use changes that arise because of changes in the albedo (see Glossary) of the surface. The effects of aviation that are in addition to the source of carbon dioxide emissions it provides arise from its influence on high cloud cover through its emissions of water vapour. As we shall see in Chapter 5 (page 91) high cloud provides a blanketing effect on the Earth's surface similar to that of greenhouse gases and therefore leads to positive radiative forcing. Many examples exist of extensive contrail formation over regions where many aircraft flights regularly occur. Aircraft may also influence cirrus cloud formation through the effect of the particles in aircraft emissions. The overall greenhouse effect of aircraft has been estimated as up to the equivalent of two or three times the effect of their carbon dioxide emissions.[18]

Details of radiative forcings as projected for the twenty-first century are presented in Chapter 6 (page 120). Chapters 5 and 6 will explain how estimates of radiative forcing can be incorporated into computer climate models so as to predict the climate change that is likely to occur because of human activities. However, before considering predictions of future climate change, it is helpful to gain perspective by looking at some of the climate changes that have occurred in the past.

Questions

1 The lifetime of a carbon dioxide molecule in the atmosphere before it is exchanged with the ocean is typically less than a year, while the time taken for an increase in carbon dioxide concentration from fossil fuel burning to diminish substantially is typically many years. Explain the reasons for this difference.

2 Estimate how much carbon dioxide you emit each year through breathing.

3 Estimate the size of your share of carbon dioxide emissions from the burning of fossil fuels.

4 A typical city in the developed world with a population of about one million produces about half a million tonnes of municipal waste each year. Suppose the waste is buried in a landfill site where the waste decays producing equal quantities of carbon dioxide and methane. Making assumptions about the likely carbon content of the waste and the proportion that eventually decays, estimate the annual production of methane. If all the methane leaks away, using the information in note 10, compare the greenhouse effect of the carbon dioxide and methane produced from the landfill site with that of the carbon dioxide produced if the waste were incinerated instead.

5 A new forest is planted containing a million trees which will mature in forty years. Estimate the amount of carbon sequestered per year by the forest.

6 Find figures for the amount of fuel used by a typical aircraft and the size of fleets of the world's airlines and airforces and estimate the carbon dioxide emitted globally each year by the world's aircraft.

7 Search for information about the ozone hole and explain why it occurs mainly in the Antarctic.

8 What are the main uses of CFCs? Suggest ways in which the emissions of CFCs to the atmosphere could be reduced more rapidly.

9 Evidence is sometimes presented suggesting that the variations in global average temperature over the last century or more can all be explained as due to variations in the energy output of the Sun. There is therefore nothing left to attribute to the increase in greenhouse gases. What is the fallacy in this argument?

10 With the use of the formula given in the text, calculate the radiative forcing due to carbon dioxide for atmospheric concentrations of 150, 280, 450, 560 and 1000 ppm.

Notes for Chapter 3

1 It is convenient to define radiative forcing as the radiative imbalance at the top of the troposphere rather than at the top of the whole atmosphere.

2 Reference sources for Table 3.1 are as follows:

Prentice, I. C. *et al.* 2001. The carbon cycle and atmospheric carbon dioxide. Chapter 3 in Houghton, J. T., Ding, Y., Griggs, D. J., Noguer, M., van der Linden, P. J., Dai, X., Maskell, K., Johnson, C. A. (eds.) *Climate Change 2001: The Scientific Basis. Contribution of Working Group I to the Third Assessment Report of the Intergovernmental Panel on Climate Change.* Cambridge: Cambridge University Press.

House, J. I. *et al.* 2003. Reconciling apparent inconsistencies in estimates of terrestrial CO_2 sources and sinks. *Tellus*, **55B**, pp. 345–63.

See also Table 1.2 in Watson, R.T., Noble, I. R., Bolin, B., Ravindranath, N. H., Verardo, D. J., Dokken, D. J. (eds.) *Land Use, Land-use Change, and Forestry*, Chapter 1. Cambridge: Cambridge University Press.

3 See Prentice *et al.*, *Climate Change 2001*.

4 Woods, J., Barkmann, W. 1993. The Plankton multiplier – positive feedback in the greenhouse. *Journal of Plankton Research*, **15**, pp. 1053–74.

5 Keeling, R. F., Piper, S. C., Heimann, M. 1996. Global and hemispheric sinks deduced from changes in atmospheric O_2 concentration. *Nature*, **381**, pp. 218–21.

6 For more details see House, J. I. *et al.* 2003. Reconciling apparent inconsistencies in estimates of terrestrial CO_2 sources and sinks. *Tellus*, **55B**, pp. 345–63.

7 See Jones, C. D. and Cox, P. M. 2001. Atmospheric science letters. doi.1006/asle. 2001.0041; the respiration rate varies approximately with a factor $2^{(T-10)/10}$.

8 Cox, P. M. *et al.* 2004. Amazon die-back under climate-carbon cycle projections for the 21st century. *Theoretical and Applied Climatology*, in press.

9 In 1992 the increase slowed to almost zero. The reason for this is not known but one suggestion is that, because of the collapse of the Russian economy, the leakage from Siberian natural gas pipelines was much reduced.

10 The ratio of the enhanced greenhouse effect of a molecule of methane compared to a molecule of carbon dioxide is known as its global warming potential (GWP); a definition of GWP is given later in the chapter. The figure of about 8 given here for the GWP of methane is for a time horizon of 100 years – see Lelieveld, J., Crutzen, P. J. *Nature*, **355**, 1992, pp. 339–41; see also Prather *et al.*, Chapter 4, and Ramaswamy *et al.*, Chapter 6 in, Houghton *et al.*, *Climate Change 2001*. The GWP is also often expressed as the ratio of the effect for unit mass of each gas in which case the GWP for methane (whose molecular mass is 0.36 of that of carbon dioxide) becomes

about 23 for the 100 year time horizon. About seventy-five per cent of the contribution of methane to the greenhouse effect is because of its direct effect on the outgoing thermal radiation. The other twenty-five per cent arises because of its influence on the overall chemistry of the atmosphere. Increased methane eventually results in small increases in water vapour in the upper atmosphere, in tropospheric ozone and in carbon dioxide, all of which in turn add to the greenhouse effect.

11 Taking into account the loss processes due to reaction with OH in the troposphere, chemical reactions and soil loss lead to a lifetime of about ten years. However, the effective lifetime of methane against a perturbation in concentration in the atmosphere (the number quoted here) is complex because it depends on the methane concentration. This is because the concentration of the radical OH (interaction with which is the main cause of methane destruction), due to chemical feedbacks, is itself dependent on the methane concentration (more details in Prather *et al.*, Chapter 4, in Houghton *et al*, *Climate Change 2001*).

12 For more detail, see *Scientific Assessment of Ozone Depletion: 1998. Global Ozone Research and Monitoring Project   Report No 44.* Geneva: World Meteorological Organization, 732 pp.

13 Prather *et al.*, Chapter 4, in Houghton *et al.*, *Climate Change 2001*.

14 More detail on this and the radiative effects of minor gases and particles can be found in Ramaswamy *et al.*, Chapter 6, in Houghton *et al. Climate Change 2001*.

15 More detail in Penner, J. E. *et al.* (eds.) 1999. *Aviation and the Global Atmosphere*. An IPCC Special Report. Cambridge: Cambridge University Press.

16 Prather *et al.*, Chapter 4, in Houghton *et al.*, *Climate Change 2001*.

17 Penner *et al.*, Chapter 5, in Houghton *et al.*, *Climate Change 2001*.

18 For more complete information of the effects of aviation see Penner, *Aviation and the Global Atmosphere*.

Chapter 4
Climates of the past

To obtain some perspective against which to view future climate change, it is helpful to look at some of the climate changes that have occurred in the past. This chapter will briefly consider climatic records and climate changes in three periods: the last hundred years, then the last thousand years and finally the last million years. At the end of the chapter some interesting recent evidence for the existence of relatively rapid climate change at various times during the past one or two hundred thousand years will be presented.

The last hundred years

The 1980s and early 1990s have brought unusually warm years for the globe as a whole (see Chapter 1) as is illustrated in Figure 4.1, which shows the global average temperature since 1860, the period for which the instrumental record is available with good accuracy and coverage. An increase over this period has taken place of about 0.6 °C (ninety-five per cent confidence limits of 0.4 to 0.8 °C). The year 1998 is very likely[1] to have been the warmest year during this period. An even more striking statistic is that each of the first eight months of 1998 was very likely the warmest of those months in the record. Although there is a distinct trend in the record, the increase is by no means a uniform one. In fact, some periods of cooling as well as warming have occurred and an obvious feature of the record is the degree of variability from year to year and from decade to decade.

A sceptic may wonder how a diagram like Figure 4.1 can be prepared and whether any reliance can be placed upon it. After all, temperature

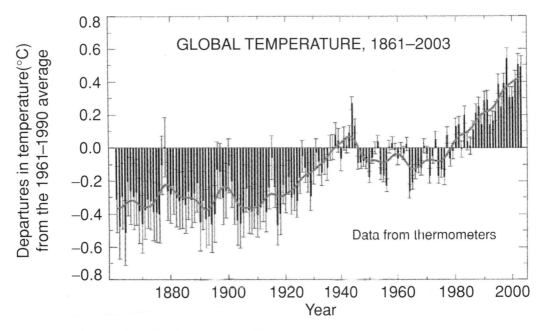

Figure 4.1 Variations of the globally averaged Earth's surface temperature over the last 140 years. The dark bars are the year by year averaged values; the grey line is a smoothed annual curve to illustrate decadal variations. Uncertainties in the data are also shown; the thin whiskers represent the 95% confidence range. The graph is based on an improved analysis of all years since the original publication in the Third IPCC Scientific Assessment. The changes are small.

varies from place to place, from season to season and from day to day by many tens of degrees. But here we are not considering changes in local temperature but in the average over the whole globe. A change of a few tenths of a degree in that average is a large change.

First of all, just how is a change in global average temperature estimated from a combination of records of changes in the near-surface temperature over land and changes in the temperature of the sea surface? To estimate the changes over land, weather stations are chosen where consistent observations have been taken from the same location over a substantial proportion of the whole 130-year period. Changes in sea surface temperature have been estimated by processing over sixty million observations from ships – mostly merchant ships – over the same period. All the observations, from land stations and from ships, are then located within a grid of squares, say 1° of latitude by 1° of longitude, covering the Earth's surface. Observations within each square are averaged; the global average is obtained by averaging (after weighting them by area) over the averages for each of the squares.

A number of research groups in different countries have made careful and independent analyses of these observations. In somewhat different ways they have made allowances for factors that could have introduced artificial changes in the records. For instance, the record at some land stations could have been affected by changes in their surroundings as these have become more urban. In the case of ships, the standard method of observation used to be to insert a thermometer into a bucket of water taken from the sea. Small changes of temperature have been shown to occur during this process; the size of the changes varies between day and night and is also dependent on several other factors including the material from which the bucket is made – over the years wooden, canvas and metal buckets have been variously employed. Nowadays, a large proportion of the observations are made by measuring the temperature of the water entering the engine cooling system. Careful analysis of the effects of these details on observations both on land and from ships has enabled appropriate corrections to be made to the record, and good agreement has been achieved between analyses carried out at different centres.

Confidence that the observed variations are real is increased by noticing that the trend and the shape of the changes are similar when different selections of the total observations are made. For instance, the separate records from the land and sea surface and from the northern and southern hemispheres are closely in accord. Further indirect indicators such as changes in borehole temperatures and sub-surface ocean temperatures, decrease in snow cover and glacier shrinkage provide independent support for the observed warming.

During the last thirty years or so observations have been available from satellites orbiting around the Earth. Their great advantage is that they automatically provide data with global coverage, which are often lacking in other data sets. The length of the record from satellites, however, is generally less than twenty years, a comparatively short period in climate terms. It has been suggested that satellite measurements of lower atmospheric temperature since 1979 are not consistent with the trend of rising temperatures in surface observations. The satellite observations do not cast doubt on the accuracy of the surface measurements – the two measurements are of different quantities. However, it is expected that trends in lower atmosphere measurements and surface measurements should be related and much careful work has been carried out studying the two records and interpreting the differences (see box).

The most obvious feature of the climate record illustrated in Figure 4.1 is that of considerable variability, not just from year to year, but from decade to decade. Some of this variability will have arisen through causes external to the atmosphere and the oceans, for instance as a result of volcanic eruptions such as those of Krakatoa in 1883 or of

Atmospheric temperature observed by satellites

Since 1979 meteorological satellites flown by the National Oceanic and Atmospheric Administration (NOAA) of the United States have carried a microwave instrument, the Microwave Sounding Unit (MSU), for the remote observation of the average temperature of the lower part of the atmosphere.

Figure. 4.2(a) shows the record of global average temperature deduced from the MSU and compares it with data from sounding instruments carried on balloons for the same region of the atmosphere, showing very good agreement for the period of overlap. The record of temperature at the surface is also added for comparison over the period from 1960 to 2000. All three measurements show similar variability, the variations at the surface tracking well with those in the lower troposphere. The plots also illustrate the difficulty of deriving accurate trends from a short period of record where there is also substantial variability. Since 1979 the trends from the three measurement sources have been carefully studied and compared. They are $0.04 \pm 0.11\,°C$ per decade and $0.03 \pm 0.10\,°C$ per decade for the satellite and balloon data respectively compared with $0.16 \pm 0.06\,°C$ per decade for the surface data. The trend in the difference of the surface and lower troposphere of $0.13 \pm 0.06\,°C$ per decade is statistically significant. This is in contrast to near-zero surface temperature trends from 1958 to 1978 when the global lower troposphere temperature warmed by about $0.03\,°C$ per decade relative to the surface. There are substantial regional variations in the differences between surface and lower tropospheric temperature trends since 1979. For instance, the differences are particularly apparent in many parts of the tropics and sub-tropics where the surface has warmed faster than the lower troposphere, while over some other regions, e.g. North America, Europe and Australia, the trends are very similar.

Why the surface and lower temperature trends show significant differences, especially over the tropical and sub-tropical oceans, is not completely understood although there are a number of possible reasons for the differences.[2] It is of course well known that the presence of increased concentrations of greenhouse gases leads to cooling of the atmosphere at higher levels (see Chapter 2). In the stratosphere therefore the temperature trends are reversed (Figure 4.2(b)) ranging from a decrease of about $0.5\,°C$ per decade in the lower stratosphere to $2.5\,°C$ per decade in the upper stratosphere.

Pinatubo in the Philippines in 1991 (the low global average temperature in 1992 and 1993, compared with neighbouring years, is almost certainly due to the Pinatubo volcano). But there is no need to invoke volcanoes or other external causes to explain all the variations in the record. Many of

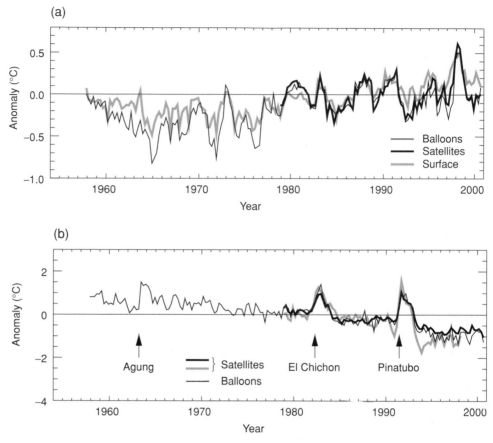

Figure 4.2 Time series of observations of global average temperature (°C) (relative to average for 1979–90) (a) for the lower troposphere based on measurements from balloons and satellites compared with measurements at the surface; (b) for the lower stratosphere from balloons and satellites.

them result from internal variations within the total climate system, for instance between different parts of the ocean. (See Chapter 5 for more details.)

The warming during the twentieth century has not been uniform over the globe. For instance, the recent warming has been greatest over Northern Hemisphere continents at mid to high latitudes. There have also been areas of cooling, for instance over some parts of the North Atlantic ocean associated with changes in ocean circulation (see Chapter 5). Some of the recent regional patterns of temperature change are related to different phases of atmosphere-ocean oscillations, such as the North Atlantic Oscillation (NAO). The positive phase of the NAO, with high pressure over the sub-tropical Atlantic and southern Europe and mild winters over northwest Europe, has tended to be dominant since the mid 1980s.

An interesting feature of the increasing temperature during the last few decades has been that, in the daily cycle of temperature, minimum temperatures over land have increased about twice as much as maximum temperatures. A likely explanation for this, in addition to the effects of enhanced greenhouse gases, is an increase in cloud cover which has been observed in many of the areas with reduced temperature range. An increase in cloud tends to obstruct daytime sunshine and tends also to reduce the escape of terrestrial radiation at night.

As might be expected the increases in temperature have led on average to increases in precipitation, although precipitation shows even more variability in both space and time than temperature. The increases have been particularly noticeable in the northern hemisphere in mid to high latitudes, often appearing particularly as increases in heavy rainfall events (see Table 4.1).

The broad features of these changes in temperature and precipitation are consistent with what is expected because of the influence of increased greenhouse gases (see Chapter 5), although there is much variability in the record that arises for reasons not associated with human activities. For instance, the particular increase from 1910 to 1940 (Figure 4.1) is too rapid to have been due to the rather small increase in greenhouse gases during that period. The particular reasons for this will be discussed in the next chapter where comparisons of observed temperatures with simulations from climate models for the whole of the twentieth century will be presented, not just as they concern the global mean but also the regional patterns of change. We conclude there that, although the expected signal is still emerging from the noise of natural variability, most of the observed warming over the last fifty years is likely to have been due to the increase in greenhouse gas concentrations.[3]

Significant cooling of the lower stratosphere (the region at altitudes between about 10 and 30 km) has been observed over the last two decades (Figure 4.2). This is to be expected both because of the decrease in the concentration of ozone (which absorbs solar radiation) and because of the increased carbon dioxide concentration which leads to increased cooling at these levels (see Chapter 2).

A further source of information regarding climate change comes from measurements of change in sea level. Over the last hundred years sea level has risen by between 10 and 20 cm. The best known contributions to this rise are from the thermal expansion of ocean water because of the global average temperature rise (estimated as up to about 7 cm)

Table 4.1 *Twentieth-century changes in the Earth's atmosphere, climate and biophysical system*

Indicator	Observed changes
Concentration indicators	
Atmospheric concentration of CO_2	280 ppm for the period 1000–1750 to 368 ppm in year 2000 ($31 \pm 4\%$ increase)
Terrestrial biospheric CO_2 exchange	Cumulative source of about 30 GtC between the years 1800 and 2000; but during the 1990s a net sink of about 14 ± 7 GtC
Atmospheric concentration of CH_4	700 ppb for the period 1000–1750 to 1750 ppb in year 2000 ($151 \pm 25\%$ increase)
Atmospheric concentration of N_2O	270 ppb for the period 1000–1750 to 316 ppb in the year 2000 ($17 \pm 5\%$ increase)
Tropospheric concentration of O_3	Increased by $35 \pm 15\%$ from the years 1750 to 2000, varies with region
Stratospheric concentration of O_3	Decreased over the years 1970 to 2000, varies with altitude and latitude
Atmospheric concentrations of HFCs, PFCs and SF_6	Increased globally over the last fifty years
Weather indicators	
Global mean surface temperature	Increased by $0.6 \pm 0.2\,°C$ over the twentieth century; land areas warned more than the oceans (*very likely*)
Northern hemisphere surface temperature	Increase over the twentieth century greater than during any other century in the last 1000 years; 1990s warmest decade of the millennium (*likely*)
Diurnal surface temperature range	Decreased over the years 1950 to 2000 over land; night-time minimum temperatures increased at twice the rate of daytime maximum temperatures (*likely*)
Hot days/heat index	Increased (*likely*)
Cold/frost days	Decreased for nearly all land areas during the twentieth century (*very likely*)
Continental precipitation	Increased by five to ten per cent over the twentieth century in the Northern Hemisphere (*very likely*), although decreased in some regions (e.g. north and west Africa and parts of the Mediterranean)
Heavy precipitation events	Increased at mid and high northern latitudes (*likely*)
Frequency and severity of drought	Increased summer drying and associated incidence of drought in a few areas (*likely*). In some regions, such as parts of Asia and Africa, the frequency and intensity of droughts have been observed to increase in recent decades

Table 4.1 (*cont.*)

Indicator	Observed changes
Biological and physical indicators	
Global mean sea level	Increased at an average annual rate of 1–2 mm during the twentieth century
Duration of ice cover of rivers and lakes	Decreased by about two weeks over the twentieth century in mid and high latitudes of the Northern Hemisphere (*very likely*)
Arctic sea-ice extent and thickness	Thinned by forty per cent in recent decades in late summer to early autumn (*likely*) and decreased in extent by ten to fifteen per cent since the 1950s in spring and summer
Non-polar glaciers	Widespread retreat during the twentieth century
Snow cover	Decreased in area by ten per cent since global observations became available from satellites in the 1960s (*very likely*)
Permafrost	Thawed, warmed and degraded in parts of the polar, sub-polar and mountainous regions
El Niño events	Became more frequent, persistent and intense during the last twenty to thirty years compared to the previous 100 years
Growing season	Lengthened by about one to four days per decade during the last forty years in the Northern Hemisphere, especially at higher latitudes
Plant and animal ranges	Shifted poleward and up in elevation for plants, insects, birds and fish
Breeding, flowering and migration	Earlier plant flowering, earlier bird arrival, earlier dates of breeding season and earlier emergence of insects in the Northern Hemisphere
Coral reef bleaching	Increased frequency, especially during El Niño events
Economic indicators	
Weather-related economic losses	Global inflation-adjusted losses rose an order of magnitude over the last forty years. Part of the observed upward trend is linked to socio-economic factors and part is linked to climatic factors

This table provides examples of key observed changes and is not an exhaustive list. It includes both changes attributable to anthropogenic climate change and those that may be caused by natural variations or anthropogenic climate change. Confidence levels (for explanation see note 1) are reported where they are explicitly assessed by the relevant Working Group of the IPCC.

Source: Table SPM-1 from IPCC 2001 Synthesis Report.

and from glaciers which have generally been retreating over the last century (estimated as up to about 4 cm). The net contribution from the Greenland and Antarctic ice caps is more uncertain but is believed to be small.

In Chapter 1, we mentioned the increasing vulnerability of human populations to climate extremes, which has brought about more awareness of recent extremes in the forms of floods, droughts, tropical cyclones and windstorms. It is therefore of great importance to know whether there is evidence of an increase in the frequency or severity of these and other extreme events. The available evidence regarding how these and other relevant parameters have changed during the twentieth century is summarised in Table 4.1 in terms of different indicators: concentrations of greenhouse gases; temperature, hydrological and storm-related indicators and biological and physical indicators. To what extent these changes are expected to continue or to intensify during the twenty-first century will be addressed in Chapter 6.

Eventually, as greenhouse gases increase further, the amount of warming is expected to become sufficiently large that it will swamp the natural variations in climate. In the meantime, the global average temperature may continue to increase or it could, because of natural variability, show periods of decrease. Over the next few years, scientists will be inspecting climate changes and climate events most carefully as they occur to see how far actual events can be related to scientific predictions especially those associated with increasing greenhouse gas emissions. Some details of these predictions will be discussed in later chapters especially Chapter 6.

The last thousand years

The detailed systematic record of weather parameters such as temperature, rainfall, cloudiness and the like presented above for the last 140 years and which covers a good proportion of the globe is not available for earlier periods. Further back, the record is more sparse and doubt arises over the consistency of the instruments used for observation. Most thermometers in use 200 years ago were not well calibrated or carefully exposed. However, many diarists and writers kept records at different times; from a wide variety of sources weather and climate information can be pieced together. Indirect sources, such as are provided by ice cores, tree rings and records of lake levels, of glacier advance and retreat, and of pollen distribution, can also yield information to assist in building up the whole climatic story. From a variety of sources, for instance, it has been possible to put together for China a systematic atlas of weather patterns covering the last 500 years.

Similarly, from direct and indirect sources, it has been possible to deduce the average temperature over the Northern Hemisphere for the last millennium (Figure 4.3). Sufficient data are not available for the same reconstruction to be carried out over the Southern Hemisphere. In Figure 4.3 it is just possible to identify the 'Medieval Warm Period' associated with the eleventh to fourteenth centuries and a relatively cool

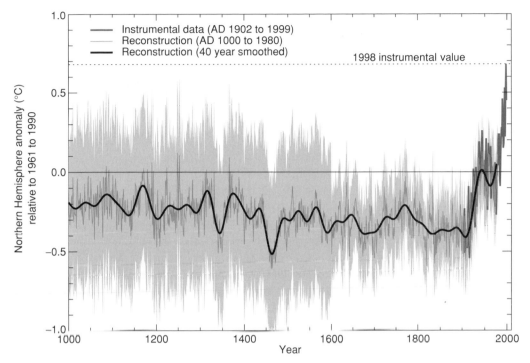

Figure 4.3 Millennial Northern Hemisphere temperature reconstruction from tree rings, corals, ice cores and historical records (for the period 1000–1980) and from instrumental data (for 1902–1999). A smoothed version of the proxy record is also shown, as are the 95% confidence limits (grey shaded).

period the 'Little Ice Age' associated with the fifteenth to nineteenth centuries. These only affected part of the Northern Hemisphere and are therefore more prominent in local records, for instance those from central England. The increase in temperature over the twentieth century is particularly striking. The 1990s are likely to have been the warmest decade of the millennium in the Northern Hemisphere and 1998 is likely to have been the warmest year.

Although there is as yet no certain explanation for the variations that occurred between 1000 and 1900, it is clear that greenhouse gases such as carbon dioxide and methane cannot have been the cause of change. For the millennium before 1800 their concentration in the atmosphere was rather stable, the carbon dioxide concentration, for instance, varying by less than three per cent. However, the combined influences of variations in volcanic activity and variations in the output of energy from the Sun can provide some part of the explanation.[4] The effect of individual volcanic eruptions can be very noticeable. For instance, one of the largest eruptions during the period was that of Tambora in Indonesia

in April 1815, which was followed in many places by two exceptionally cold years: 1816 was described in New England and Canada as the
'year without a summer'. Although the effect on the climate even of
an eruption of the magnitude of Tambora only lasts a few years, variations in *average* volcanic activity have a longer term effect. Regarding
the Sun's output, although accurate direct measurements are not available (apart from those made during the last two decades from satellite
instruments), other evidence suggests that the solar output could have
varied significantly in the past. For instance, compared with its value
today it may have been somewhat lower (by a few tenths of a watt per
square metre) during the Maunder Minimum in the seventeenth century
(a period when almost no sunspots were recorded; see also box on p. 138).
There is no need, however, to invoke volcanoes or variations in solar
output as the cause of all the climate variations over this period. As
with the shorter-term changes mentioned earlier, such variations of climate can arise naturally from internal variations within the atmosphere
and the ocean and in the two-way relationship – coupling – between
them.

The millennial record of Figure 4.3 is particularly important because it provides an indication of the range and character of climate
variability that arises from natural causes. As we shall see in the next
chapter, climate models also provide some information on natural climate variability. Careful assessments of these observational and model
results confirm that natural variability (the combination of internal variability and naturally forced, e.g. by volcanoes or change in solar output)
is unlikely to explain the warming in the latter half of the twentieth
century.[5]

The past million years

To go back before recorded human history, scientists have to rely on
indirect methods to unravel much of the story of the past climate. A
particularly valuable information source is the record stored in the ice that
caps Greenland and the Antarctic continent. These ice caps are several
thousands of metres thick. Snow deposited on their surface gradually
becomes compacted as further snow falls, becoming solid ice. The ice
moves steadily downwards, eventually flowing outwards at the bottom of
the ice-sheet. Ice near the top of the layer will have been deposited fairly
recently; ice near the bottom will have fallen on the surface many tens
or hundreds of thousands of years ago. Analysis of the ice at different
levels can, therefore, provide information about the conditions prevailing
at different times in the past.

Deep cores have been drilled out of the ice at several locations in both Greenland and Antarctica. At Russia's Vostok station in east Antarctica, for instance, drilling has been carried out for over twenty years. The longest and most recent core reached a depth of over 3.5 km; the ice at the bottom of the hole fell as snow on the surface of the Antarctic continent over 400 000 years ago.

Small bubbles of air are trapped within the ice. Analysis of the composition of that air shows what was present in the atmosphere for the

Paleoclimate reconstruction from isotope data

The isotope ^{18}O is present in natural oxygen at a concentration of about 1 part in 500 compared with the more abundant isotope ^{16}O. When water evaporates, water containing the lighter isotope is more easily vaporised, so that water vapour in the atmosphere contains less ^{18}O compared with sea water. Similar separation occurs in the process of condensation when ice crystals form in clouds. The amount of separation between the two oxygen isotopes in these processes depends on the temperatures at which evaporation and condensation occur. Measurements on snowfall in different places can be used to calibrate the method; it is found that the concentration of ^{18}O varies by about 0.7 of a part per thousand for each degree of change in average temperature at the surface. Information is therefore available in the ice cores taken from polar ice caps concerning the variation in atmospheric temperature in polar regions during the whole period when the ice core was laid down.

Since the ice caps are formed from accumulated snowfall which contains less ^{18}O compared with sea water, the concentration of ^{18}O in water from the oceans provides a measure of the total volume of the ice in the ice caps; it changes by about one part in 1000 between the maximum ice extent of the ice ages and the warm periods in between. Information about the ^{18}O content of ocean water at different times is locked up in corals and in cores of sediment taken from the ocean bottom, which contain carbonates from fossils of plankton and small sea creatures from past centuries and millennia. Measurements of radioactive isotopes, such as the carbon isotope ^{14}C, and correlations with other significant past events enable the corals and sediment cores to be dated. Since the separation between the oxygen isotopes which occurs as these creatures are formed also depends on the temperature of the sea water (although the dependence is weaker than the other dependencies considered above) information is also available about the distribution of ocean surface temperature at different times in the past.[6]

time at which the ice was formed – gases such as carbon dioxide or methane. Dust particles that may have come from volcanoes or from the sea surface are also contained within the ice. Further information is provided by analysis of the ice itself. Small quantities of different oxygen isotopes and of the heavy isotope of hydrogen (deuterium) are contained in the ice. The ratios of these isotopes that are present depend sensitively on the temperatures at which evaporation and condensation took place for the water in the clouds from which the ice originated (see box above). These in turn are dependent on the average temperature near the surface of the Earth. A temperature record for the polar regions can therefore be constructed from analyses of the ice cores. The associated changes in global average temperature are estimated to be about half the changes in the polar regions.

Such a reconstruction from a Vostok core for the temperature and the carbon dioxide content is shown in Figure 4.4 for the past 160 000 years, which includes the last major ice age that began about 120 000 years ago and began to come to an end about 20 000 years ago. It also demonstrates the close connections that exist between temperature and carbon dioxide concentrations. Similar close correlation is found with the methane concentration. Note from Figure 4.4 the likely growth of atmospheric carbon dioxide during the twenty-first century, taking it to levels that are unlikely to have been exceeded during the past twenty million years.

Data from ice cores can take us back 400 000 years or so over four ice age cycles during which the correlations between temperature and carbon dioxide concentrations shown in Figure 4.4 are repeated.[7] To go further back, over the past million years, the composition of ocean sediments can be investigated to yield information. Fossils of plankton and other small sea creatures deposited in these sediments also contain different isotopes of oxygen. In particular the amount of the heavier isotope of oxygen (^{18}O) compared with the more abundant isotope (^{16}O) is sensitive both to the temperature at which the fossils were formed and to the total volume of ice in the world's ice caps at the time of the fossils' formation (see box above). For instance, from oxygen isotope and other data we can deduce that the sea level at the last glacial maximum, 20 000 years ago, was about 120 m lower than today.

From the variety of paleoclimate data available, variations in the volume of ice in the ice caps can be reconstructed over the greater part of the last million years (Figure 4.5(c)). In this record six or seven major ice ages can be identified with warmer periods in between, the period between these major ice ages being approximately 100 000 years. Other cycles are also evident in the record.

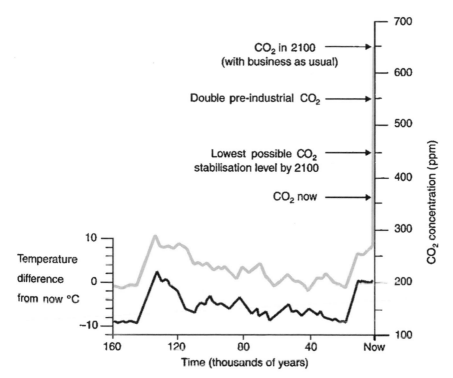

Figure 4.4 Variations over the last 160 000 years of polar temperature and atmospheric carbon dioxide concentrations derived from the Vostok ice core from Antarctica. It is estimated that the variation of global average temperature is about half that in the polar regions. Also shown is the current carbon dioxide concentration of about 370 ppm and the likely rise during the twenty-first century under various projections of its growth.

The most obvious place to look for the cause of regular cycles in climate is outside the Earth, in the Sun's radiation. Has this varied in the past in a cyclic way? So far as is known the output of the Sun itself has not changed to any large extent over the last million years or so. But because of variations in the Earth's orbit, the distribution of solar radiation has varied in a more or less regular way during the last millennium.

Three regular variations occur in the orbit of the Earth around the Sun (Figure 4.5(a)). The Earth's orbit, although nearly circular, is actually an ellipse. The eccentricity of the ellipse (which is related to the ratio between the greatest and the least diameters) varies with a period of about 100 000 years; that is the slowest of the three variations. The Earth also spins on its own axis, the axis of spin being tilted with respect to the axis of the Earth's orbit, the angle of tilt varying between $21.6°$ and $24.5°$ (currently it is $23.5°$) with a period of about 41 000 years.

The third variation is of the time of year when the Earth is closest to the Sun (the Earth's perihelion). The time of perihelion moves through the months of the year with a period of about 23 000 years (see also Figure 5.19); in the present configuration, the Earth is closest to the Sun in January.

As the Earth's orbit changes its relationship to the Sun, although the total quantity of solar radiation reaching the Earth varies very little, the distribution of that radiation with latitude and season over the Earth's surface changes considerably. The changes are especially large in polar regions where the variations in summer sunshine, for instance, reach about ten per cent (Figure 4.5(b)). James Croll, a British scientist, first pointed out in 1867 that the major ice ages of the past might be linked with these regular variations in the seasonal distribution of solar radiation reaching the Earth. His ideas were developed in 1920 by Milutin Milankovitch, a climatologist from Yugoslavia, whose name is usually linked with the theory. Inspection by eye of the relationship between the variations of polar summer sunshine and global ice volume shown in Figure 4.5 suggests a significant connection. Careful study of the correlation between the two curves confirms this and demonstrates that sixty per cent of the variance in the climatic record of global ice volume falls close to the three frequencies of regular variations in the Earth's orbit, thus providing support for the Milankovitch theory.

More careful study of the relationship between the ice ages and the Earth's orbital variations shows that the size of the climate changes is larger than might be expected from forcing by the radiation changes alone. Other processes that enhance the effect of the radiation changes (in other words, positive feedback processes) have to be introduced to explain the climate variations. One such feedback arises from the changes in carbon dioxide influencing atmospheric temperature through the greenhouse effect, illustrated by the strong correlation observed in the climatic record between average atmospheric temperature and carbon dioxide concentration (Figure 4.4). Such a correlation does not, of course, prove the existence of the greenhouse feedback; in fact part of the correlation arises because the atmospheric carbon dioxide concentration is itself influenced, through biological feedbacks (see Chapter 3), by factors related to the average global temperature. However, as we shall see in Chapter 5, climates of the past cannot be modelled successfully without taking the greenhouse feedback into account.[8]

An obvious question to ask is when, on the Milankovitch theory, is the next ice age due? It so happens that we are currently in a period of relatively small solar radiation variations and the best projections for the long term are of a longer than normal interglacial period leading to the beginning of a new ice age perhaps in 50 000 years' time.[9]

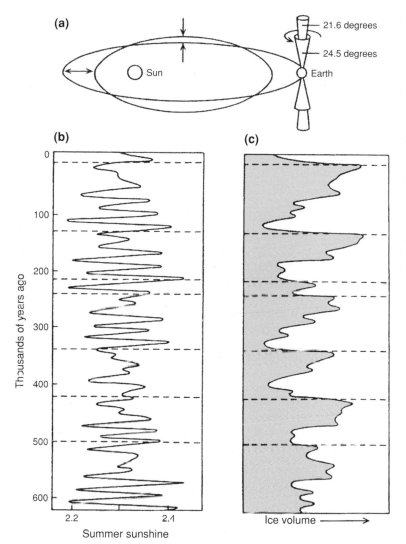

Figure 4.5 Variations in the Earth's orbit (a), in its eccentricity, the orientation of its spin axis (between 21.6° and 24.5°) and the longitude of perihelion (i.e. the time of year when the Earth is closest to the Sun, see also Figure 5.19), cause changes in the average amount of summer sunshine (in millions of joules per square metre per day) near the poles (b). These changes appear as cycles in the climate record in terms of the volume of ice in the ice caps (c).

How stable has past climate been?

The major climate changes considered so far in this chapter have taken place relatively slowly. The growth and recession of the large polar ice-sheets between the ice ages and the intervening warmer interglacial periods have taken on average many thousands of years. However, the ice core records such as those in Figures 4.4 and 4.6 show evidence of large and relatively rapid fluctuations. Ice cores from Greenland provide more detailed evidence of these than those from Antarctica. This is because at the summit of the Greenland ice cap, the rate of accumulation of snow

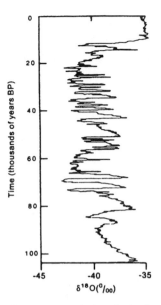

Figure 4.6 Variations in Arctic temperature over the past 100 000 years as deduced from oxygen isotope measurements (in terms of δ^{18}O) from the 'Summit' ice core in Greenland. The quantity δ^{18}O plotted in Figures 4.6 and 4.7 is the difference (in parts per thousand) between the ^{18}O/^{16}O ratio in the sample and the same ratio in a laboratory standard. The overall shape of the record is similar to that from the Vostok ice core shown in Figure 4.4 but much more detail is apparent in the 'Summit' record's stable period over the last 8000 years. A change of five parts per thousand in δ^{18}O in the ice core corresponds to about a 7 °C change in temperature.

has been higher than that at the Antarctica drilling locations. For a given period in the past, the relevant part of the Greenland ice core is longer and more detail of variations over relatively short periods is therefore available.

The data show that the last 8000 years have been unusually stable compared with earlier epochs. In fact, as judged from the Vostok (Figure 4.4) and the Greenland records (Figure 4.6) this long stable period in the Holocene is a unique feature of climate during the past 420 000 years. It has been suggested that this had profound implications for the development of civilisations.[10] Model simulations (see Chapter 5) indicate that the detail of long-term changes during the Holocene is consistent with the influence of orbital forcing (Figure 4.5).

It is also interesting to inspect the rate of temperature change during the recovery period from the last glacial maximum about 20 000 years ago and compare it with recent temperature changes. The data indicate an average warming rate of about 0.2 °C per century between 20 000 and 10 000 years before present (BP) over Greenland, with lower rates for other regions. Compare this with a temperature rise during the twentieth century of about $0.6\,°C$ and the rates of change of a few degrees centigrade per century projected to occur during the twenty-first century because of human activities (see Chapter 6).

The ice core data (Figures 4.6) demonstrate that a series of rapid warm and cold oscillations called Dansgaard–Oeschger events punctuated the last glaciation. Comparison between the results from ice cores drilled at different locations within the Greenland ice cap confirm the details up to about 100 000 years ago. Comparison with data from Antarctica suggests that the fluctuations of temperature over Greenland (perhaps up to $16\,°C$) have been larger than those over Antarctica. Similar large and relatively rapid variations are evident from North Atlantic deep sea sediment cores.

Another particularly interesting period of climatic history, more recently, is the Younger Dryas event (so called because it was marked by the spread of an arctic flower, *Dryas octopetala*), which occurred over a period of about 1500 years between about 12 000 and 10 700 years ago. For 6000 years before the start of this event the Earth had been warming up after the end of the last ice age. But then during the Younger Dryas period, as demonstrated from many different sources of paleoclimatic data, the climate swung back again into much colder conditions similar to those at the end of the last ice age (Figure 4.7). The ice core record shows that at the end of the event, 10 700 years ago, the warming in the Arctic of about $7\,°C$ occurred over only about fifty years and was associated with decreased storminess (shown by a dramatic fall in the amount of dust in the ice core) and an increase of precipitation of about fifty per cent.

Two main reasons for these rapid variations in the past have been suggested. One reason particularly applicable to ice age conditions is that, as the ice-sheets over Greenland and eastern Canada have built up, major break-ups have occurred from time to time, releasing massive numbers of icebergs into the North Atlantic in what are called Heinrich events. The second possibility is that the ocean circulation in the North Atlantic region has been strongly affected by injections of fresh water from the melting of ice. At present the ocean circulation in this region is strongly influenced by cold salty water sinking to deep ocean levels because its saltiness makes it dense; this sinking process is part of the 'conveyor belt' which is the major feature of the circulation of deep ocean water around the world (see Figure 5.18). Large quantities of fresh water from the melting of ice would make the water less salty, preventing it from sinking and thereby altering the whole Atlantic circulation.

This link between the melting of ice and the ocean circulation is a key feature of the explanation put forward by Professor Wallace

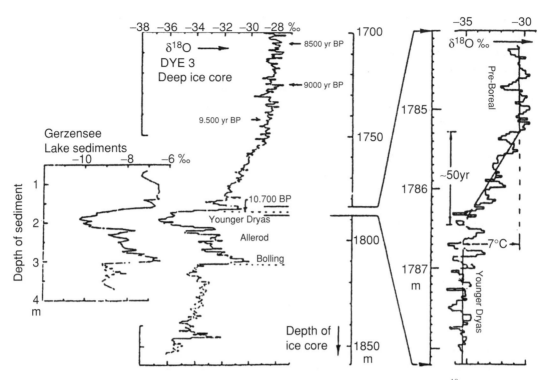

Figure 4.7 Records of the variations of the oxygen isotope $\delta^{18}O$ from lake sediments from Lake Gerzen in Switzerland and from the Greenland ice core 'Dye 3' showing the 'Younger Dryas' event and its rapid end about 10 700 years ago. Dating of the ice core was by counting the annual layers down from the surface; dating of the lake sediment was by the ^{14}C method. A change of five parts in a thousand in $\delta^{18}O$ in the ice core corresponds to about a 7 °C change in temperature.

Broecker for the Younger Dryas event.[11] As the great ice-sheet over north America began to melt at the end of the last ice age, the melt water at first drained through the Mississippi into the Gulf of Mexico. Eventually, however, the retreat of the ice opened up a channel for the water in the region of the St Lawrence river. This influx of fresh water into the North Atlantic reduced its saltiness, thus, Broecker postulates, cutting off the formation of deep water and that part of the ocean 'conveyor belt'.[12] Warm water was therefore prevented from flowing northward, resulting in a reversal to much colder conditions. The suggestion is also that a reversal of this process with the starting up of the Atlantic 'conveyor belt' could lead to a sudden onset of warmer conditions.

Although debate continues regarding the details of the Younger Dryas event, there is considerable evidence from paleodata, especially those from ocean sediments, for the main elements of the Broecker explanation which involve the deep ocean circulation. It is also clear from paleodata that large changes have occurred at different times in the past in the formation of deep water and in the deep ocean circulation. Chapter 3 mentioned the possibility of such changes being induced by global warming through the growth of greenhouse gas concentrations. Our perspective regarding the possibilities of future climate change needs to take into account the rapid climate changes that have occurred in the past.

Having now in these early chapters set the scene, by describing the basic science of global warming, the greenhouse gases and their origins and the current state of knowledge regarding past climates, I move on in the next chapter to describe how, through computer models of the climate, predictions can be made about what climate change can be expected in the future.

Questions

1   Given that the sea level at the end of the last glacial maximum was 120 m lower than that today, estimate the volume of ice in the ice-sheets that covered the northern parts of the American and Euroasian continents.

2   How much energy would be required to melt the volume of ice you have calculated in question 1? Compare this with the extra summer sunshine north of latitude 60° which might have been available between 18 000 and 6000 years before the present according to the data in Figure 4.5. Does your answer support the Milankovitch theory?

3   It is sometimes suggested that the large reserves of fossil fuels on Earth should be preserved until the onset of the next ice age is closer so that some of its impact can be postponed. From what you know of the greenhouse effect and of the behaviour of carbon dioxide in the atmosphere and the oceans, consider the influences that human burning of the known reserves of fossil fuels – see Figure 11.2 – could have on the onset of the next ice age.

Notes for Chapter 4

1   In the IPCC 2001 Report, expressions of certainty such as 'very likely' were related so far as possible to quantitative statement of confidence, *virtually certain* (greater than ninety-nine per cent chance that a result is true), *very likely* (ninety to ninety-nine per cent chance), *likely* (sixty-six to ninety per cent chance), *medium likelihood* (thirty-three to sixty-six per cent chance), *unlikely* (ten to thirty-three per cent chance), *very unlikely* (one to ten per cent chance), *exceptionally unlikely* (less than one per cent chance).

2   Further information in Folland, C. K. *et al.* (Chapter 2) and Mitchell, J. F. B. *et al.* (Chapter 12). In Houghton, J. T., Ding, Y., Griggs, D. J., Noguer, M., van der Linden, P. J., Dai, X., Maskell, K., Johnson, C. A. (eds.) *Climate Change 2001: The Scientific Basis. Contribution of Working Group I to*

   the Third Assessment Report of the Intergovernmental Panel on Climate Change. Cambridge: Cambridge University Press.

3 From Summary for policymakers, Houghton *et al.*, *Climate Change 2001: The Scientific Basis.*

4 See, for instance, Crowley, T. J. 2000. Causes of climate change over the past 1000 years. *Science*, **289**, 270–7.

5 More information in Mitchell, J. F. B. *et al.* (Chapter 12). In Houghton *et al. Climate Change 2001.*

6 For more information about paleoclimates and how they are investigated see, for instance, Crowley, T. L. and North, G. R. 1991. *Paleoclimatology.* Oxford: Oxford University Press.

7 Folland, C. K. *et al.* (Chapter 2). Figure 2.22. In Houghton *et al.*, *Climate Change 2001.*

8 See also Raynaud, D. *et al.*, 1993. The ice core record of greenhouse gases. *Science*, **259**, 926–34.

9 Berger, A. and Loutre, M. F. 2002. *Science*, **297**, 1287–8.

10 Petit, J. R. *et al.* 1999. *Nature*, **399**, 429–36.

11 Broecker, W. S. and Denton, G. H. 1990. What drives glacial cycles? *Scientific American*, **262**, 43–50.

12 More information in Chapter 5, see especially Figure 5.18.

Chapter 5
Modelling the climate

Chapter 2 looked at the greenhouse effect in terms of a simple radiation balance. That gave an estimate of the rise in the average temperature at the surface of the Earth as greenhouse gases increase. But any change in climate will not be distributed uniformly everywhere; the climate system is much more complicated than that. More detail in climate change prediction requires very much more elaborate calculations using computers. The problem is so vast that the fastest and largest computers available are needed. But before computers can be set to work on the calculation, a model of the climate must be set up for them to use.[1] A model of the weather as used for weather forecasting will be used to explain what is meant by a numerical model on a computer, followed by a description of the increase in elaboration required to include all parts of the climate system in the model.

Modelling the weather

An English mathematician, Lewis Fry Richardson, set up the first numerical model of the weather. During his spare moments while working for the Friends' Ambulance Unit (he was a Quaker) in France during the First World War he carried out the first numerical weather forecast. With much painstaking calculation with his slide-rule, he solved the appropriate equations and produced a six-hour forecast. It took him six months – and then it was not a very good result. But his basic methods, described in a book published in 1922,[2] were correct. To apply his methods to real forecasts, Richardson imagined the possibility of a very large concert hall filled with people, each person carrying out part of the calculation,

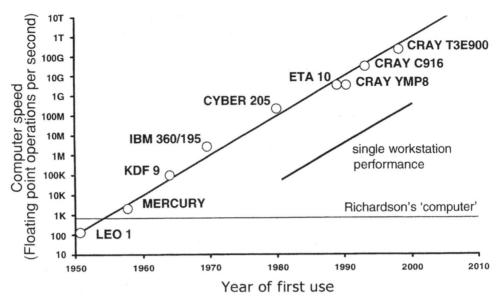

Figure 5.1 Illustrating the growth of computer power available at major forecasting centres. The computers are those used by the UK Meteorological Office for numerical weather prediction research, since 1965 for operational weather forecasting and most recently for research into climate prediction. Richardson's computer refers to his dream of a large 'human' computer mentioned at the beginning of the chapter.

so that the integration of the numerical model could keep up with the weather. But he was many years before his time! It was not until some forty years later that, essentially using Richardson's methods, the first operational weather forecast was produced on an electronic computer. Computers more than one million times faster than the one used for that first forecast (Figure 5.1) now run the numerical models that are the basis of all weather forecasts.

Numerical models of the weather and the climate are based on the fundamental mathematical equations that describe the physics and dynamics of the movements and processes taking place in the atmosphere, the ocean, the ice and on the land. Although they include empirical information, they are not based to any large degree on empirical relationships – unlike numerical models of many other systems, for instance in the social sciences.

Setting up a model of the atmosphere for a weather forecast (see Figure 5.2) requires a mathematical description of the way in which energy from the Sun enters the atmosphere from above, some being reflected by the surface or by clouds and some being absorbed at the surface or in the atmosphere (see Figure 2.6). The exchange of energy and water vapour between the atmosphere and the surface must also be described.

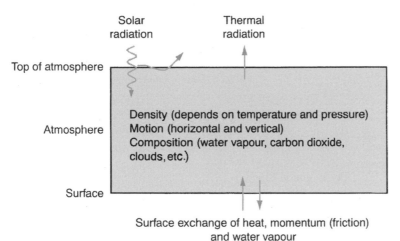

Solar radiation   Thermal radiation

Top of atmosphere

Atmosphere

Density (depends on temperature and pressure)
Motion (horizontal and vertical)
Composition (water vapour, carbon dioxide, clouds, etc.)

Surface

Surface exchange of heat, momentum (friction)
and water vapour

Figure 5.2 Schematic illustrating the parameters and physical processes involved in atmospheric models.

Water vapour is important because of its associated latent heat (in other words, it gives out heat when it condenses) and also because the condensation of water vapour results in cloud formation, which modifies substantially the interaction of the atmosphere with the incoming energy from the Sun. Variations in both these energy inputs modify the atmospheric temperature structure, causing changes in atmospheric density (since warmed gases expand and are therefore less dense). It is these density changes that drive atmospheric motions such as winds and air currents, which in their turn alter and feed back on atmospheric density and composition. More details of the model formulation are given in the box below.

To forecast the weather for several days ahead a model covering the whole globe is required; for example, the southern hemisphere circulation today will affect northern hemisphere weather within a few days and vica versa. In a global forecasting model, the parameters (i.e. pressure, temperature, humidity, wind velocity and so on) that are needed to describe the dynamics and physics (listed in the box below) are specified at a grid of points (Figure 5.3) covering the globe. A typical spacing between points in the horizontal would be 100 km and about 1 km in the vertical; typically there would be twenty levels or so in the model in the vertical. The fineness of the spacing is limited by the power of the computers currently available.

Having set up the model, to generate a forecast from the present, it is started off from the atmosphere's current state and then the equations are integrated forward in time (see box below) to provide new descriptions of the atmospheric circulation and structure up to six or more days ahead. For a description of the atmosphere's current state, data from a wide variety of sources (see box below) have to be brought together and fed into the model.

Setting up a numerical atmospheric model

A numerical model of the atmosphere contains descriptions, in appropriate computer form and with necessary approximations, of the basic dynamics and physics of the different components of the atmosphere and their interactions.[3] When a physical process is described in terms of an algorithm (a process of step-by-step calculation) and simple parameters (the quantities that are included in a mathematical equation), the process is said to have been parameterised.

The dynamical equations are:

- The horizontal momentum equations (Newton's Second Law of Motion). In these, the horizontal acceleration of a volume of air is balanced by the horizontal pressure gradient and the friction. Because the Earth is rotating, this acceleration includes the Coriolis acceleration. The 'friction' in the model mainly arises from motions smaller than the grid spacing, which have to be parameterised.
- The hydrostatic equation. The pressure at a point is given by the mass of the atmosphere above that point. Vertical accelerations are neglected.
- The continuity equation. This ensures conservation of mass.

The model's physics consists of:

- The equation of state. This connects the quantities of pressure, volume and temperature for the atmosphere.
- The thermodynamic equation (the law of conservation of energy).
- Parameterisation of moist processes (such as evaporation, condensation, formation and dispersal of clouds).
- Parameterisation of absorption, emission and reflection of solar radiation and of thermal radiation.
- Parameterisation of convective processes.
- Parameterisation of exchange of momentum (in other words, friction), heat and water vapour at the surface.

Most of the equations in the model are differential equations, which means they describe the way in which quantities such as pressure and wind velocity change with time and with location. If the rate of change of a quantity such as wind velocity and its value at a given time are known, then its value at a later time can be calculated. Constant repetition of this procedure is called integration. Integration of the equations is the process whereby new values of all necessary quantities are calculated at later times, providing the model's predictive powers.

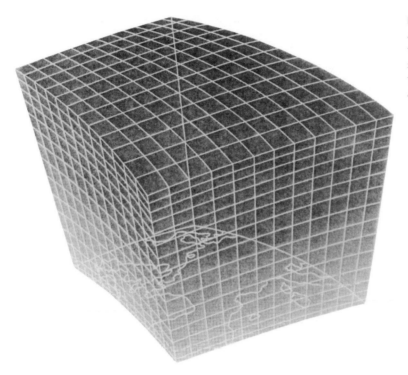

Figure 5.3 Illustration of a model grid. The levels in the vertical are not equally spaced; the top level is typically at about 30 km in altitude.

Since computer models for weather forecasting were first introduced, their forecast skill has improved to an extent beyond any envisaged by those involved in the development of the early models. As improvements have been made in the model formulation, in the accuracy or coverage of the data used for initialisation (see box above) or in the resolution of the model (the distance between grid points), the resulting forecast skill has increased. For instance, for the British Isles, three-day forecasts of surface pressure today are as skilful on average as two-day forecasts of ten years ago, as can be seen from Figure 5.5.

When looking at the continued improvement in weather forecasts, the question obviously arises as to whether the improvement will continue or whether there is a limit to the predictability we can expect. Because the atmosphere is a partially chaotic system (see box below), even if perfect observations of the atmospheric state and circulation could be provided, there would be a limit to our ability to forecast the detailed state of the atmosphere at some time in the future. In Figure 5.6 current forecast skill is compared with the best estimate of the limit of the forecast skill for the British Isles (similar results would be obtained with any other mid-latitude situation) with a perfect model and near perfect data. According to that estimate, the limit of significant future skill is about 20 days ahead.

Data to initialise the model

At a major global weather forecasting centre, data from many sources are collected and fed into the model. This process is called initialisation. Figure 5.4 illustrates some of the sources of data for the forecast beginning at 1200 hours Universal Time (UT) on 1 July 1990. To ensure the timely receipt of data from around the world a dedicated communication network has been set up, used solely for this purpose. Great care needs to be taken with the methods for assimilation of the data into the model as well as with the data's quality and accuracy.

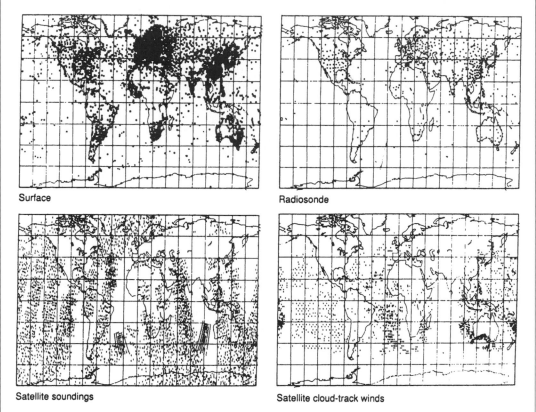

Surface

Radiosonde

Satellite soundings

Satellite cloud-track winds

Figure 5.4 Illustrating some of the sources of data for input into the UK Meteorological Office global weather forecasting model on a typical day. Surface observations are from land observing stations (manned and unmanned), from ships and from buoys. Radiosonde balloons make observations at up to 30 km altitude from land and from ship-borne stations. Satellite soundings are of temperature and humidity at different atmospheric levels deduced from observations of infrared or microwave radiation. Satellite cloud-track winds are derived from observing the motion of clouds in images from geostationary satellites.

Forecast skill varies considerably between different weather situations or patterns. In other words some situations are more 'chaotic' (in the technical sense in which that word is used – see box below) than others. One way of identifying the skill that might be achieved in a

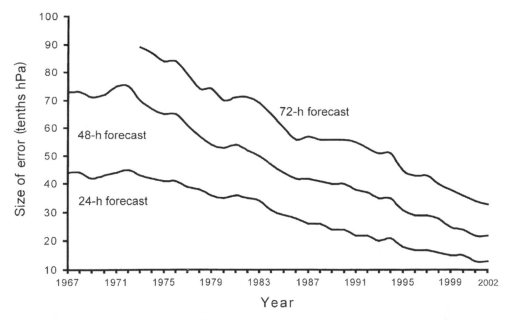

Figure 5.5 Errors (root mean square differences of forecasts of surface pressure compared with analyses) of UK Meteorological Office forecasting models for the north Atlantic and Western Europe since 1966 for 24-h, 48-h and 72-h forecasts. Note that 1 hPa = 1 mbar.

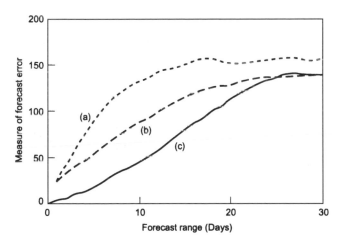

Figure 5.6 Illustrating potential improvements in forecast skill if there were better data or a better model. The ordinate (vertical axis) is a measure of the error of model forecasts (it is the root mean square differences of forecasts of the 500 hPa height field compared with analyses). Curve (a) is the error of 1990 UK Meteorological Office forecasts as a function of forecast range. Curve (b) is an estimate showing how, with the same initial data, the error would be reduced if a perfect model could be used. Curve (c) is an estimate showing the further improvement which might be expected if near-perfect data could be provided for the initial state. After a sufficiently long period, all the curves approach a saturation value of the average root mean square difference between any forecasts chosen at random.

Weather forecasting and chaos[4]

The science of chaos has developed rapidly since the 1960s (when a meteorologist Edward Lorenz was one of its pioneers) along with the power of electronic computers. In this context, chaos is a term with a particular technical meaning. A chaotic system is one whose behaviour is so highly sensitive to the initial conditions from which it started that precise future prediction is not possible. Even quite simple systems can exhibit chaos under some conditions. For instance, the motion of a simple pendulum (Figure 5.7) can be 'chaotic' under some circumstances, and, because of its extreme sensitivity to small disturbances, its detailed motion is not then predictable.

A condition for chaotic behaviour is that the relationship between the quantities which govern the motion of the system be non-linear; in other words, a description of the relationship on a graph would be a curve rather than a straight line.[5] Since the appropriate relationships for the atmosphere are non-linear it can be expected to show chaotic behaviour. This is illustrated in Figure 5.6, which shows the improvement in predictability that can be expected if the data describing the initial state are improved. However, even with virtually perfect initial data, the predictability in terms of days ahead only moves from about six days to about twenty days, because the atmosphere is a chaotic system.

For the simple pendulum not all situations are chaotic. Not surprisingly, therefore, in a system as complex as the atmosphere, some occasions are more predictable than others. A good illustration of an occasion with particular sensitivity to the initial data and to the way in which the data were assimilated into the model is provided by the Meteorological Office forecasts for the storm which hit southeast England in the early hours of Friday, 16 October 1987. During this storm gusts of over 90 knots (170 km h^{-1}) were recorded and approximately fifteen million trees were blown down. Although as early as the previous Sunday forecasts had given good early warning of a storm of unusual severity, the model forecasts available during 15 October gave much poorer guidance than earlier forecasts and failed to predict the intensity or the correct track of the storm. The question was raised at the time as to whether the numerical models used were capable of the accurate prediction of such an exceptional event. Figure 5.8 shows that, using all the data which could have been available and better assimilation procedures in the model, a good forecast of the event can be achieved.

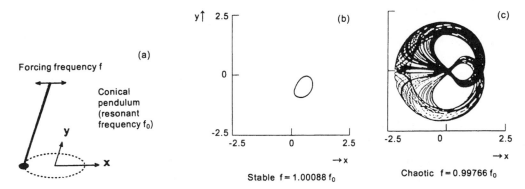

Figure 5.7 (a) Illustrating a simple pendulum consisting of a bob at the end of a string of length 10 cm attached to a point of suspension which is moved with a linear oscillatory forcing motion at frequencies near the pendulum's resonance frequency f_0. (b) and (c) show plots of the bob's motion on a horizontal plane, the scale being in centimetres. (b) For a forcing frequency just above f_0 the motion of the bob settles down to a simple, regular pattern. (c) For a forcing frequency just below f_0 the bob shows 'chaotic' motion (although contained within a given region) which varies randomly and discontinuously as a function of the initial conditions.

Weather forecasting and chaos *continued*

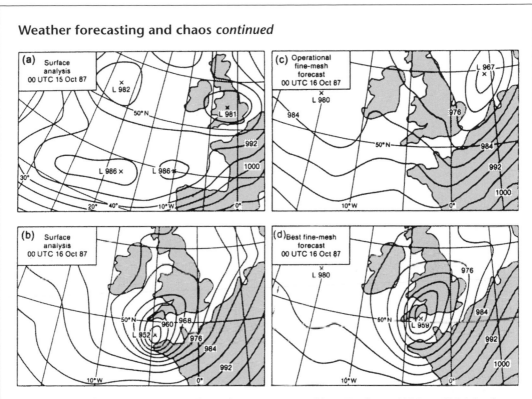

Figure 5.8 Analyses and forecasts of the surface pressure in millibars (1 mbar = 100 Pa = 1hPa) for the storm which passed over southern England on 15/16 October 1987. The fully developed storm is shown in (b) about 4 hours before it reached the London area; (a) is the situation twenty-four hours earlier from which the storm developed. (c) and (d) are twenty-four-hour forecasts with the Meteorological Office operational fine-mesh model, (c) being the one available at the time and (d) being one produced after the event using more complete data and better assimilation procedures.

given situation is to employ *ensemble* forecasting in which an ensemble of forecasts is run from a cluster of initial states that are generated by adding to an initial state small perturbations that are within the range of observational or analysis errors. The forecasts provided from the means of such ensembles show significant improvement compared with individual forecasts. Further, ensemble forecasts where the spread amongst the ensemble is low possess more skill that those where the spread in the ensemble is comparatively high.

Seasonal forecasting

So far short-term forecasts of detailed weather have been considered. After twenty days or so they run out of skill. What about further into the future? Although we cannot expect to forecast the weather in detail, is there any possibility of predicting the average weather, say, a few months ahead? As this section shows, it is possible for some parts of the world, because of the influence of the distribution of ocean surface temperatures on the atmosphere's behaviour. For seasonal forecasting it is no longer the initial state of the atmosphere about which detailed knowledge is required. Rather, we need to know the conditions at the surface and how they might be changing.

In the tropics, the atmosphere is particularly sensitive to sea surface temperature. This is not surprising because the largest contribution to the heat input to the atmosphere is due to evaporation of water vapour from the ocean surface and its subsequent condensation in the atmosphere, releasing its latent heat. Because the saturation water vapour pressure increases rapidly with temperature – at higher temperatures more water can be evaporated before the atmosphere is saturated – evaporation from the surface and hence the heat input to the atmosphere is particularly large in the tropics.

It is during El Niño events in the east tropical Pacific (see Figure 5.9) that the largest variations are found in ocean temperature. Anomalies in the circulation and rainfall in all tropical regions and to a lesser extent at

Figure 5.9 Changes in sea surface temperature 1871–2002, relative to the 1961–90 average, for the eastern tropical Pacific off Peru.

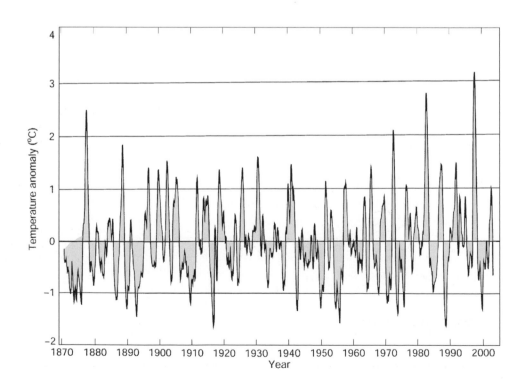

Forecasting for the African Sahel region

The Sahel region of Africa forms a band about 500 km wide along the southern edge of the Sahara Desert that gets most of its rainfall during northern hemisphere summer (particularly July to September). This region is known for its prolonged periods of drought, which can have a devastating impact on the local economy, an example of which occurred in the 1970s and 1980s (Figure 5.10). The droughts have been linked to fluctuations in several patterns of sea surface temperature (SST). Figure 5.11 illus-trates the difference in pattern of the mean world-wide SST between the five wettest and five driest Sahel years since 1901. This and other patterns form the basis of the statistical and atmosphere global circulation model seasonal forecasts that have been made by the UK Meteorological Office for the Sahel since 1986.[8] Long-lead forecasts are made in May for the July–September rainfall season and updated in early July.

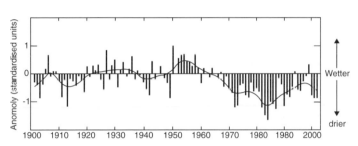

Figure 5.10 Annual rainfall differences from a long-term average (anomalies) for the Sahel in standardised units (the standard deviation from the average). The smoothed curve is approximately a decadal average.

A common difficulty with the forecasts has been that changes in the SST patterns occur between the time of the forecast and the main rainfall season. To overcome this problem, an atmosphere–ocean coupled model (see 'The climate system', next section) is being used to forecast changes in SST in addition to changes in atmospheric circulation. Forecast skill has improved although remaining modest (correlation between observed and modelled mean rainfall is about 0.37), in particular major changes to drier conditions are being picked up.[9] Such models, additionally coupled to variations in the characteristics of land surface vegetation and soil which have also been shown to be important,[10] provide the greatest promise for seasonal forecasts in the Sahel and in many other regions of the world.

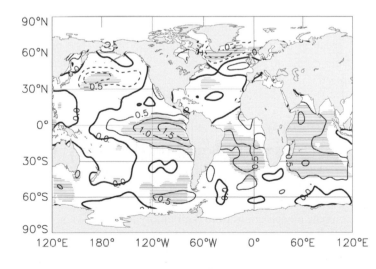

Figure 5.11 Mean difference in July–September SST (in °C) observed during the five driest years in the Sahel since 1900 (1972, 1982, 1983, 1984, 1987) and during the five wettest years (1927, 1950, 1953, 1954, 1958). Shaded areas show the most consistently (statistically significant) different regions.

mid latitudes are associated with these El Niño events (see Figure 1.4). A good test of the atmospheric models described above is to run them with an El Niño sequence of sea surface temperatures and see whether they are able to simulate these climate anomalies. This has now been done with a number of different atmospheric models; they have shown considerable skill in the simulation of many of the observed anomalies, especially those in the tropics and sub-tropics.[6]

Because of the large heat capacity of the oceans, anomalies of ocean surface temperature tend to persist for some months. The possibility therefore exists, for regions where there is a strong correlation between weather and patterns of ocean surface temperature, of making forecasts of climate (or average weather) some weeks or months in advance. Such seasonal forecasts have been attempted especially for regions with low rainfall; for instance, for north east Brazil and for the Sahel region of sub-Saharan Africa, a region where human survival is very dependent on the marginal rainfall (see box below).

The ability to make accurate seasonal forecasts significantly in advance depends critically on being able to forecast changes in ocean surface temperature. To do that requires understanding of, and the ability to model, the ocean circulation and the way it is coupled to the atmospheric circulation. Because the largest changes in ocean surface temperature occur in the tropics and because there are reasons to suppose that the ocean may be more predictable in the tropics than elsewhere, most emphasis on the prediction of ocean surface temperature has been placed in tropical regions, in particular on the prediction of the El Niño events themselves (see box below).

Later on in this chapter the coupling of atmospheric models and ocean models is described. For the moment it will suffice to say that, using coupled models together with detailed observations of both atmosphere and ocean in the Pacific region, significant skill in the prediction of El Niño events up to a year in advance has been achieved[7] (see also Chapter 7).

The climate system

So far the forecasting of detailed weather over a few days and of average weather for a month or so, up to perhaps a season ahead, has been described in order to introduce the science and technology of modelling, and also because some of the scientific confidence in the more elaborate climate models arises from their ability to describe and forecast the processes involved in day-to-day weather.

Climate is concerned with substantially longer periods of time, from a few years to perhaps a decade or longer. A description of the climate

A simple model of the El Niño

El Niño events are good examples of the strong coupling which occurs between the circulations of ocean and atmosphere. The stress exerted by atmospheric circulation – the wind – on the ocean surface is a main driver for the ocean circulation. Also, as we have seen, the heat input to the atmosphere from the ocean, especially that arising from evaporation, has a big influence on the atmospheric circulation.

A simple model of an El Niño event that shows the effect of different kinds of wave motions that can propagate within the ocean is illustrated in Figure 5.12. In this model a wave in the ocean, known as a Rossby wave, propagates westwards from a warm anomaly in ocean surface temperature near the equator. When it reaches the ocean's western boundary it is reflected as a different sort of wave, known as a Kelvin wave, which travels eastward. This Kelvin wave cancels and reverses the sign of the original warm anomaly, so triggering a cold event. The time taken for this half-cycle of the whole El Niño process is determined by the speed with which the waves propagate in the ocean; it takes about two years. It is essentially driven by ocean dynamics, the associated atmospheric changes being determined by the patterns of ocean surface temperature (and in turn reinforcing those patterns) which result from the ocean dynamics. Expressed in terms of this simple model, some of the characteristics of the El Niño process appear to be essentially predictable.

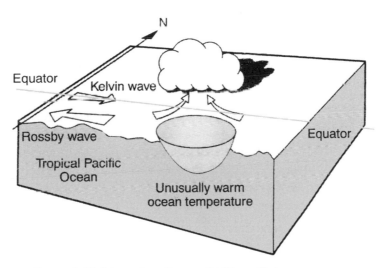

Figure 5.12 Schematic to illustrate El Niño oscillation

over a period involves the averages of appropriate components of the weather (for example, temperature and rainfall) over that period together with the statistical variations of those components. In considering the effect of human activities such as the burning of fossil fuels, changes in climate over periods of decades up to a century or two ahead must be predicted .

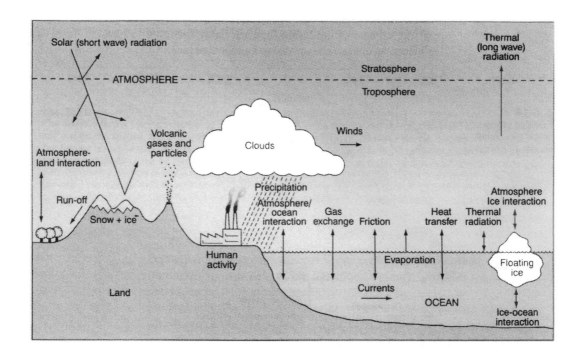

Figure 5.13 Schematic of the climate system.

Since we live in the atmosphere the variables commonly used to describe climate are mainly concerned with the atmosphere. But climate cannot be described in terms of atmosphere alone. Atmospheric processes are strongly coupled to the oceans (see above); they are also coupled to the land surface. There is also strong coupling to those parts of the Earth covered with ice (the cryosphere) and to the vegetation and other living systems on the land and in the ocean (the biosphere). These five components – atmosphere, ocean, land, ice and biosphere – together make up the climate system (Figure 5.13).

Feedbacks in the climate system

Chapter 2 considered the rise in global average temperature which would result from the doubling of the concentration of atmospheric carbon dioxide assuming that no other changes occurred apart from the increased temperature at the surface and in the lower atmosphere. The rise in temperature was found to be 1.2 °C. However, it was also established that, because of feedbacks (which may be positive or negative) associated with the temperature increase, the actual rise in global average temperature was likely to be approximately doubled to about 2.5 °C. This section lists the most important of these feedbacks.

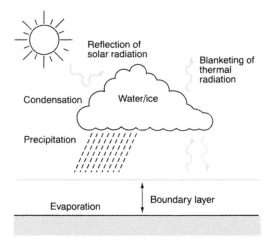

Figure 5.14 Schematic of the physical processes associated with clouds.

Water vapour feedback

This is the most important.[11] With a warmer atmosphere more evaporation occurs from the ocean and from wet land surfaces. On average, therefore, a warmer atmosphere will be a wetter one; it will possess a higher water vapour content. Since water vapour is a powerful greenhouse gas, on average a positive feedback results of a magnitude that models estimate to approximately double the increase in the global average temperature that would arise with fixed water vapour.[12]

Cloud-radiation feedback

This is more complicated as several processes are involved. Clouds interfere with the transfer of radiation in the atmosphere in two ways (Figure 5.14). Firstly, they reflect a certain proportion of solar radiation back to space, so reducing the total energy available to the system. Secondly, they act as blankets to thermal radiation from the Earth's surface in a similar way to greenhouse gases. By absorbing thermal radiation emitted by the Earth's surface below, and by themselves emitting thermal radiation, they act to reduce the heat loss to space by the surface.

   The effect that dominates for any particular cloud depends on the cloud temperature (and hence on the cloud height) and on its detailed optical properties (those properties which determine its reflectivity to solar radiation and its interaction with thermal radiation). The latter depend on whether the cloud is of water or ice, on its liquid or solid

Cloud radiative forcing

A concept helpful in distinguishing between the two effects of clouds mentioned in the text is that of cloud radiative forcing. Take the radiation leaving the top of the atmosphere above a cloud; suppose it has a value R. Now imagine the cloud to be removed, leaving everything else the same; suppose the radiation leaving the top of the atmosphere is now R'. The difference $R'-R$ is the cloud radiative forcing. It can be separated into solar radiation and thermal radiation. Values of the cloud radiative forcing deduced from satellite observations are illustrated in Figure 5.15. On average, it is found that clouds tend slightly to cool the Earth-atmosphere system.

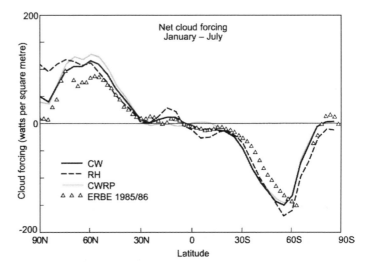

Figure 5.15 The cloud radiative forcing is made up of a solar radiation and a thermal radiation component, which generally act in opposite senses, each typically of magnitude between 50 and 100 W.m^{-2}. The average net forcing is shown here for the period January to July as a function of latitude as observed from satellites in two years (1985–6) of the Earth Radiative Budget Experiment (ERBE) and also as simulated in a climate model with different schemes of cloud formulation (a simple threshold relative humidity scheme (RH); a scheme that includes cloud water as a separate variable (CW); as CW but also with cloud radiative properties dependent on the cloud water content – thick clouds with a large water content are more reflective than thin clouds (CWRP)).There is encouraging agreement between the models' results and observations, but also differences that need to be understood. It is through comparisons of this kind that further elucidation of cloud radiation feedback will be achieved.

Table 5.1 *Estimates of global average temperature changes under different assumptions about changes in greenhouse gases and clouds*

Greenhouse gases	Clouds	Change (in °C) from current average global surface temperature of 15 °C
As now	As now	0
None	As now	−32
None	None	−21
As now	None	4
As now	As now but +3% high cloud	0.3
As now	As now but +3% low cloud	−1.0
Doubled CO₂ concentration otherwise as now	As now (no additional cloud feedback)	1.2
Doubled CO₂ concentration + best estimate of feedbacks	Cloud feedback included	2.5

water content (how thick or thin it is) and on the average size of the cloud particles. In general for low clouds the reflectivity effect wins so they tend to cool the Earth-atmosphere system; for high clouds, by contrast, the blanketing effect is dominant and they tend to warm the system. The overall feedback effect of clouds, therefore, can be either positive or negative (see box below).

Climate is very sensitive to possible changes in cloud amount or structure, as can be seen from the results of models discussed in later chapters. To illustrate this, Table 5.1 shows that the hypothetical effect on the climate of a small percentage change in cloud cover is comparable with the expected changes due to a doubling of the carbon dioxide concentration.

Ocean-circulation feedback

The oceans play a large part in determining the existing climate of the Earth; they are likely therefore to have an important influence on climate change due to human activities.

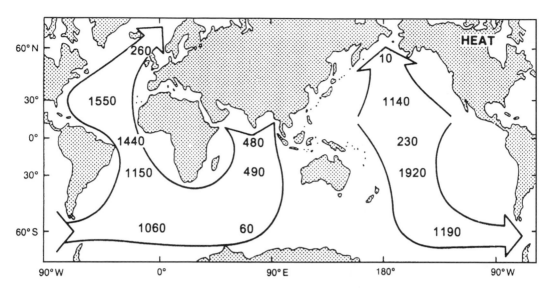

Figure 5.16 Estimates of transport of heat by the oceans. Units are terawatts (10^{12} W or million million watts). Note the linkages between the oceans and that some of the heat transported by the north Atlantic originates in the Pacific.

The oceans act on the climate in four important ways. Firstly, there are close interactions between the ocean and the atmosphere; they behave as a strongly coupled system. As we have already noted, evaporation from the oceans provides the main source of atmospheric water vapour which, through its latent heat of condensation in clouds, provides the largest single heat source for the atmosphere. The atmosphere in its turn acts through wind stress on the ocean surface as the main driver of the ocean circulation.

Secondly, they possess a large heat capacity compared with the atmosphere, in other words a large quantity of heat is needed to raise the temperature of the oceans only slightly. In comparison, the entire heat capacity of the atmosphere is equivalent to less than three metres depth of water. That means that in a world that is warming, the oceans warm much more slowly than the atmosphere. We experience this effect of the oceans as they tend to reduce the extremes of atmospheric temperature. For instance, the range of temperature change both during the day and seasonally is much less at places near the coast than at places far inland. The oceans therefore exert a dominant control on the rate at which atmospheric changes occur.

Thirdly, through their internal circulation the oceans redistribute heat throughout the climate system. The total amount of heat transported from the equator to the polar regions by the oceans is similar to that transported by the atmosphere. However, the regional distribution of that transport is very different (Figure 5.16). Even small changes in the regional heat transport by the oceans could have large implications for climate change. For instance, the amount of heat transported by the north Atlantic Ocean is over 1000 terawatts (1 terawatt = 1 million million watts = 10^{12} watts). To give an idea of how large this is, we can note that a large power station puts out about 1000 million (10^9) watts and the total amount of commercial energy produced globally is about twelve terawatts. To put it further in context, considering the region of the north Atlantic Ocean between north west Europe and Iceland, the heat input (Figure 5.16) carried by the ocean circulation is of similar magnitude to that reaching the ocean surface there from the incident solar radiation. Any accurate simulation of likely climate change, therefore, especially of its regional variations, must include a description of ocean structure and dynamics.

Ice-albedo feedback

An ice or snow surface is a powerful reflector of solar radiation (the albedo is a measure of its reflectivity). As some ice melts, therefore, at the warmer surface, solar radiation which had previously been reflected back to space by the ice or snow is absorbed, leading to further increased warming. This is another positive feedback which on its own would increase the global average temperature rise due to doubled carbon dioxide by about twenty per cent.

Four feedbacks have been identified, all of which play a large part in the determination of climate, especially its regional distribution. It is therefore necessary to introduce them into climate models. Because the global models allow for regional variation and also include the important non-linear processes in their formulation, they are able in principle to provide a full description of the effect of these feedbacks. They are, in fact, the only tools available with this potential capability. It is to a description of climate prediction models that we now turn.

Models for climate prediction

For models to be successful they need to include an adequate description of the feedbacks we have listed. The water vapour feedback and its regional distribution depend on the detailed processes of evaporation, condensation and advection (the transfer of heat by horizontal air flow) of water vapour, and on the way in which convection processes (responsible for showers and thunderstorms) are affected by higher surface temperatures. All these processes are already well included in weather forecasting models, and water vapour feedback has been very thoroughly studied.[13] The most important of the others are cloud radiation feedback and ocean-circulation feedback. How are these incorporated into the models?

For modelling purposes, clouds divide into two types – *layer* clouds present on scales larger than the grid size and *convective* clouds generally on smaller scales than a grid box. For the introduction of layer clouds, early weather-forecasting and climate models employed comparatively simple schemes. A typical scheme would generate cloud at specified levels whenever the relative humidity exceeded a critical value, chosen for broad agreement between model-generated cloud cover and that observed from climatological records. More recent models parameterise the processes of condensation, freezing, precipitation and cloud formation much more completely. They also take into account detailed cloud properties (e.g. water droplets or ice crystals and droplet number and size) that enable their radiative properties (e.g. their reflectivity and transmissivity) to be specified sufficiently well for the influence of clouds on the atmosphere's overall energy budget to be properly described. The most sophisticated models also include allowance for the effect of aerosols on cloud properties – denoted the indirect aerosol effect in Figure 3.8. The effects of convective clouds are incorporated as part of the model's scheme for the parametrisation of convection.

The amount and sign (positive or negative) of the average cloud-radiation feedback in a particular climate model is dependent on many aspects of the model's formulation as well as on the particular scheme used for the description of cloud formation. Different climate models, therefore, can show average cloud-radiation feedback that can be either positive or negative; further, the feedback can show substantial regional variation. For instance, models differ in their treatment of low cloud such that in some models the amount of low cloud increases with increased greenhouse gases, with other models it decreases. Uncertainty regarding cloud-radiation feedback is the main reason for the wide uncertainty range in what is called *climate sensitivity* (see Chapter 6) or the change in global average surface temperature due a doubling of carbon dioxide concentration.

The second important feedback is that due to the effects of the ocean circulation. Compared with a global atmospheric model for weather forecasting, the most important elaboration of a climate model is the inclusion of the effects of the ocean. Early climate models only included the ocean very crudely; they represented it by a simple slab some fifty or one hundred metres deep, the approximate depth of the 'mixed layer' of

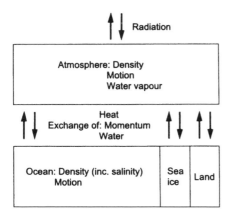

Figure 5.17 Component elements and parameters of a coupled atmosphere–ocean model including the exchanges at the atmosphere–ocean interface.

ocean which responds to the seasonal heating and cooling at the Earth's surface. In such models, adjustments had to be made to allow for the transport of heat by ocean currents. When running the model with a perturbation such as increased carbon dioxide, it was not possible to make allowance for any changes in that transport which might occur. Such models therefore possess severe limitations.

For an adequate description of the influence of the ocean it is necessary to model the ocean circulation and its coupling to the atmospheric circulation. Figure 5.17 shows the ingredients of such a model. For the atmospheric part of the model, in order to accommodate long runs on available computers, the size of the grid has to be substantially larger, typically 300 km in the horizontal. Otherwise it is essentially the same as the global model for weather forecasting described earlier. The formulation of the dynamics and physics of the ocean part of the model is similar to that of its atmospheric counterpart. The effects of water vapour are of course peculiar to the atmosphere, but the salinity (the salt content) of the oceans has to be included as a parameter together with its considerable effects on the water density. Because dynamical systems, e.g. largescale eddies in the oceans, are of smaller scale than their atmospheric counterparts, the grid size of the ocean component is typically about half that of the atmospheric component. On the other hand, because changes in the ocean are slower, the time step for model integration can be greater for the ocean component.

At the ocean–atmosphere interface there is exchange of heat, water and momentum (exchange of momentum leads to friction) between the two fluids. The importance of water in the atmosphere and its influence on the atmospheric circulation have already been shown. The distribution of fresh water precipitated from the atmosphere as rain or snow also has a large influence on the ocean's circulation through its effect on the distribution of salt in the ocean, which in turn affects the ocean density. It is not surprising, therefore, to find that the 'climate' described by the model is quite sensitive to the size and the distribution of water exchanges at the interface.

Before the model can be used for prediction it has to be run for a considerable time until it reaches a steady 'climate'. The 'climate' of the model, when it is run unperturbed by increasing greenhouse gases, should be as close as possible to the current actual climate. If the exchanges are not correctly described, this will not be the case. Getting the exchanges right in the model has proved to be difficult. Until recently, many coupled models introduced artificial adjustments to the fluxes (known as flux adjustments) so as to ensure that the model's 'climate' is as identical as possible to the current climate. However, the ocean component of the model has been improved especially through introducing higher resolution (150 km or less), so the need for such adjustments has diminished and many models are now able to provide an adequate description of the climate with no such adjustments.

Before leaving the oceans, there is a particular feedback which should be mentioned between the hydrological cycle and the deep ocean circulation (see box below). Changes in rainfall, by altering the ocean salinity, can interact with the ocean circulation. This could affect the climate, particularly of the North Atlantic region; it may also have been responsible for some dramatic climate changes in the past (see Chapter 4).

The most important feedbacks belong to the atmospheric and the ocean components of the model. They are the largest components, and, because they are both fluids and have to be dynamically coupled together, their incorporation into the model is highly demanding. However, another feedback to be modelled is the ice-albedo feedback, which arises from the variations of sea-ice and of snow.

Sea-ice covers a large part of the polar regions in the winter. It is moved about by both the surface wind and the ocean circulation. So that the ice-albedo feedback can be properly described, the growth, decay and dynamics of sea-ice have to be included in the model. Land ice is also included, essentially as a boundary condition – a fixed quantity – because its coverage changes little from year to year. However, the model needs to show whether there are likely to be changes in ice volume, even though these are small, in order to find out their effect on sea level (Chapter 7 considers the impacts of sea-level change).

Interactions with the land surface must also be adequately described. The most important properties for the model are land surface wetness or, more precisely, soil moisture content (which will determine the amount

The ocean's deep circulation

For climate change over periods up to a decade, only the upper layers of the ocean have any substantial interaction with the atmosphere. For longer periods, however, links with the deep ocean circulation become important. The effects of changes in the deep circulation are of particular importance

Experiments using chemical tracers, for instance those illustrated in Figure 5.20 (see next box), have been helpful in indicating the regions where strong coupling to the deep ocean occurs. To sink to the deep ocean, water needs to be particularly dense, in other words both cold and salty. There are two main regions where such dense water sinks down to the deep ocean, namely in the north Atlantic Ocean (in the Greenland sea between Scandinavia and Greenland and the Labrador sea

west of Greenland) and in the region of Antarctica. Salt-laden deep water formed in this way contributes to a deep ocean circulation that involves all the oceans (Figure 5.18) and is known as the *thermohaline circulation* (THC).

In Chapter 4 we mentioned the link between the THC and the melting of ice. Increases in the ice melt can lead to the ocean surface water becoming less salty and therefore less dense. It will not sink so readily, the deep water formation will be inhibited and the THC is weakened. In Chapter 6, the link between the THC and the hydrological (water) cycle in the atmosphere is mentioned. Increased precipitation in the north Atlantic region, for instance, can lead to a weakening of the THC.

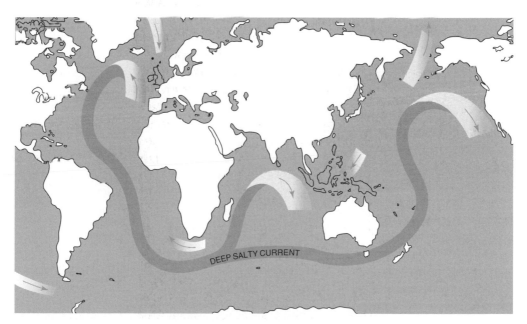

DEEP SALTY CURRENT

Figure 5.18 Deep water formation and circulation. The deep salty current originates in the north Atlantic. Northward flowing water near the surface that is unusually salty becomes cooler and even saltier through evaporation, so increasing its density and causing it to sink.

of evaporation) and albedo (reflectivity to solar radiation). The models keep track of the changes in soil moisture through evaporation and precipitation. The albedo depends on soil type, vegetation, snow cover and surface wetness.

Validation of the model

In discussing various aspects of modelling we have already indicated how some validation of the components of climate models may be carried out. The successful predictions of weather-forecasting models provide validation of important aspects of the atmospheric component, as do the simulations mentioned earlier in the chapter of the connections between sea surface temperature anomalies and precipitation patterns in some parts of the world. Various tests have also been carried out of the ocean component of climate models; for instance, through comparisons between the simulation and observation of the movement of chemical tracers (see box below).

Once a comprehensive climate model has been formulated it can be tested in three main ways. Firstly, it can be run for a number of years of simulated time and the climate generated by the model compared in detail to the current climate. For the model to be seen as a valid one, the average distribution and the seasonal variations of appropriate parameters such as surface pressure, temperature and rainfall have to compare well with observation. In the same way, the variability of the model's climate must be similar to the observed variability. Climate models that are currently employed for climate prediction stand up well to such comparisons.[14]

Secondly, models can be compared against simulations of past climates when the distribution of key variables was substantially different than at present; for example, the period around 9000 years ago when the configuration of the Earth's orbit around the Sun was different (see Figure 5.19). The perihelion (minimum Earth–Sun distance) was in July rather than in January as it is now; also the tilt of the Earth's axis was slightly different from its current value (24° rather than 23.5°). Resulting from these orbital differences (see Chapter 4), there were significant differences in the distribution of solar radiation throughout the year. The incoming solar energy when averaged over the northern hemisphere was about seven per cent greater in July and correspondingly less in January.

When these altered parameters are incorporated into a model, a different climate results. For instance, northern continents are warmer in summer and colder in winter. In summer a significantly expanded low

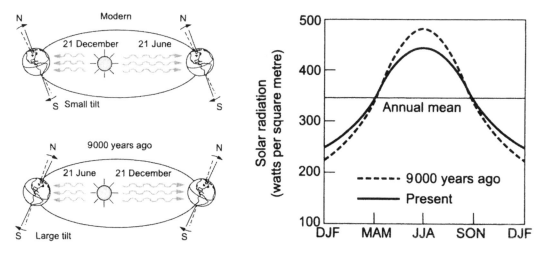

Figure 5.19 Changes in the Earth's elliptical orbit from the present configuration to 9000 years ago and (right hand side) changes in the average solar radiation during the year over the northern hemisphere.

pressure region develops over north Africa and south Asia because of the increased land–ocean temperature contrast. The summer monsoons in these regions are strengthened and there is increased rainfall. These simulated changes are in qualitative agreement with paleoclimate data; for example, these data provide evidence for that period (around 9000 years ago) of lakes and vegetation in the southern Sahara about 1000 km north of the present limits of vegetation.

The accuracy and the coverage of data available for these past periods are limited. However, the model simulations for 9000 years ago, described above, and those for other periods in the past have demonstrated the value of such studies in the validation of climate models.

A third way in which models can be validated is to use them to predict the effect of large perturbations on the climate. Good progress is being achieved with the prediction of El Niño events and the associated climate anomalies up to a year ahead (see earlier in the chapter). Other short-term perturbations are due to volcanic eruptions, the effects of which were mentioned in Chapter 1. Several climate models have been run in which the amount of incoming solar radiation has been modified to allow for the effect of the volcanic dust from Mount Pinatubo, which erupted in 1991 (Figure 5.21). Successful simulation of some of the regional anomalies of climate which followed that eruption, for instance the unusually cold winters in the Middle East and the mild winters in western Europe, has also been achieved by the models.[15]

Modelling of tracers in the ocean

A test that assists in validating the ocean component of the model is to compare the distribution of a chemical tracer as observed and as simulated by the model. In the 1950s radioactive tritium (an isotope of hydrogen) released in the major atomic bomb tests entered the oceans and was distributed by the ocean circulation and by mixing. Figure 5.20 shows good agreement between the observed distribution of tritium (in tritium units) in a section of the western north Atlantic Ocean about a decade after the major bomb tests and the distribution as simulated by a twelve-level ocean model. Similar comparisons have been made more recently of the measured uptake of one of the freons CFC-11, whose emissions into the atmosphere have increased rapidly since the 1950s, compared with the modelled uptake.

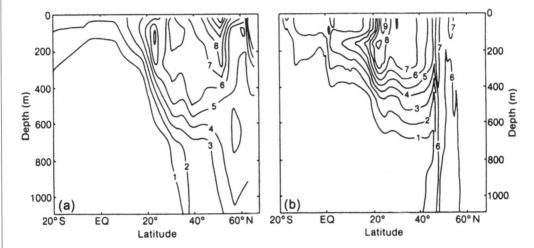

Figure 5.20 The tritium distribution in a section of the western north Atlantic Ocean approximately one decade after the major atomic bomb tests, as observed in the GEOSECS programme (a) and as modelled (b).

In these three ways, which cover a range of timescales, confidence has been built in the ability of models to predict climate change due to human activities.

Comparison with observations

More than fifteen centres in the world located in ten countries are currently running climate models of the kind we have described in which the circulations of the atmosphere and the ocean are fully coupled together. Some of these models have been employed to simulate the climate of the last 150 years allowing for variations in aspects of natural forcing (e.g. solar variations and volcanoes) and the increases in the concentrations of greenhouse gases and aerosols.

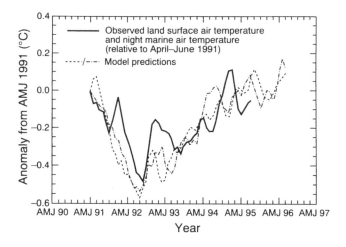

Figure 5.21 The predicted and observed changes in global land and ocean surface air temperature after the eruption of Mount Pinatubo, in terms of three-month running averages from April to June 1991 to March to May 1995.

An example of such simulations is shown in Figure 5.22, where the observed record of global average surface air temperature is compared with model simulations taking into account in turn natural forcings, anthropogenic forcings (i.e. the increase in greenhouse gases and aerosols) and the combination of natural and anthropogenic forcings. Although the simulations in Figure 5.22 are based on one model, similar results have been obtained from many models.

Three interesting features of Figure 5.22 can be noted. Firstly, that the inclusion of anthropogenic forcings provides a plausible explanation for a substantial part of the observed temperature changes over the last century (especially for the latter part of the century), but that the best match with observations occurs when both natural and anthropogenic factors are included. In particular it is likely that changes in solar output and the comparative absence of volcanic activity were important variations in natural forcing factors during the first part of the twentieth century. Secondly, the model simulations show variability up to a tenth of a degree Celsius or more over periods of a few years up to decades. This variability is due to internal exchanges in the model between different parts of the climate system, and is not dissimilar to that which appears in the observed record. Thirdly, due to the slowing effect of the oceans on climate change, the warming observed or modelled so far is less than would be expected if the climate system were in equilibrium under the amount of radiative forcing due to the current increase in greenhouse gases and aerosols.

There remains, however, a large amount of natural variability in both the observations and the simulations and much debate has taken place over the last decade or more about the strength of the evidence that global warming due to the increase in greenhouse gases has actually

Simulated annual global mean surface temperatures

(a) Natural

(b) Anthropogenic

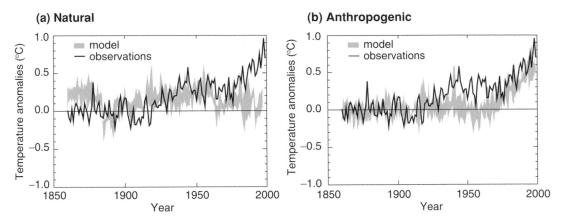

Figure 5.22 Annual global
mean surface temperatures
simulated by a climate model
compared with observations
for the period 1860–2000.
The simulations in (a) were
done with only natural
forcings – solar variation and
volcanic activity; in (b) with
only anthropogenic forcings –
greenhouse gases and
sulphate aerosols; and in (c)
with both natural and
anthropogenic forcings
combined. The simulations are
shown in a band that covers
the results from four runs with
the same model and therefore
illustrates the range of natural
variability within the model.

(c) All forcings

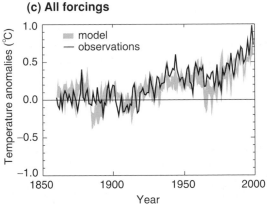

been observed in the climate record. In other words has the 'signal' that
can be attributed to global warming risen sufficiently above the 'noise'
of natural variability? The Intergovernmental Panel on Climate Change
(IPCC) has been much involved in this debate.

The IPCC's first Report published in 1990[16] made a carefully worded
statement to the effect that, although the size of the observed warming
is broadly consistent with the predictions of climate models, it is also of
similar magnitude to natural climate variability. An unequivocal statement that anthropogenic climate change had been detected could not
therefore be made at that time. By 1995 more evidence was available
and the IPCC 1995 Report[17] therefore reached the cautious conclusion
as follows.

Our ability to quantify the human influence on global climate is currently limited because the expected signal is still emerging from the noise of natural climate variability, and because there are uncertainties in key factors. These include the magnitude and patterns of long term natural variability and the time-evolving pattern of forcing by, and response to, changes in the concentrations of greenhouse gases and aerosols, and land surface changes. Nevertheless, the balance of evidence suggests a discernible influence on global climate.

Since 1995 a large number of studies have addressed the problems of *detection* and *attribution*[18] of climate change. Better estimates of natural variability have been made, especially using models, and the conclusion reached that the warming over the last one hundred years is very unlikely to be due to natural variability alone.[19] In addition to studies using globally averaged parameters, there have been detailed statistical studies using pattern correlations based on optimum detection techniques applied to both model results and observations. For example, Figure 5.23 shows a comparison between simulated and observed estimates of zonal mean temperature change as a function of altitude. Taking these studies

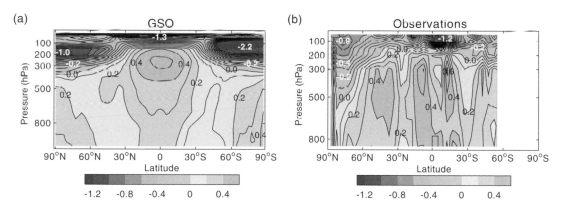

Figure 5.23 Simulated and observed zonal mean temperature change as a function of latitude and height. The changes plotted are the differences between the 1986–1995 decadal average and the twenty-year average from 1961 to 1980. The contour interval is 0.1 °C. (a) simulated changes taking into account increases in carbon dioxide, sulphate aerosols and the effect of observed changes in stratospheric ozone; (b) observed changes. A common pattern of stratospheric cooling and tropospheric warming is evident in the observations and both model experiments; the stratospheric cooling is partially due to the increase of carbon dioxide (see also Figure 4.2(b)) and the reduction in ozone (see Chapter 3).

into account, the conclusion reached in the IPCC 2001 Report[20] is as follows.

> In the light of new evidence and taking into account the remaining uncertainties, most of the observed warming over the last 50 years is likely to have been due to the increase in greenhouse gas concentrations.

Confidence having been established in climate models in the ways we have outlined in the last two sections, these models can now be used to generate projections of the likely climate change in the future due to human activities. Details of such projections will be presented in the next chapter.

Before leaving comparison with observations, I should mention some recent work relating to the warming of the ocean that adds further confirmation to the picture that has been presented. In Chapter 2, the effect of an increase of greenhouse gases was expressed in terms of radiative forcing or, in other words, a net input of heat energy into the earth-atmosphere system. Most of this extra energy is stored in the ocean. The amount of this extra energy has been estimated from measurements of the temperature increase in the ocean at different locations and depths down to 3 km. From 1957 to 1994 the estimate is $(19 \pm 9) \times 10^{22}$ joules, the uncertainty relating mostly to inadequate sampling of some large ocean areas.[21] This amount is a little over 0.5% of the solar energy received by the Earth over this period. Within the limits of uncertainty, it agrees well with the estimates of radiative forcing presented earlier and also with model estimates of ocean heat uptake.[22]

Is the climate chaotic?

Throughout this chapter the implicit assumption has been made that climate change is predictable and that models can be used to provide predictions of climate change due to human activities. Before leaving this chapter I want to consider whether this assumption is justified.

The capability of the models themselves has been demonstrated so far as weather forecasting is concerned. They also possess some skill in seasonal forecasting. They can provide a good description of the current climate and its seasonal variations. Further, they provide predictions which on the whole are reproducible and which are reasonably consistent between different models. But, it might be argued, this consistency could be a property of the models rather than of the climate. Is there any other evidence to support the view that the climate is predictable?

A good place to look for further evidence is in the record of climates of the past, presented in Chapter 4. Correlation between the Milankovitch cycles in the Earth's orbital parameters and the cycles of climate change over the past half million years (see Figures 4.4 and 5.19) provides strong evidence to substantiate the Earth's orbital variations as the main factor responsible for the triggering of climate change. The nature of the feedbacks which control the very different amplitudes of response to the three orbital variations still need to be understood. Some $60 \pm 10\%$ of the variance in the record of global average temperature from paleontological sources over the past million years occurs close to frequencies identified in the Milankovitch theory. The existence of this surprising amount of regularity suggests that the climate system is not strongly chaotic so far as these large changes are concerned, but responds in a largely predictable way to the Milankovitch forcing.

This Milankovitch forcing arises from changes in the distribution of solar radiation over the Earth because of variations in the Earth's orbit. Changes in climate as a result of the increase of greenhouse gases are also driven by changes in the radiative regime at the top of the atmosphere. These changes are not dissimilar in kind (although different in distribution) from the changes that provide the Milankovitch forcing. It can be argued therefore that the increases in greenhouse gases will also result in a largely predictable response.

Regional climate modelling

The simulations we have so far described in this chapter are with global circulation models (GCM) that typically possess a horizontal resolution (grid size) of around 300 km – the size being limited primarily by the availability of computer power. Weather and climate on scales large compared with the grid size are described reasonably well. However, at scales comparable with the grid size, described as the regional scale,[23] the results from global models possess serious limitations. The effects of forcings and circulations that exist on the regional scale need to be properly represented. For instance, patterns of precipitation depend critically on the major variations in orography and surface characteristics that occur on this scale (see Figure 5.24). Patterns generated by a global model therefore will be a poor representation of what actually occurs on the regional scale.

To overcome these limitations regional modelling techniques have been developed.[24] That most readily applicable to climate simulation and prediction is the *Regional Climate Model* (RCM). A model covering an appropriate region at a horizontal resolution of say 25 or 50 km can

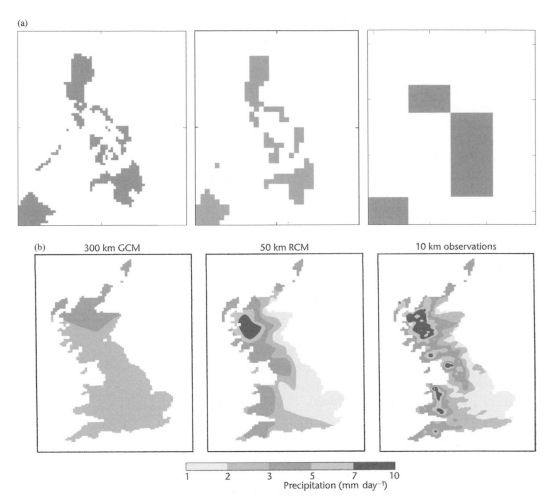

Figure 5.24 (a) Representation of the Philippines in RCMs with resolutions of 25 km and 50 km and in a GCM with 400 km resolution; (b) patterns of present-day winter precipitation over Great Britain, (i) as simulated with a 300 km resolution global model, (ii) with 50 km resolution regional model, (iii) as observed with 10 km resolution.

be 'nested' in a global model. The global model provides information about the response of the global circulation to large-scale forcings and the evolution of boundary information for the RCM. Within the region, physical information, for instance concerning forcings, is entered on the scale of the regional grid and the evolution of the detailed circulation is developed within the RCM. The RCM is able to account for forcings on smaller scales than are included in the GCM (e.g. due to topography or land cover inhomogeneity, see Figure 5.24) and can also simulate atmospheric circulations and climate variables on these smaller scales.

A limitation of the regional modelling technique we have described is that, although the global model provides the boundary inputs for the RCM, the RCM provides no interaction back on to the global model. As larger computers become available it will be possible to run global models at substantially increased resolution so that this limitation becomes less serious; at the same time RCMs will acquire an ability to deal with detail on even smaller scales. Some examples of regional model simulations are given in Chapter 6 (Figure 6.10).

Another technique is that of *Statistical Downscaling* that has been widely employed in weather forecasting. This uses statistical methods to relate large scale climate variables (or 'predictors') to regional or local variables. The predictors from a global circulation climate model can be fed into the statistical model to estimate the corresponding regional climate characteristics. The advantage of this technique is that it can easily be applied. Its disadvantage from the point of view of simulating climate change is that it is not possible to be sure how far the statistical relations apply to a climate-changed situation.

The future of climate modelling

Very little has been said in this chapter about the biosphere. Chapter 3 referred to comparatively simple models of the carbon cycle which include chemical and biological processes and simple non-interactive descriptions of atmospheric processes and ocean transport. The large three-dimensional global circulation climate models described in this chapter contain a lot of dynamics and physics but no interactive chemistry or biology. As the power of computers increases, global dynamical and physical circulation models that couple in the biological and chemical processes that make up the carbon cycle and the chemistry of other gases are now being developed. Before very long we can expect that models will be available that are fully interactive and comprehensive in their inclusion of dynamical, physical, chemical and biological processes in the atmosphere, the ocean and on the land.

Climate modelling continues to be a rapidly growing science. Although useful attempts at simple climate models were made with early computers it is only during the last ten years or so that computers have been powerful enough for coupled atmosphere–ocean models to be employed for climate prediction and that their results have been sufficiently comprehensive and credible for them to be taken seriously by policy makers. The climate models which have been developed are probably the most elaborate and sophisticated of computer models developed in any area of natural science. Further, climate models that describe the natural science of climate are now being coupled with socio-economic information in integrated assessment models (see box in Chapter 9, page 237).

As the power of computers increases it becomes more possible to investigate the sensitivity of models by running a variety of ensembles that include different initial conditions, model parameterisations and formulations. A particularly interesting project[25] involves thousands of computer users around the world in running state-of-the-art climate prediction models on their home, school or work computers. By collating data from thousands of models it will generate the world's largest climate modelling prediction experiment.

A great deal remains to be done to narrow the uncertainty of model predictions. The first priorities must be to improve the modelling of clouds and the description in the models of the ocean–atmosphere interaction. Larger and faster computers are required to tackle this problem, especially to enable the resolution of the model grid to be increased, as well as more sophisticated model physics and dynamics. Much more thorough observations of all components of the climate system are also necessary, so that more accurate validation of the model formulations can be achieved. Further, regional climate modelling techniques will develop rapidly as they are applied to a wide variety of situations. Very substantial national and international programmes are underway to address all these issues.

Questions

1   Make an estimate of the speed in operations per second of Richardson's 'people' computer. Do you agree with the estimate in Figure 5.1?

2   If the spacing between the grid points in a model is 100 km and there are twenty levels in the vertical, what is the total number of grid points in a global model? If the distance between grid points in the horizontal is halved, how much longer will a given forecast take to run on the computer?

3   Take your local weather forecasts over a week and describe their accuracy for twelve, twenty-four and forty-eight hours ahead.

4   Estimate the average energy received from the Sun over a square region of the ocean surface, one side of the square being a line between northern Europe and Iceland. Compare with the average transport of energy into the region by the North Atlantic Ocean (Figure 5.16).

5   Take a hypothetical situation in which a completely absorbing planetary surface at a temperature of 280 K is covered by a non-absorbing and non-emitting atmosphere. If a cloud which is non-absorbing in the visible part of the spectrum but completely absorbing in the thermal infrared is present above the surface, show that its equilibrium temperature will be 235 K ($=280/2^{0.25}$ K).[26] Show also that if the cloud reflects fifty per cent of

solar radiation, the rest being transmitted, the planet's surface will receive the same amount of energy as when the cloud is absent. Can you substantiate the statement that the presence of low clouds tends to cool the Earth while high clouds tend towards warming of it?

6   Associated with the melting of sea-ice which results in increased evaporation from the water surface, additional low cloud can appear. How does this affect the ice-albedo feedback? Does it tend to make it more or less positive?

7   Work out the total energy received by the Earth from the Sun over the thirty-seven-year period from 1957 to 1994; allow for that lost by reflection and scattering to space. What precise proportion of this is: (1) the total radiative forcing over this period due to increased greenhouse gases (see for instance Figure 3.8) and (2) the energy absorbed by the ocean as derived by Levitus *et al.* (page 106)? Comment on your results.

8   It is sometimes argued that weather and climate models are the most sophisticated and soundly based models in natural science. Compare them (e.g. in their assumptions, their scientific basis, their potential accuracy, etc.) with other computer models with which you are familiar both in natural science and social science (e.g. models of the economy).

Notes for Chapter 5

1   Further information regarding the subject of this chapter can be found in the following texts:

Houghton, J. T. 1991. The Bakerian Lecture, 1991: The predictability of weather and climate. *Philosophical Transactions of the Royal Society, London, A*, **337**, pp. 521–71.

Houghton, J. T. 2002. *The Physics of Atmospheres*, third edition. Cambridge: Cambridge University Press.

Houghton, J. T., Jenkins, G. J., Ephraums, J. J. (eds.) 1990. *Climate Change: the IPCC Scientific Assessments*. Cambridge: Cambridge University Press.

Houghton, J. T., Callander, B. A., Varney, S. K. (eds.) 1992. *Climate Change 1992: the Supplementary Report to the IPCC Scientific Assessments*. Cambridge: Cambridge University Press.

Houghton, J. T., Meira Filho, L. G., Callander, B. A., Harris, N., Kattenberg, A., Maskell, K. (eds.) 1996. *Climate Change 1995: the Science of Climate Change*. Cambridge: Cambridge University Press.

Houghton, J. T., Ding, Y., Griggs, D. J., Noguer, M., van der Linden,P. J., Dai, X., Maskell, K., Johnson, C. A. (eds.) 2001. *Climate Change 2001: the Scientific Basis*. Cambridge: Cambridge University Press.

McGuffie, K., Henderson-Sellers, A. 1997. *A Climate Modelling Primer*, second edition. New York, Wiley.

Trenberth, K. E. (ed.) 1992. *Climate System Modelling*. Cambridge: Cambridge University Press.

2   Richardson, L. F. 1922. *Weather Prediction by Numerical Processes.* Cambridge: Cambridge University Press. Reprinted by Dover, 1965.

3   For more details see, for instance, Houghton, *The Physics of Atmospheres.*

4   For more detail see:
Chapter 13 in Houghton, *The Physics of Atmospheres.*
Palmer, T. N. 1993. A nonlinear perspective on climate change. *Weather*, **48**, pp. 314–26.
Palmer, T. N. 1999. *Journal of Climatology*, **12**, pp. 575–91.

5   An equation such as $y = ax + b$ is linear; a plot of y against x is a straight line. Examples of non-linear equations are $y = ax^2 + b$ or $y + xy = ax + b$; plots of y against x for these equations would not be straight lines. In the case of the pendulum, the equations describing the motion are only approximately linear for very small angles from the vertical where the sine of the angle is approximately equal to the angle; at larger angles this approximation becomes much less accurate and the equations are non-linear.

6   More detail in McAvaney, B. J. *et al.* 2001. In Houghton, J. T., Ding, Y., Griggs, D. J., Noguer, M., van der Linden, P. J., Dai, X., Maskell, K., Johnson, C. A. (eds.) *Climate Change 2001: The Scientific Basis. Contribution of Working Group I to the Third Assessment Report of the Intergovernmental Panel on Climate Change.* Cambridge: Cambridge University Press, Chapter 8.

7   See, for instance:
Cane, M. A. 1992. In Trenberth, K. E. (ed.) *Climate System Modelling.* Cambridge: Cambridge University Press, pp. 583–616.
McAvaney *et al.*, Chapter 8, in Houghton, *Climate Change 2001.*
Federov, A. V. *et al.* 2003. How predictable is El Niño? *Bulletin of the American Meteorological Society*, **84**, pp. 911–919.

8   Folland, C. K., Owen, J., Ward, M. N., Colman, A. 1991. Prediction of seasonal rainfall in the Sahel region using empirical and dynamical methods. *Journal of Forecasting*, **10**, pp. 21–56.

9   Information from Folland, C., Hadley Centre, UK.

10 Xue, Y. 1997. Biospheric feedback on regional climate in tropical north Africa. *Quarterly Journal of the Royal Meteorological Society*, **123**, pp. 1483–1515.

11 Associated with water vapour feedback is also *lapse rate feedback* which occurs because, associated with changes of temperature and water vapour content in the troposphere, are changes in the average lapse rate (the rate of fall of temperature with height). Such changes lead to this further feedback, which is generally much smaller in magnitude than water vapour feedback but of the opposite sign, i.e. negative instead of positive. Frequently, when overall values for water vapour feedback are quoted the lapse rate feedback has been included. For more details see Houghton, *The Physics of Atmospheres.*

12 Stocker, T. F. *et al*. 2001. Physical climate processes and feedbacks. In Houghton, J. T., Ding, Y., Griggs, D. J., Noguer, M., van der Linden, P. J., Dai, X., Maskell, K., Johnson, C. A. (eds.) *Climate Change 2001: The Scientific Basis. Contribution of Working Group I to the Third Assessment Report of the Intergovernmental Panel on Climate Change.* Cambridge: Cambridge University Press, Section 7.2.1.1.

13 Lindzen, R. S. 1990. Some coolness concerning global warming. *Bulletin of the American Meteorological Society*, **71**, pp. 288–99. In this paper, Lindzen queries the magnitude and sign of the feedback due to water vapour, especially in the upper troposphere, and suggests that it could be much less positive than predicted by models and could even be slightly negative. Much has been done through observational and modelling studies to investigate the likely magnitude of water vapour feedback. More detail can be found in Stocker, T. F. *et al*. 2001. Physical climate processes and feedbacks. In Houghton, *Climate Change 2001*, Chapter 7. The conclusion of that chapter, whose authors include Linzden, is that 'the balance of evidence favours a positive clear-sky water vapour feedback of a magnitude comparable to that found in simulations.

14 For more details see McAvaney, B. J. *et al*. 2001. Chapter 8 in Houghton, *Climate Change 2001*.

15 Graf, H.-E. *et al*, 1993. Pinatubo eruption winter climate effects. model versus observations. *Climate Dynamics*, **9**, pp. 61–73.

16 See Policymakers summary. In Houghton, J. T., Jenkins, G. J., Ephraums, J. J. (eds.) 1990. *Climate Change: the IPCC Scientific Assessment.* Cambridge: Cambridge University Press.

17 See Summary for policymakers. In Houghton, J. T., Meira Filho, L. G., Callander, B. A., Harris, N., Kattenberg, A., Maskell, K. (eds.) 1996. *Climate Change 1995: the Science of Climate Change.* Cambridge: Cambridge University Press.

18 Detection is the process demonstrating that an observed change is significantly different (in a statistical sense) than can be explained by natural variability. Attribution is the process of establishing cause and effect with some defined level of confidence, including the assessment of competing hypotheses.

19 For this and other information about detection and attribution studies see Mitchell, J. F. B., Karoly, D. J. 2001. Detection of climate change and attribution of causes. In Houghton, *Climate Change 2001*, Chapter 12.

20 Summary for policymakers. In Houghton, *Climate Change 2001*.

21 See Levitus, S. *et al*. 2000. *Science*, **287**, pp. 2225–9; and Levitus, S. *et al*. 2001. *Science*, **292**, pp. 267–70.

22 Gregory, J. *et al*. 2002. *Journal of Climate*, **15**, 3117–21.

23 The regional scale is defined as describing the range of 10^4 to 10^7 km^2. The upper end of the range (10^7 km^2) is often described as a typical sub-continental scale. Circulations at larger than the sub-continental scale are on the planetary scale.

24 For more information see Giorgi, F., Hewitson, B. 2001. Regional climate information – evaluation and projections. In Houghton, *Climate Change 2001*, Chapter 10.

25 See www.climateprediction.net

26 Hint: recall Stefan's blackbody radiation law that the energy emitted is proportional to the fourth power of the temperature.

Chapter 6
Climate change in the twenty-first century and beyond

The last chapter showed that the most effective tool we possess for the prediction of future climate change due to human activities is the climate model. This chapter will describe the predictions of models for likely climate change during the twenty-first century. It will also consider other factors that might lead to climate change and assess their importance relative to the effect of greenhouse gases.

Emission scenarios

A principal reason for the development of climate models is to learn about the detail of the likely climate change this century and beyond. Because model simulations into the future depend on assumptions regarding future anthropogenic emissions of greenhouse gases, which in turn depend on assumptions about many factors involving human behaviour, it has been thought inappropriate and possibly misleading to call the simulations of future climate so far ahead 'predictions'. They are therefore generally called 'projections' to emphasise that what is being done is to explore likely future climates which arise from a range of assumptions regarding human activities.

A starting point for any projections of likely climate change into the future is a set of descriptions of likely future global emissions of greenhouse gases. These will depend on a variety of assumptions regarding human behaviour and activities, including population, economic growth, energy use and the sources of energy generation. As was mentioned in Chapter 3, such descriptions of future emissions are called *scenarios*.

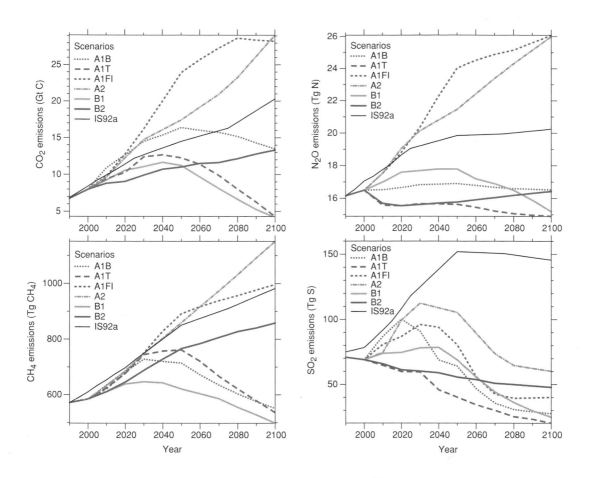

Figure 6.1 Anthropogenic emissions of carbon dioxide, methane, nitrous oxide and sulphur dioxide for the six illustrative SRES scenarios, A1B, A2, B1 and B2, A1F1 and A1T. For comparison the IS 92a scenario is also shown.

A wide range of scenarios was developed by the IPCC in a Special Report on Emission Scenarios (SRES)[1] in preparation for its 2001 Report (see box below). It is these scenarios that have been used in developing the projections of future climate presented in this chapter. In addition, because it has been widely used in modelling studies, results are also presented using a scenario (IS 92a) taken from a set developed by the IPCC in 1992 and widely described as representative of 'business-as-usual'.[2] Details of these scenarios are presented in Figure 6.1.

The SRES scenarios include estimates of greenhouse gas emissions resulting from all sources including land-use change. Estimates in the different scenarios begin from the current values for land-use change including deforestation (see Table 3.1). Assumptions in different scenarios vary, from continued deforestation, although reducing as less forest remains available for clearance, to substantial afforestation leading to an increased carbon sink.

The emission scenarios of the Special Report on Emission Scenarios (SRES)

The SRES scenarios are based on a set of four different story lines within each of which a family of scenarios has been developed – leading to a total of thirty-five scenarios.[3]

A1 Storyline

The A1 storyline and scenario family describes a future world of very rapid economic growth, a global population that peaks in mid-century and declines thereafter, and the rapid introduction of new and more efficient technologies. Major underlying themes are convergence among regions, capacity building and increased cultural and social interactions, with a substantial reduction in regional differences in per capita income. The A1 scenario family develops into three groups that describe alternative directions of technological change in the energy system. The three groups are distinguished by their technological emphasis: fossil fuel intensive (A1FI), non-fossil fuel energy sources (A1T), or a balance across all sources (A1B) – where balance is defined as not relying too heavily on one particular energy source, on the assumption that similar improvement rates apply to all energy-supply and end-use technologies.

A2 Storyline

The A2 storyline and scenario family describes a very heterogeneous world. The underlying theme is self-reliance and preservation of local identities. Fertility patterns across regions converge very slowly, which results in a continuously increasing population. Economic development is primarily regionally oriented and per capita economic growth and technological change more fragmented and slower than other story lines.

B1 Storyline

The B1 storyline and scenario family describes a convergent world, with the same global population that peaks in mid century and declines thereafter as in the A1 storyline, but with rapid change in economic structures towards a service and information economy, with reductions in material intensity and the introduction of clean and resource-efficient technologies. The emphasis is on global solutions to economic, social and environmental sustainability, including improved equity, but without additional climate-related initiatives.

B2 Storyline

The B2 storyline and scenario family describes a world in which the emphasis is on local solutions to economic, social and environmental sustainability. It is a world with a continuously increasing global population, at a rate lower than in A2, intermediate levels of economic development and less rapid and more diverse technological change than in the B1 and A1 storylines. While the storyline is also oriented towards environmental protection and social equity, it focuses on local and regional levels.

From the total set of thirty-five scenarios, an illustrative scenario was chosen for each of the six scenario groups A1B, A1FI, A1T, A2, B1 and B2. All should be considered equally sound. It is mostly for this set of six illustrative scenarios that data are presented in this chapter.

The SRES scenarios do not include additional climate initiatives, which means that no scenarios are included that explicitly assume implementation of the United Nations Framework Convention on Climate Change or the emissions targets of the Kyoto Protocol.

The next stage in the development of projections of climate change is to turn the emission profiles of greenhouse gases into greenhouse gas concentrations (Figure 6.2) and then into radiative forcing (Table 6.1 and Figure 6.4(a)). The methods by which these are done are described in Chapter 3, where the main sources of uncertainty are also mentioned. For the carbon dioxide concentration scenarios these uncertainties, especially those concerning the magnitude of the climate feedback from the terrestrial biosphere (see box on page 40), amount to a range of about −10% to +30% in 2100 for each profile.[4]

For most scenarios, emissions and concentrations of the main greenhouse gases increase during the twenty-first century. However, despite the increases projected in fossil fuel burning – very large increases in some cases – emissions of sulphur dioxide (Figure 6.1) and hence the concentrations of sulphate particles are expected to fall substantially because of the spread of policies to abate the damaging consequences of air pollution and 'acid-rain' deposition to both humans and ecosystems.[5] The influence of sulphate particles in tending to reduce the warming due to increased greenhouse gases is therefore now projected to be much less than for projections made in the mid 1990s (see the IS 92a scenario for sulphur dioxide in Figure 6.1). The other anthropogenic sources of particles in the atmosphere listed in Figure 3.8 will also contribute small amounts of positive or negative radiative forcing during the twenty-first century.[6]

Model projections

Results which come from the most sophisticated coupled atmosphere–ocean models of the kind described in the last chapter provide fundamental information on which to base climate projections. However, because they are so demanding on computer time only a limited number of results from such models are available. Many studies have also therefore been carried out with simpler models. Some of these, while possessing a full description of atmospheric processes, only include a simplified description of the ocean; these can be useful in exploring regional change. Others, sometimes called energy balance models (see box on page 121), drastically simplify the dynamics and physics of both atmosphere and ocean and are useful in exploring changes in the global average response with widely different emission scenarios. Results from simplified models need to be carefully compared with those from the best coupled atmosphere–ocean models and the simplified models 'tuned' so that, for the particular parameters for which they are being employed, agreement with the more complete models is as close as possible. The projections presented in the next sections depend on results from all these kinds of models.

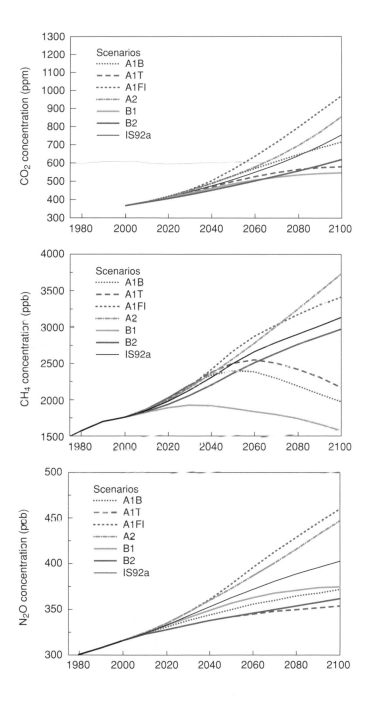

Figure 6.2 Atmospheric concentrations of carbon dioxide, methane and nitrous oxide resulting from the six illustrative SRES scenarios and from the IS 92a scenario. Uncertainties for each profile, especially those due to possible carbon feedbacks, have been estimated as from about −10% to +30% in 2100

In order to assist comparison between models, experiments with many models have been run with the atmospheric concentration of carbon dioxide doubled from its pre-industrial level of 280 ppm. The global average temperature rise under steady conditions of doubled carbon dioxide

Table 6.1 *Radiative forcing (W m^{-2}), globally averaged, for greenhouse gases from the year 1750 to 2000 and from SRES scenarios to 2050 and 2100*

Greenhouse gas	Year		A1B	A1T	A1FI	A2	B1	B2	IS 92a
CO_2	2000	1.46							
	2050		3.36	3.08	3.70	3.36	2.92	2.83	3.12
	2100		4.94	3.85	6.61	5.88	3.52	4.19	4.94
CH_4	2000	0.48							
	2050		0.70	0.73	0.78	0.75	0.52	0.68	0.73
	2100		0.56	0.62	0.99	1.07	0.41	0.87	0.91
N_2O	2000	0.15							
	2050		0.25	0.23	0.33	0.32	0.27	0.23	0.29
	2100		0.31	0.26	0.55	0.51	0.32	0.29	0.40
O_3(trop)	2000	0.35							
	2050		0.59	0.72	1.01	0.78	0.39	0.63	0.67
	2100		0.50	0.46	1.24	1.22	0.19	0.78	0.90
Halocarbons	2000	0.34							
	2050	0.49							
	2100	0.57							

Data from Ramaswamy, V. *et al.* 2001. Radiative forcing of climate change. In Houghton, *Climate Change 2001*, Chapter 6. Data selected from Tables 6.1 and 6.4. For 2050 and 2100 for the halocarbons, all scenarios make the same assumptions.

concentration has become known as the climate sensitivity.[7] The Intergovernmental Panel on Climate Change (IPCC) in its 1990 Report gave a 'best estimate' of 2.5 °C for the climate sensitivity; it also considered that it was unlikely to lie outside the range of 1.5 °C to 4.5 °C, a range that encompasses the results of the best coupled atmosphere–ocean general circulation models (AOGCMs). The IPCC 1995 and 2001 Reports have confirmed these values. Reasons why there remains this range of uncertainty in the estimates from climate models were explained in Chapter 5. The projections presented in this chapter follow the IPCC 2001 Assessment.[8]

Projections of global average temperature

When information of the kind illustrated in Figures 6.1, 6.2 and 6.4(a) is incorporated into simple or more complex models, projections of climate change can be made. As we have seen in earlier chapters, a useful proxy

Simple climate models

In Chapter 5 a detailed description was given of general circulation models (GCMs) of the atmosphere and the ocean and of the way in which they are coupled together (in AOGCMs) to provide simulations of the current climate and of climate perturbed by anthropogenic emissions of greenhouse gases. These models provide the basis of our projections of the detail of future climate. However, because they are so elaborate, they take a great deal of computer time so that only a few simulations can be run with these large coupled models.

To carry out more simulations under different future emission profiles of greenhouse gases or of aerosols or to explore the sensitivity of future change to different parameters (for instance, parameters describing the feedbacks in the atmosphere which largely define the climate sensitivity), extensive use has been made of simple climate models.[9] These simpler models are 'tuned' so as to agree closely with the results of the more complex AOGCMs in cases where they can be compared.

The most radical simplification in the simpler models is to remove one or more of the dimensions so that the quantities of interest are averaged over latitude circles (in two-dimensional models) or over the whole globe (in one-dimensional models). Such models can, of course, only simulate latitudinal or global averages – they can provide no regional information.

Figure 6.3 illustrates the components of such a model in which the atmosphere is contained within a 'box' with appropriate radiative inputs and outputs. Exchange of heat occurs at the land surface (another 'box') and the ocean surface. Within the ocean allowance is made for vertical diffusion and vertical circulation. Such a model is appropriate for simulating changes in global average surface temperature with increasing greenhouse gases or aerosols. When exchanges of carbon dioxide across the interfaces between the atmosphere, the land and the ocean are also included, the model can be employed to simulate the carbon cycle.

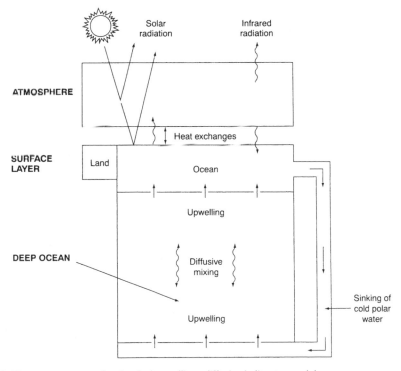

Figure 6.3 The components of a simple 'upwelling–diffusion' climate model.

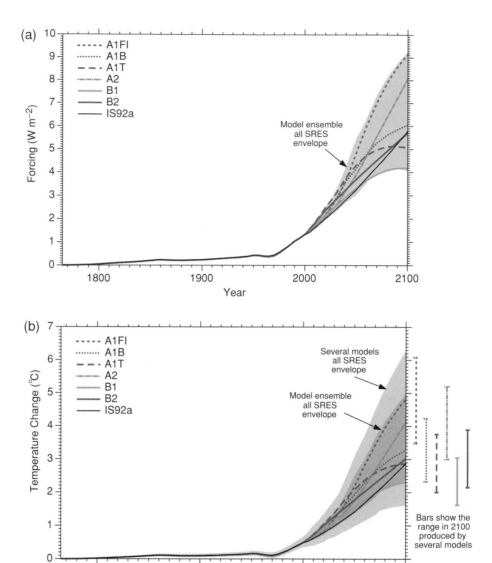

Figure 6.4 Simple model results. (a) Anthropogenic globally averaged radiative forcing, based on historical information about greenhouse gas and aerosol concentrations to the year 2000 (see Figure 3.8) and the SRES scenarios to the year 2100. The shading shows the envelope of forcing that encompasses the full range of thirty-five SRES scenarios. (b) Historic anthropogenic global mean temperature change and future changes for the SRES scenarios and the IS 92a scenario calculated using a simple climate model tuned to seven AOGCMs (with climate sensitivity in the range 1.7 to 4.2 °C). The darker shading represents the envelope of the full set of thirty-five SRES scenarios using the average of the model results (mean climate sensitivity is 2.8 °C). The lighter shading is the envelope including all seven model projections (the range of model results for each scenario is also shown by the bars on the right hand side).

for climate change that has been widely used is the change in global average temperature.

The projected rise in global average temperature due to the increase in greenhouse gases and aerosols from pre-industrial times is illustrated in Figure 6.4(b). It shows an increase of about 0.6 °C up to the year 2000 and an increase ranging from about 2 °C to about 6 °C by 2100 – that wide range resulting from the very large uncertainty regarding future emissions and also from the uncertainty that remains regarding the feed-backs associated with the climate response to the changing atmospheric composition (as described in Chapter 5).[10]

Compared with the temperature changes normally experienced from day to day and throughout the year, changes of between 2 °C and 6 °C may not seem very large. But, as was pointed out in Chapter 1, it is in fact a large amount when considering *globally averaged* temperature. Compare it with the 5 °C or 6 °C change in global average temperature that occurs between the middle of an ice age and the warm period in between ice ages. The changes projected for the twenty-first century are from one-third to a whole ice age in terms of the degree of climate change!

The rate of change of global average temperature projected for the twenty-first century is in the range of 0.15 °C to 0.6 °C per decade. These might seem small rates of change; most people would find it hard to detect a change in temperature of a fraction of a degree. But remembering again that these are global averages, such rates of change become very large. Indeed, they are much larger than any rates of change the climate has experienced for at least the past 10 000 years as inferred from paleoclimate data. As we shall see in the next chapter, the ability of both humans and ecosystems to adapt to climate change depends critically on the rate of change.

The changes in global average temperature shown in Figure 6.4(b) from the IPCC 2001 Report are substantially greater than those shown in the IPCC 1995 Report. The main reason for the difference is the much smaller aerosol emissions in the SRES scenarios compared with the IS 92 scenarios. This is illustrated by the temperature in 2100 shown in Figure 6.4(b) for the IS 92a scenario which is similar to that for the SRES B2 scenario even though the carbon dioxide emissions at that date for IS 92a are fifty per cent greater than those for B2.

In many of the modelling studies of climate change, the situation of doubled pre-industrial atmospheric carbon dioxide has often been in-troduced as a benchmark especially to assist in comparisons between different model projections and their possible impacts. Since the pre-industrial concentration was about 280 ppm, doubled carbon dioxide is about 560 ppm. From the curves in Figure 6.2 this is likely to occur sometime in the second half of the twenty-first century, depending on the scenario. But, other greenhouse gases are also increasing and contributing to the radiative forcing. So as to achieve an overall picture more easily, it is often convenient to convert other greenhouse gases to equivalent amounts of carbon dioxide, in other words to amounts of carbon dioxide that would give the same radiative forcing.[11] The information in Table 6.1 enables the conversion to be carried out. For instance, the increases in the greenhouse gases other than carbon dioxide (including ozone) to date produce about eighty per cent of the radiative forcing due to the increase in carbon dioxide to date (see Figure 3.8). This proportion will drop substantially during the next few decades as the growth in carbon dioxide becomes more dominant in nearly all scenarios. Referring to Figure 6.4(a), and noting that doubled carbon dioxide produces a radiative forcing of about 3.7 W m^{-2}, it can be seen that doubling of the equivalent carbon dioxide amount from pre-industrial times will occur between 2040 and 2070 depending on the scenario.

Now, in Figure 6.4(a), select one of the scenarios, say A1B, and note that the radiative forcing for this scenario reaches that equivalent to doubled carbon dioxide (i.e. 3.7 W m^{-2}) in about 2040. Then move to Figure 6.4(b) and note that in 2040 the temperature rise is about $1.7 \,^{\circ}\text{C}$. This is only just over half the $2.8 \,^{\circ}\text{C}$ (the value for climate sensitivity used for the results presented in Figure 6.4(b) see figure caption) that would be expected for doubled carbon dioxide under steady conditions. As was shown in Chapter 5, this difference occurs because of the slowing effect of the oceans on the temperature rise. But this means that, as the carbon dioxide concentration continues to increase, at any given time there exists a commitment to further significant temperature rise which has not been realised at that time.

Regional patterns of climate change

So far we have been presenting global climate change in terms of likely increases in global average surface temperature that provide a useful overall indicator of the magnitude of climate change. In terms of regional implications, however, a global average conveys rather little information. What is required is spatial detail. It is in the regional or local changes that the effects and impacts of global climate change will be felt.

With respect to regional change, it is important to realise that, because of the way the atmospheric circulation operates and the interactions that govern the behaviour of the whole climate system, climate change over the globe will not be at all uniform. We can, for instance, expect substantial differences between the changes over large land masses and over the ocean; land possesses a much smaller thermal capacity and so can respond more quickly. Listed below are some of the broad features on the continental scale that characterise the projected temperature changes; more detailed patterns are illustrated in Figure 6.5(a). Reference to Chapter 4 indicates that many of these characteristics are already being found in the observed record of the last few decades.

- Generally greater surface warming of land areas than of the oceans typically by about forty per cent compared with the global average, greater than this in northern high latitudes in winter (associated with reduced sea ice and snow cover) and less than forty per cent in south and southeast Asia in summer and in southern South America in winter.
- Minimum warming around Antarctica and in the northern North Atlantic which is associated with deep oceanic mixing in those areas.
- Little warming over the Arctic in summer.
- Little seasonal variation of the warming in low latitudes or over the southern circumpolar ocean.
- A reduction in diurnal temperature range over land in most seasons and most regions; night time lows increase more than daytime highs.

So far we have been presenting results solely for atmospheric temperature change. An even more important indicator of climate change is precipitation. With warming at the Earth's surface there is increased evaporation from the oceans and also from many land areas leading on average to increased atmospheric water vapour content and therefore also on average to increased precipitation. Since the water-holding capacity of the atmosphere increases by about 6.5% per degree celsius,[12] the increases in precipitation as surface temperature rises can be expected to be substantial. In fact, model projections indicate increases in precipitation broadly related to surface temperature increases of about three per cent per degree celsius.[13] Further, since the largest component of the energy input to the atmospheric circulation comes from the release of latent heat as water vapour condenses, the energy available to the atmosphere's circulation will increase in proportion to the atmospheric water content. A characteristic therefore of anthropogenic climate change due to the increase of greenhouse gases will be a more intense hydrological cycle. The likely effect of this on precipitation extremes will be discussed in the next section.

In Figure 6.5(b) are shown the projected changes in the distribution of precipitation. Although on average precipitation increases there are large regional variations and large areas where there are likely to be decreases in average precipitation and also changes in its seasonal distribution. For instance, at high northern latitudes there are large increases in winter and over south Asia in summer. Southern Europe, Central America, southern Africa and Australia are likely to have drier summers.

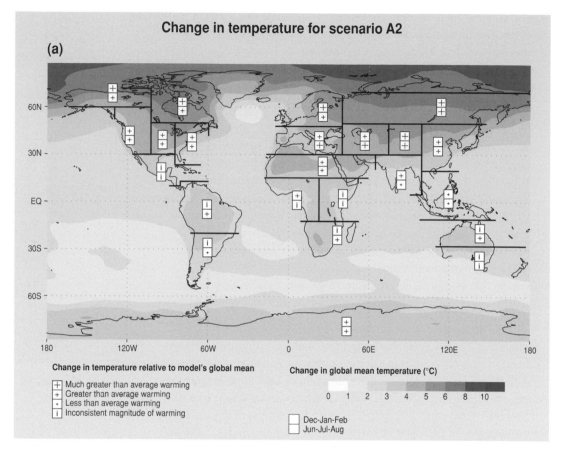

Figure 6.5 Projections for the SRES scenario A2 for the period 2071–2100 relative to 1961–1990 from an ensemble of nine different ocean–atmosphere general circulation models. (a) The annual mean change of temperature in °C shown by the shading. The boxes show an analysis of inter-model consistency in regional relative warming (i.e. warming relative to each model's global average warming) for winter and summer seasons. Regions are classified as showing either agreement on warming in excess of forty per cent above the global mean annual average (*much greater than average warming*), agreement on warming greater or less than the global mean annual average (*greater* or *less than average warming*) or disagreement amongst models on the magnitude of regional relative warming (*inconsistent magnitude of warming*). A consistent result from at least seven of the nine models is defined as being necessary for agreement. The global mean average warming of the models used are in the range 1.2 to 4.5 °C for this scenario. (b) The boxes show an analysis of inter-model consistency in regional precipitation change for winter and summer seasons. Regions are classified as showing agreement on change of greater than +20% (*large increase*), between +5% and +20% (*small increase*) and more than −20% (*large decrease*), or disagreement (*inconsistent sign*). A consistent result from at least seven of the nine models is defined as being necessary for agreement.

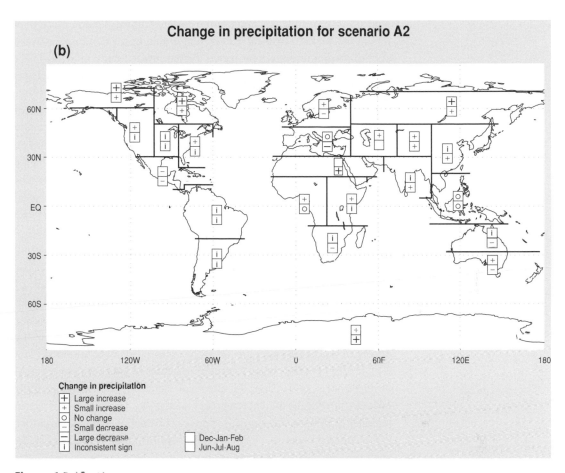

Figure 6.5 (*Cont.*).

Much natural climate variability occurs because of changes in, or oscillations between, persistent climatic patterns or regimes. The Pacific, North Atlantic Anomaly (PNA – that is dominated by high pressure over the eastern Pacific and western North America and which tends to lead to very cold winters in the eastern United States), the North Atlantic Oscillation (NAO – that has a strong influence on the character of the winters in north west Europe) and the El Niño events mentioned in Chapter 5 are examples of such regimes. Important components of climate change in response to the forcing due to the increase in greenhouse gases can be expected to be in the form of changes in the intensity or the frequency of established climate patterns illustrated by these regimes.[14] There is little consistency at the present time between models regarding projections of many of these patterns. However, recent trends in the tropical Pacific for the surface temperature to become more El Niño-like (see Table 4.1 on page 62–3), with the warming in the eastern tropical Pacific more than that in the western tropical Pacific and with a corresponding eastward shift of precipitation, are projected to continue by many models. There is also evidence that warming associated with increasing greenhouse gas concentrations will cause an intensification of the Asian summer monsoon and an increase of variability in its precipitation. The influence of increased greenhouse gases on these major climate regimes, especially the El Niño, is an important and urgent area of research.

A complication in the interpretation of patterns of climate change arises because of the differing influence of atmospheric aerosols as compared with that of greenhouse gases. Although in the projections based on the SRES scenarios the influence of aerosols is much less than in those based on the IS 92 scenarios published by the IPCC in its 1995 Report,[15] their projected radiative forcing is still significant. When considering global average temperature and its impact on, for instance, sea level rise (see Chapter 7) it is appropriate in the projections to use the values of globally averaged radiative forcing. The negative radiative forcing from sulphate aerosol, for instance, then becomes an offset to the positive forcing from the increase in greenhouse gases. However, because the effects of aerosol forcing are far from uniform over the globe (Figure 3.7), it is not correct, when considering climate change and its regional characteristics, to consider the effects of increasing aerosol as a simple offset to those of the increase in greenhouse gases. The large variations in regional forcing due to aerosols produce substantial regional variations in the climate response. Detailed regional information from the best climate models needs to be employed to assess the climate change under different assumptions about the increases in both greenhouse gases and aerosols.

Changes in climate extremes

The last section looked at the likely regional patterns of climate change. Can anything be said about likely changes in the frequency or intensity of climate extremes in the future? It is, after all, not the changes in average climate that are generally noticeable, but the extremes of climate – the droughts, the floods, the storms and the extremes of temperature in very cold or very warm periods – which provide the largest impact on our lives (see Chapter 1).

The most obvious change we can expect in extremes is a large increase in the number of extremely warm days (Figure 6.6) coupled with a decrease in the number of extremely cold days. A number of model projections show a generally decreased daily variability of surface air temperature in winter and increased daily variability in summer in

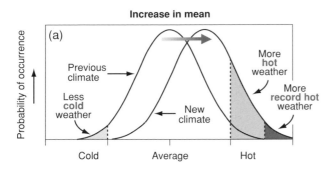

Increase in mean

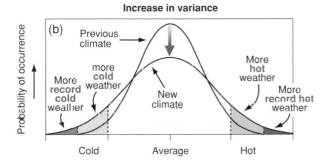

Increase in variance

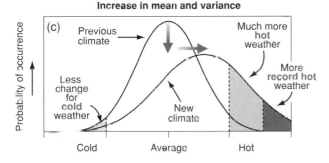

Increase in mean and variance

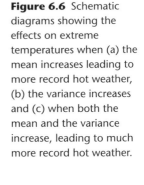

Figure 6.6 Schematic diagrams showing the effects on extreme temperatures when (a) the mean increases leading to more record hot weather, (b) the variance increases and (c) when both the mean and the variance increase, leading to much more record hot weather.

Northern Hemisphere land areas, suggesting that the situation in Figure 6.6(c) could apply in these areas. An example of this can be found in the box in Chapter 7 on page 177.

However, the changes that are likely to lead to most impact are those connected with the hydrological cycle. In the last section it was explained that in a warmer world with increased greenhouse gases, average precipitation increases and the hydrological cycle becomes more intense.[16] Consider what might occur in regions of increased rainfall. Often in such regions with the more intense hydrological cycle the larger amounts of rainfall will come from increased convective activity: more really heavy showers and more intense thunderstorms. The result of a study with an Australian climate model of the effect on rainfall amounts of doubling

Figure 6.7 Changes in the frequency of occurrence of different daily rainfall amounts with doubled carbon dioxide as estimated by a CSIRO model in Australia.

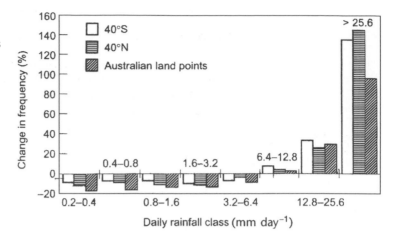

the carbon dioxide concentration is shown in Figure 6.7. The number of days with large rainfall amounts (greater than 25 mm day^{-1}) doubled. The probability of conditions leading to floods would also at least have doubled. Similar results (fewer rainy days, higher maximum daily rainfalls for a given mean rainfall rate) have been obtained from many other climate models. For instance, a recent modelling study (Figure 6.8) has shown that if atmospheric carbon dioxide concentration is doubled from its pre-industrial value, the probability of extreme seasonal precipitation in winter is likely to increase substantially over large areas of central and northern Europe and likely to decrease over parts of the Mediterranean and north Africa. The increase in parts of central Europe is such that the return period of extreme rainfall events would decrease by about a factor of five (e.g. from fifty years to ten years). Similar results have been obtained in a study of major river basins around the world.[17]

Note also from Figure 6.7 that the number of days with lighter rainfall events (less than 6 mm/day) is expected to decrease in the globally warmed world. This is because, with the more intense hydrological cycle, a greater proportion of the rainfall will fall in the more intense events and, furthermore, in regions of convection the areas of downdraught become drier as the areas of updraught become more moist. In many areas with relatively low rainfall, the rainfall will tend to become less – which has implications for the likelihood of drought.

Take, for instance, the likelihood of drought in regions where the average summer rainfall falls by perhaps twenty per cent as is likely to occur, for instance, in southern Europe (Figure 6.5(b)). The likely result of such a drop in rainfall is not that the number of rainy days will remain the same, with less rain falling each time; it is more likely that there will be substantially fewer rainy days and considerably more chance of

Greenhouse/control

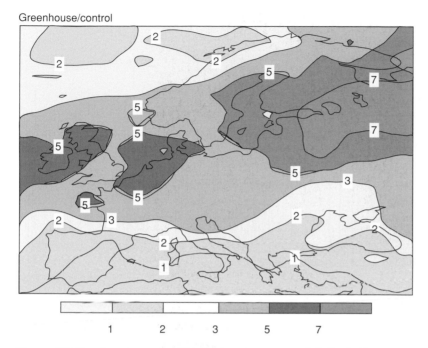

1   2   3   5   7

Figure 6.8 The changing probability of extreme season precipitation in Europe in winter as estimated from an ensemble of nineteen runs with a climate model starting from slightly different initial conditions. The figure shows the ratio of probabilities of extreme precipitation events in the years sixty-one to eighty of eighty-year runs that assumed an increase of carbon dioxide concentration of one per cent per year (hence doubling in about seventy years) compared with control runs with no change in carbon dioxide.

prolonged periods of no rainfall at all (see Figure 6.9). In other words, much more likelihood of drought. Further, the higher temperatures will lead to increased evaporation reducing the amount of moisture available at the surface – thus adding to the drought conditions. The proportional increase in the likelihood of drought is much greater than the proportional decrease in average rainfall.

Thus in the warmer world of increased greenhouse gases, different places will experience more frequent droughts and floods – we noted in Chapter 1 that these are the climate extremes which cause the greatest problems and we will be considering the impacts in more detail in the next chapter.

What about other climate extremes, intense storms, for instance? How about hurricanes and typhoons, the violent rotating cyclones that are found over the tropical oceans and which cause such devastation when they hit land? The energy for such storms largely comes from the latent heat of the water which has been evaporated from the warm

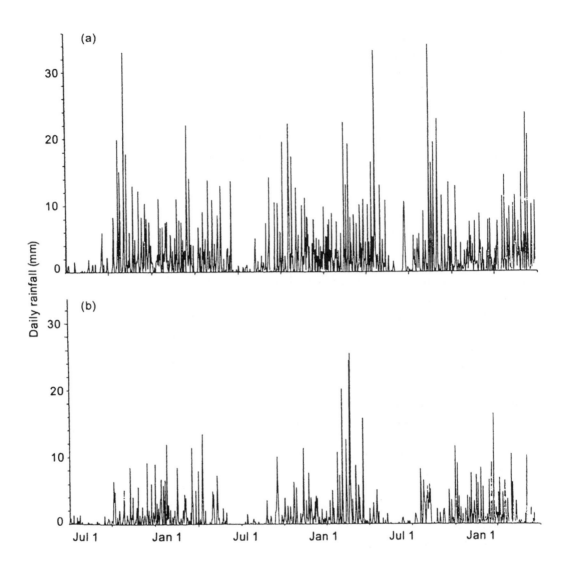

Figure 6.9 Daily rainfall for Italy for a three-year period simulated by a climate model, (a) for the current climate situation and (b) for the climate if the equivalent carbon dioxide concentration increased by a factor of four from its pre-industrial value (predicted for instance to occur before 2100 under the A2 scenario).

ocean surface and which condenses in the clouds within the storm, releasing energy. It might be expected that warmer sea temperatures would mean more energy release, leading to more frequent and intense storms. However, ocean temperature is not the only parameter controlling the genesis of tropical storms; the nature of the overall atmospheric flow is also important. Further, although based on limited data, observed variations in the intensity and frequency of tropical cyclones show no clear trends in the last half of the twentieth century. Models can take all the relevant factors into account but, because of the relatively large size of their grid, they are unable to simulate very reliably the detail of relatively small disturbances such as tropical cyclones. There is no consistent evidence from model projections of changes in the frequency of tropical cyclones or their areas of formation. However, model projections[18] and theoretical studies suggest that, if carbon dioxide concentration is doubled, the peak wind intensities will tend to increase by perhaps five or ten per cent and the mean and peak precipitation intensities by twenty to thirty per cent.

Regarding storms at mid latitudes, the various factors that control their incidence are complex. Two factors tend to an increased intensity of storms. The first, as with tropical storms, is that higher temperatures, especially of the ocean surface, tend to lead to more energy being available. The second factor is that the larger temperature contrast between land and sea, especially in the Northern Hemisphere, tends to generate steeper temperature gradients, which in turn generate stronger flow and greater likelihood of instability. The region around the Atlantic seaboard of Europe is one area where such increased storminess might be expected, a result that some model projections have shown. However, such a picture may well be too simple; other models suggest changes in storm tracks that result in very different changes in some regions and there is little overall consistency between model projections.

For some other extremes such as very small-scale phenomena (e.g. tornadoes, thunderstorms, hail and lightning) that cannot be simulated in global models, although they may have important impacts, there is currently insufficient information to assess recent trends, and understanding is inadequate to make firm projections.

Table 6.2 summarises the state of knowledge regarding the likely future incidence of extreme events. Although general indications of trends can be given, there have been few projections with quantitative estimates of likely changes in the frequency or intensity of extreme events. In many research centres, work is under way on more detailed studies of the influence of increased greenhouse gases on extreme events and changes in climate variability.

Regional climate models

Most of the likely changes that we have presented have been on the scale of continents. Can more specific information be provided about change for smaller regions? In Chapter 5 we referred to the limitation of global circulation models (GCM) in the simulation of changes on the regional scale[19] arising from the coarse size of their horizontal grid – typically 300 km or more. Also in Chapter 5 we introduced the regional climate model (RCM) that typically possesses a resolution of 50 km and can be 'nested' in a global circulation model. Examples are shown in Figures 6.10 and 7.8 of the improvement achieved by RCMs in the simulation of extremes

Table 6.2 *Estimates of confidence in observed and projected changes in extreme weather and climate events*

Confidence in observed changes (latter half of the twentieth century)	Changes in phenomenon	Confidence in projected changes (during the twenty-first century)
Likely[b]	Higher maximum temperatures and more hot days over nearly all land areas	Very likely
Very likely	Higher minimum temperatures, fewer cold days and frost days over nearly all land areas	Very likely
Very likely	Reduced diurnal temperature range over most land areas	Very likely
Likely, over many areas	Increase of heat index[a] over land areas	Very likely, over most areas
Likely, over many Northern Hemisphere mid to high latitude land areas	More intense precipitation events	Very likely, over most areas
Likely, in a few areas	Increased summer continental drying and associated risk of drought	Likely, over most mid-latitude continental interior (lack of projections consistent in other areas)
Not observed in the few analyses available	Increase in tropical cyclone peak wind intensities	Likely, over some areas
Insufficient data for assessment	Increase in tropical cyclone mean and peak precipitation intensities	Likely, over some areas

[a]Heat index is a combination of temperature and humidity that measures effects on human comfort.

[b]See Note 1 of Chapter 4 for explanation of likely, very likely, etc.

Source: Table SPM-1 from Summary for policymakers. In Houghton, J. T., Ding, Y., Griggs, D. J., Noguer, M., van der Linden, P. J., Dai, X., Maskell, K., Johnson, C. A. (eds.) *Climate Change 2001: The Scientific Basis. Contribution of Working Group I to the Third Assessment Report of the Intergovernmental Panel on Climate Change*. Cambridge: Cambridge University Press.

and in providing regional detail that in many cases (especially for precipitation) shows substantial disagreement with the averages provided by a GCM.

Regional models are providing a powerful tool for the investigation of detail in patterns of climate change. In the next chapter the importance of such detail will be very apparent in studies that assess the impacts of climate change. However, it is important to realise that, even if the models were perfect, because much greater natural variability is apparent in local climate than in climate averaged over continental or larger scales, projections on the local and regional scale are bound to be more uncertain than those on larger scales.

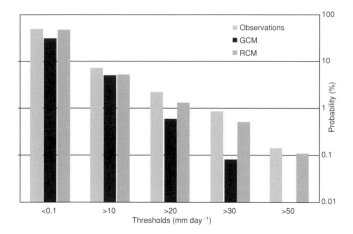

Figure 6.10 Example of simulations showing the probability of winter days over the Alps with different daily rainfall thresholds, as observed, simulated by a 300-km resolution GCM and by a 50-km resolution RCM. The RCM shows much better agreement with observations especially for higher thresholds.

Longer-term climate change

Most of the projections of future climate that have been published cover the Twenty-first century. For instance, the curves plotted in Figure 6.2 extend to the year 2100. They illustrate what is likely to occur if fossil fuels continue to provide most of the world's energy needs during that period.

From the beginning of the industrial revolution until 2000 the burning of fossil fuels released approximately 600 Gt of carbon in the form of carbon dioxide into the atmosphere. Under the SRES A1B scenario it is projected that a further 1500 Gt will be released by the year 2100. As Chapter 11 will show, the reserves of fossil fuels in total are sufficient to enable their rate of use to continue to grow well beyond the year 2100. If that were to happen the global average temperature would continue to rise and could, in the twenty-second century, reach very high levels, perhaps up to 10 °C higher than today (see Chapter 9). The associated changes in climate would be correspondingly large and could well be irreversible.

A further longer-term effect that may become important during this century is that of positive feedbacks on the carbon cycle due to climate change. This was mentioned in Chapter 3 (see box on page 40) and the +30% uncertainty in 2100 in the atmospheric concentrations of carbon dioxide shown in Figure 6.2 was introduced to allow for it. Taking this effect into account would add nearly a further degree Celsius to the projected increase in global average temperature in 2100 at the top end of the range shown in Figure 6.4. Some of the further implications of this feedback will be considered in Chapter 10 on page 255–259.

Especially when considering the longer term, there is also the possibility of surprises – changes in the climate system that are unexpected.

The discovery of the 'ozone hole' is an example of a change in the atmosphere due to human activities, which was a scientific 'surprise'. By their very nature such 'surprises' cannot, of course, be foreseen. However, there are various parts of the system which are, as yet, not well understood, where such possibilities might be looked for[20]; for instance, in the deep ocean circulation (see box in Chapter 5, page 99) or in the stability of the major ice sheets. In the next section we shall look in more detail at the first of these possibilities; the second will be addressed in the section of the next chapter entitled 'How much will sea level rise?'.

Changes in the ocean thermohaline circulation

The ocean thermohaline circulation (THC) was introduced in the box on page 99 in Chapter 5, where Figure 5.18 illustrates the deep ocean currents that transport heat and freshwater between all the world's oceans. Also mentioned in the box was the influence on the THC in past epochs of the input of large amounts of fresh water from ice melt to the region in the north Atlantic between Greenland and Scandinavia where the main source region for the THC is located.

With climate change due to increasing greenhouse gases we have seen that the precipitation will increase substantially especially at high latitudes (Figure 6.5(b)), leading to additional fresh water input to the oceans. The dense salty water in the north Atlantic source region for the THC will become less salty and therefore less dense. As a result the THC will weaken and less heat will flow northward from tropical regions to the north Atlantic. All coupled ocean–atmosphere GCMs show this occurring; an example is shown in Figure 6.11(a), which indicates a weakening of about twenty per cent by 2100. Although there is disagreement between the models as to the extent of weakening, all model projections of the pattern of temperature change under increasing greenhouse gases show less warming in the region of the north Atlantic (Figure 6.5(a)) – but none show actual cooling occurring in this region during the twenty-first century. In the longer term, some models show the THC actually cutting off completely after two or three centuries of increasing greenhouse gases. Figure 6.11(b) illustrates the effect that cut-off would have on the pattern of surface temperature around the globe. Note the severe cooling that would occur in the north Atlantic and north west Europe and the small compensating warming in the Southern hemisphere. Intense research is being pursued – both observations and modelling – to elucidate further likely changes in the thermohaline circulation and their possible impact.

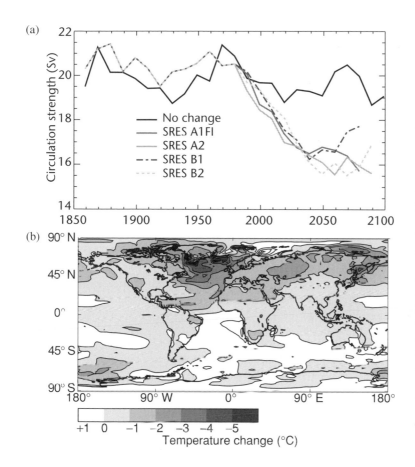

(a)

(b)

Figure 6.11 (a) Change in the strength of the thermohaline circulation (THC) in the north Atlantic as simulated by the Hadley Centre climate model for different SRES scenarios. The unit of circulation is the Sverdrup, $10^6\,m^3\,s^{-1}$. (b) Changes in the surface air temperature, relative to the present day, twenty years after the hypothetical collapse of the thermohaline circulation as simulated by the Hadley Centre climate model. To obtain the temperature distribution for a real situation, the changes due to increased greenhouse gases at the time of collapse would have to be added to these changes resulting from THC collapse.

Other factors that might influence climate change

So far climate change due to human activities has been considered. Are there other factors, external to the climate system, which might induce change? Chapter 4 showed that it was variations in the incoming solar energy as a result of changes in the Earth's orbit which triggered the ice ages and the major climate changes of the past. These variations are, of course, still going on; what influence are they having now?

Over the past 10 000 years, because of these orbital changes, the solar radiation incident at 60 °N in July has decreased by about 35 W m^{-2}, which is quite a large amount. But over one hundred years the change is only at most a few tenths of a watt per square metre, which is much less than the changes due to the increases in greenhouse gases (remember that doubling carbon dioxide alters the thermal radiation, globally averaged, by about 4 W m^{-2} – see Chapter 2). Looking to the future and the effect of the Earth's orbital variations, over at least the next

Does the Sun's output change?

Some scientists have suggested that all climate variations, even short-term ones, might be the result of changes in the Sun's energy output. Such suggestions are bound to be somewhat speculative because the only direct measurements of solar output that are available are those since 1978, from satellites outside the disturbing effects of the Earth's atmosphere. These measurements indicate a very constant solar output, changing by about 0.1% between a maximum and a minimum in the cycle of solar magnetic activity indicated by the number of sunspots.

It is known from astronomical records and from measurements of radioactive carbon in the atmosphere that this solar sunspot activity has, from time to time over the past few thousand years, shown large variations. Of particular interest is the period known as the Maunder Minimum in the seventeenth century when very few sunspots were recorded.[21] Studies of the recent measurements of solar output correlated with other indicators of solar activity, when extrapolated to this earlier period, suggest that the Sun was a little less bright in the seventeenth century, perhaps by about 0.4% or about 1 W m^{-2} in the average solar energy incident on the Earth's surface. This reduction in solar energy may have been a cause of the cooler period at that time known as the 'Little Ice Age'. Careful studies have estimated that since 1850 the maximum variations in the solar energy incident on the Earth's surface are unlikely to be greater than about 0.5 W m^{-2} (Figure 6.12). This is about the same as the change in the energy regime at the Earth's surface due to about a ten years' increase in greenhouse gases at the current rate.

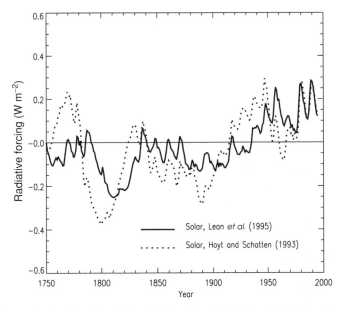

Figure 6.12 Radiative forcing due to variations in the energy input from the Sun as estimated by Lean *et al.* (1995) and by Hoyt and Schatten (1993).

50 000 years the solar radiation incident in summer on the polar regions will be unusually constant so that the present interglacial is expected to last for an exceptionally long period.[22] Suggestions therefore that the current increase of greenhouse gases might delay the onset of the next ice age are unfounded.

These orbital changes only alter the *distribution* of incoming solar energy over the Earth's surface; the total amount of energy reaching the Earth is hardly affected by them. Of more immediate interest are suggestions that the actual energy output of the Sun might change with time. As we mentioned in Chapter 3 (see Figure 3.8) and as is described in the box above, such changes, if they occur, are estimated to be much smaller than changes in the energy regime at the Earth's surface due to the increase in greenhouse gases.

There have also been suggestions of plausible indirect mechanisms whereby effects on the Sun might influence climate on Earth. Changes in solar ultraviolet radiation will influence atmospheric ozone and hence might affect climate. There is a possibility that the galactic cosmic ray flux, modified by the varying Sun's magnetic field, could influence cloudiness and hence climate. From studies of these possible mechanisms there is as yet insufficient evidence of significant climate effects.[23]

Another influence on climate comes from volcanic eruptions. Their effects, lasting typically a few years, are relatively short-term compared with the much longer-term effects of the increase of greenhouse gases. The recent large volcanic eruption of Mount Pinatubo in the Philippines which occurred in June 1991 has already been mentioned (Figure 5.21). Estimates of the change in the net amount of radiation (solar and thermal) at the top of the atmosphere resulting from this eruption are of about 0.5 W m^{-2}. This perturbation lasts for about two or three years while the major part of the dust settles out of the atmosphere; the longer-term change in radiation forcing, due to the minute particles of dust which last in the stratosphere for somewhat longer, is much smaller.

To summarise this chapter:

- The increase in greenhouse gases is by far the largest of the factors which can lead to climate change during the twenty-first century.
- The likely climate changes for a range of scenarios of greenhouse gas emissions have been described in terms of global average temperature and in terms of regional change of temperature and precipitation and the occurrence of extremes.
- The rate of change is likely to be larger than any the Earth has seen at any time during the past 10 000 years.

- The changes that are likely to have the greatest impact will be changes in the frequencies, intensities and locations of climate extremes, especially droughts and floods.
- Sufficient fossil fuel reserves are available to provide for continuing growth in fossil fuel emissions of carbon dioxide well into the twenty-second century. If this occurred the climate change could be very large indeed and have unpredictable features or 'surprises'.

The next chapter will look at the impact of such changes on sea level, on water, on food supplies and on human health. Later chapters of the book will then suggest what action might be taken to slow down and eventually to terminate the rate of change.

Questions

1 Suggest, for Figure 6.6, an appropriate temperature scale for a place you know. Define what might be meant by a very hot day and estimate the percentage increase in the probability of such days if the average temperature increases by 1, 2 and 4 °C.

2 From Figure 6.9, compare the maximum length of periods in the summer with less than 1 mm, 2 mm, 5 mm of total rainfall under normal climate conditions and under conditions with increased carbon dioxide.

3 It is stated in the text describing extremes that in convective regions, with global warming, as the updraughts become more moist the downdraughts tend to be drier. Why is this?

4 Look at the assumptions underlying the full range of SRES emission scenarios in the IPCC 2001 Report. Would you want to argue that some of the scenarios are more likely to occur than the others? Which (if any) would you designate as the most likely scenario?

5 It is sometimes suggested that northwest Europe could become colder in the future while most of the rest of the world becomes warmer. What could cause this and how likely do you think it is to occur?

6 How important do you consider it is to emphasise the possibility of 'surprises' when presenting projections of likely future climate change?

7 Estimate the effect on the projected carbon dioxide concentrations for 2100 shown in Figure 6.2, the projected radiative forcing for 2100 shown in Figure 6.4(a) and the projected temperatures for 2100 shown in Figure 6.4(b), of assuming the climate feedback on the carbon cycle illustrated in Figure 3.5 (note: first turn the accumulated atmospheric carbon in Figure 3.5 into an atmospheric concentration).

Notes for Chapter 6

1 Nakicenovic, N. *et al.* 2000. *Special Report on Emission Scenarios (SRES). A Special Report of the IPCC*. Cambridge: Cambridge University Press.

2   For details of IS 92a see Leggett, J., Pepper, W. J., Swart, R. J. 1992. Emission scenarios for the IPCC: an update. In Houghton, J. T., Callender, B. A., Varney, S. K. (eds.) *Climate Change 1992: the Supplementary Report to the IPCC Assessments*. Cambridge: Cambridge University Press, pp. 69–95. Small modifications have been made to the IS 92a scenario to take into account developments in the Montreal Protocol.

3   This summary is based on the Summary of SRES in the Summary for policymakers. In Houghton, *Climate Change 2001*, p. 18.

4   The +30% amounting to an addition of between 200 and 300 ppm to the carbon dioxide concentration in 2100 (see box on carbon feedbacks on page 40).

5   The World Energy Council Report *Energy for Tomorrow's World*. 1993. London: World Energy Council. In its most likely scenario this report projects that global sulphur emissions in 2020 will be almost the same as in 1990, although with a different distribution (more over Asia but less over Europe and North America). An extension of this study (*Global Energy Perspectives to 2050 and Beyond*. London: World Energy Council, 1995) projects at 2050 global sulphur emissions that are little more than half the 1990 levels.

6   Detailed values listed Ramaswamy, V. *et al.* 2001. Technical summary. In Houghton, *Climate Change 2001*, Chapter 6.

7   Note that because the response of global average temperature to the increase of carbon dioxide is logarithmic in the carbon dioxide concentration, the increase of global average temperature for doubling of carbon dioxide concentration is the same whatever the concentration that forms the base for the doubling, e.g. doubling from 280 ppm or from 360 ppm produces the same rise in global temperature. For a discussion of 'climate sensitivity' see Cubasch, U., Meehl, G. A. *et al.* 2001. In Houghton, *Climate Change 2001*, Chapter 9.

8   See Houghton, *Climate Change 2001*.

9   See Harvey, D. D. 1997. An introduction to simple climate models used in the IPCC Second Assessment Report. In *IPCC Technical Paper 2*. Geneva: IPCC.

10 Note that the uncertainty ranges in Figure 6.4 do not include those that arise from lack of knowledge concerning climate feedbacks on the carbon cycle (see box in Chapter 3 on page 40).

11 The assumption that greenhouse gases may be treated as equivalent to each other is a good one for many purposes. However, because of the differences in their radiative properties, accurate modelling of their effect should treat them separately. More details of this problem are given in Gates, W. L. *et al.* 1992. Climate modelling, climate prediction and model validation. In Houghton, J. T., Callander, B. A., Varney, S. K. (eds.) *Climate Change 1992: the Supplementary Report to the IPCC Scientific Assessments*. Cambridge: Cambridge University Press, pp. 171–5.

12 Related through the Clausius Clapeyron equation, $e^{-1} \, de/dT = L/RT^2$, where e is the saturation vapour pressure at temperature T, L the latent heat of evaporation and R the gas constant.

13   Allen, M. R., Ingram, W. J. 2002. *Nature*, **419**, pp. 224–32.

14   For more on this see Palmer, T. N. 1993. *Weather*, **48**, pp. 314–25; and Palmer, T. N. 1999. *Journal of Climate*, **12**, pp. 575–91.

15   Kattenberg, A. *et al*. 1996. Climate models – projections of future climate. In Houghton, J. T., Meira Filho, L. G., Callander, B. A., Harris, N., Kattenberg, A., Maskell, K. (eds.) *Climate Change 1995: the Science of Climate Change*. Cambridge: Cambridge University Press, Chapter 6.

16   For a more detailed discussion of the effect of global warming on the hydrological cycle, see Allen, M. R., Ingram, W. J. 2002. *Nature*, **419**, pp. 224–32.

17   Milly, P. C. D. *et al*. 2002. Increasing risk of great floods in a changing climate. *Nature*, **415**, pp. 514–17.

18   More information in Georgi, F., Hewitson, B. *et al*., Regional climate information: evaluation and projections. Chapter 10 in Houghton, *Climate Change 2001*, Chapter 10.

19   For definition of continental and regional scales see Note 23 in Chapter 5.

20   See also Table 7.4.

21   Studies of other stars are providing further information, see Nesme-Ribes, E. *et al*. 1996. *Scientific American*, August, pp. 31–6.

22   Berger, A., Loutre, M. F. 2002. *Science*, **297**, pp. 1287–8.

23   For a review and assessment of these mechanisms see Ramaswamy, V. *et al*. 2001. In Houghton, *Climate Change 2001*, Chapter 6, Section 6.11.2.

Chapter 7
The impacts of climate change

The last two chapters have detailed the climate change that we can expect during the twenty-first century because of human activities in terms of temperature and rainfall. To be useful to human communities, these details need to be turned into descriptions of the impact of climate change on human resources and activities. The questions to which we want answers are: how much will sea level rise and what effect will that have?; how much will water resources be affected?; what will be the impact on agriculture and food supply?; will natural ecosystems suffer damage and how will human health be affected? This chapter considers these questions.[1]

A complex network of changes

In outlining the character of the likely climate change in different regions of the world, the last chapter showed that it is likely to vary a great deal from place to place. For instance, in some regions precipitation will increase, in other regions it will decrease. Not only is there a large amount of variability in the character of the likely change, there is also variability in the sensitivity (for definition see box below) of different systems to climate change. Different ecosystems, for instance, will respond very differently to changes in temperature or precipitation.

There will be a few impacts of the likely climate change that will be positive so far as humans are concerned. For instance, in parts of Siberia or northern Canada increased temperature will tend to lengthen the growing season with the possibility in these regions of growing a greater variety of crops. In some places, increased carbon dioxide will aid the growth of some types of plants leading to increased crop yields.

Sensitivity, adaptive capacity and vulnerability: some definitions[2]

Sensitivity is the degree to which a system is affected, either adversely or beneficially, by climate-related stimuli. These encompass all the elements of climate change, including mean climate characteristics, climate variability, and the frequency and magnitude of extremes. This may be direct (e.g. a change in crop yield in response to a change in the mean, range or variability of temperature) or indirect (e.g. damage caused by an increase in the frequency of coastal flooding due to sea level rise).

Adaptive capacity is the ability of a system to adjust to climate change (including climate variability and extremes), to moderate potential damage, to take advantage of opportunities or to cope with the consequences.

Vulnerability is the degree to which a system is susceptible to, or unable to cope with, adverse effects of climate change, including climate variability and extremes. Vulnerability is a function of the character, magnitude and rate of climate change and also the extent to which a system is exposed, its sensitivity and its adaptive capacity.

Both the magnitude and the rate of climate change are important in determining the sensitivity, adaptability and vulnerability of a system.

However, because, over centuries, human communities have adapted their lives and activities to the present climate, most changes in climate will tend to produce an adverse impact. If the changes occur rapidly, quick and possibly costly adaptation to a new climate will be required by the affected community. An alternative might be for the affected community to migrate to a region where less adaptation would be needed – a solution which has become increasingly difficult or, in some cases, impossible in the modern crowded world.

As we consider the questions posed at the start of this chapter, it will become clear that the answers are far from simple. It is relatively easy to consider the effects of a particular change (in say, sea level or water resources) supposing nothing else changes. But other factors will change. Some adaptation, for both ecosystems and human communities, may be relatively easy to achieve; in other cases, adaptation may be difficult, very costly or even impossible. In assessing the effects of global warming and how serious they are, allowance must be made for response and adaptation. The likely costs of adaptation also need to be put alongside the costs of the losses or impacts connected with global warming.

Sensitivity, adaptive capacity and vulnerability (see box above) vary a great deal from place to place and from country to country. In particular, developing countries, especially the least developed countries, have less capacity to adapt than developed countries, which contributes to the relative high vulnerability to damaging effects of climate change in developing countries.

The assessment of the impacts of global warming is also made more complex because global warming is not the only human-induced environmental problem. The loss of soil and its impoverishment (through poor agricultural practice), the over-extraction of groundwater and the damage due to acid rain are examples of environmental degradations on local or regional scales that are having a substantial impact now.[3] If they are not corrected they will tend to exacerbate the negative impacts likely to arise from global warming. For these reasons, the various effects of climate change so far as they concern human communities and their activities will be put in the context of other factors that might alleviate or exacerbate their impact.

The assessment of climate change impacts, adaptations and vulnerability draws on a wide range of physical, biological and social science disciplines and consequently employs a large variety of methods and tools. It is therefore necessary to integrate information and knowledge from these diverse disciplines; the process is called *Integrated Assessment* (see box in Chapter 9 on page 237).

The following paragraphs will look at various impacts in turn and then bring them together in a consideration of the overall impact.

How much will sea level rise?

There is plenty of evidence for large changes in sea level during the Earth's history. For instance, during the warm period before the onset of the last ice age, about 120 000 years ago, the global average temperature was a little warmer than today (Figure 4.4). Average sea level then was about 5 or 6 m higher than it is today. When ice cover was at its maximum towards the end of the ice age, some 18 000 years ago, sea level was over 100 m lower than today, sufficient, for instance, for Britain to be joined to the continent of Europe.

It is often thought that the main cause of these sea level changes was the melting or growth of the large ice-sheets that cover the polar regions. It is certainly true that the main reason for the drop in sea level 18 000 years ago was the amount of water locked up in the large extension of the polar ice-sheets. In the northern hemisphere these extended in Europe as far south as southern England and in North America to south of the Great Lakes. It must also be true that the main reason for the 5 or 6 m higher sea level during the last warm interglacial period was a reduction in the Antarctic or Greenland ice-sheets. But changes over shorter periods are largely governed by other factors that combine to produce a significant effect on the average sea level.

During the twentieth century observations show that the average sea level rose by between 10 and 20 cm.[4] The largest contribution to this rise (about one-third) is from thermal expansion of ocean water; as the oceans warm the water expands and the sea level rises (see box below). Other significant contributions come from the melting of glaciers and as a result of long-term adjustments that are still occurring because of the removal of the major ice-sheets 20 000 years or so ago. The contributions from the ice caps of Greenland and Antarctica are believed to be small. A further contribution to sea level change of uncertain magnitude arises from changes in terrestrial storage of water, for instance from the growth of reservoirs or irrigation.

Thermal expansion of the oceans

A large component of sea level rise is due to thermal expansion of the oceans. Calculation of the precise amount of expansion is complex because it depends critically on the water temperature. For cold water the expansion for a given change of temperature is small. The maximum density of sea water occurs at temperatures close to $0\,°C$; for a small temperature rise at a temperature close to $0\,°C$, therefore, the expansion is negligible. At $5\,°C$ (a typical temperature at high latitudes), a rise of $1\,°C$ causes an increase of water volume of about 1 part in 10 000 and at $25\,°C$ (typical of tropical latitudes) the same temperature rise of $1\,°C$ increases the volume by about 3 parts in 10 000. For instance, if the top 100 m of ocean (which is approximately the depth of what is called the mixed layer) was at $25\,°C$, a rise to $26\,°C$ would increase its depth by about 3 cm.

A further complication is that not all the ocean changes temperature at the same rate. The mixed layer fairly rapidly comes into equilibrium with changes induced by changes in the atmosphere. The rest of the ocean changes comparatively slowly (the whole of the top kilometre will, for instance, take many decades to warm); some parts may not change at all. Therefore, to calculate the sea level rise due to thermal expansion – its global average and its regional variations – it is necessary to employ the results of an ocean climate model, of the kind described in Chapter 5.

Various contributions to the likely sea level rise in the twenty-first century can be identified. Again the largest is from the thermal expansion of ocean water. The other main contribution comes from the melting of glaciers. If all glaciers outside Antarctica and Greenland were to melt, the rise in sea level would be about 50 cm (between 40 and 60 cm). Substantial glacier retreat has occurred in recent decades adding an estimated 2–4 cm to the sea level rise in the twentieth century. Modelling the effect of climate change on the behaviour of glaciers is, however, complex. The growth or decay of a glacier depends on the balance between the amount of snowfall on it, especially in winter, and the amount of melting in the summer. Both winter snowfall and average summer temperature are important, and both must be taken into account in future projections of the rate of glacier melting.

The average sea level rise during the twenty-first century for each of the Special Report on Emission Scenarios (SRES) has been calculated by adding up the various contributions. Those due to thermal expansion (typically about sixty per cent of the total) and land ice changes (typically about twenty-five per cent of the total) were calculated using a simple climate model calibrated separately for each of seven coupled atmosphere–ocean general circulation models (AOGCMs) (as in the calculations for changes in global average temperature in Figure 6.4). The relatively small contributions from changes in permafrost, the effect of sediment deposition and the long-term adjustment of the ice-sheets to past climate change were then added. The results are shown in Figure 7.1, where it will be seen that the uncertainties in the estimates are substantial. Apart from the uncertainties inherent in the emissions scenarios, there is the uncertainty in the actual temperature rise (and hence in the contribution from thermal expansion) depending on the value chosen for the climate sensitivity (Figure 6.4). Different models also give substantially different estimates of the amount of sea level rise due to the melt from glaciers and small ice caps. The total range of uncertainty by 2100 is from about 10 cm to 90 cm.

The projections in Figure 7.1 apply to the next 100 years. During that period, because of the slow mixing that occurs throughout much of the oceans, only a small part of the ocean will have warmed significantly. Sea level rise resulting from global warming will therefore lag behind temperature change at the surface (Figure 7.2). During the following centuries, as the rest of the oceans gradually warm, sea level will continue to rise at about the same rate, even if the average temperature at the surface were to be stabilised.

The estimates of average sea level rise in Figure 7.1 provide a general guide as to what can be expected during the twenty-first century. Sea level rise, however, will not be uniform over the globe. The effects of thermal expansion in the oceans will vary considerably with location. Further, movements of the land occurring for natural reasons due, for instance, to tectonic movements or because of human activities (for instance, the

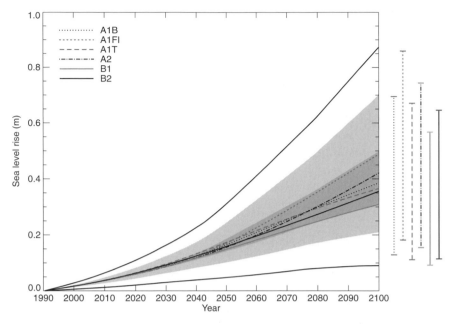

Figure 7.1 Global average sea level rise 1990–2100 for the SRES scenarios. Each of the six lines identified in the key is the average of the AOGCMs for one of the six illustrative SRES scenarios. The region in dark shading shows the range of the average of AOGCMs for all thirty-five SRES scenarios. The region in light shading shows the range of all AOGCMs for all thirty-five SRES scenarios. The region delineated by the outermost lines shows the range of all AOGCMs and scenarios including uncertainties in land ice changes, permafrost changes and sediment deposition. Note that this range does not allow for uncertainty relating to ice-dynamical changes in the West Antarctic ice-sheet (see text). The bars at the right show the range in 2100 of all AOGCMs for the six illustrative scenarios.

Figure 7.2 Estimate of sea level rise under a scenario with increasing greenhouse gases until the year 2030, at which time it is assumed that greenhouse gases are stabilised so that there is no further radiative forcing of the climate. An additional rise in sea level occurs during the remainder of the century as the increase in temperature penetrates to more of the ocean. The rise will continue at about the same rate for the following centuries as the rest of the ocean warms.

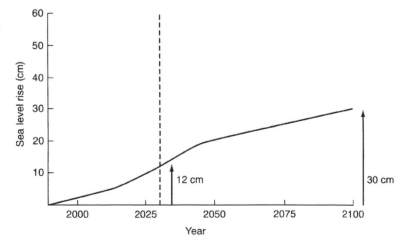

removal of groundwater) can have comparable effects to the rate of sea level rise arising from global warming. At any given place, all these factors have to be taken into account in determining the likely value of future sea level rise.

It is interesting and perhaps surprising that the net contribution expected from changes in the Antarctic and Greenland ice-sheets is small. For both ice-sheets there are two competing effects.[5] In a warmer world, there is more water vapour in the atmosphere which leads to more snowfall. But there is also more ablation (erosion by melting) of the ice around the boundaries of the ice-sheets where melting of the ice and calving of icebergs occur during the summer months. For Antarctica, the estimates are that accumulation is greater than ablation, leading to a small net growth. For Greenland, ablation is greater than accumulation. For the two taken together, under current conditions, the net effect is about zero, although there is considerable uncertainty in that estimate.

If we look further into the future, however, larger changes in the ice-sheets may begin to occur. The Greenland ice-sheet is the more vulnerable; its complete melting will cause a sea level rise of about 7 m. Model studies of the ice-sheet show that, with a temperature rise of more than 3 °C, ablation will significantly overtake accumulation and meltdown of the ice cap will begin. Figure 7.3 illustrates the rate of sea

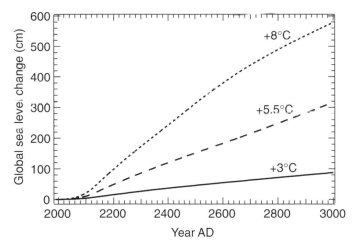

Figure 7.3 Global average sea level changes due to the response of the Greenland ice-sheet to three climate warming scenarios during the third millennium. The labels on the curves refer to the mean annual temperature rise over Greenland by 3000 AD as predicted by a two-dimensional climate and ocean model forced by greenhouse gas concentration rises until 2130 and kept constant after that. Note that projected temperature rises over Greenland are generally greater than those for global averages (by a factor of 1.2 to 3.1 for the AOGCMs used in generating Figure 6.4).

level rise over the next millennium that might be expected with different levels of temperature change at Greenland. A warming of 5.5 °C, for instance, if sustained for 1000 years, would be likely to result in about 3 m of sea level rise.

The portion of the Antarctic ice-sheet that is of most concern is that in the west of Antarctica (around 90° W longitude); its disintegration would result in about a 6-m sea level rise. Because a large portion of it is grounded well below sea level it has been suggested that rapid ice discharge could occur if the surrounding ice shelves are weakened. Although studies are far from conclusive, current ice dynamic models do not indicate that rapid disintegration is likely and suggest that, over the next millennium, the contribution of the west Antarctic ice-sheet to sea level rise will be less than 3 m.

The impacts of sea level rise

A rise in average sea level of 10 cm by 2030 and about half a metre by the end of the next century (typical values from Figure 7.1) may not seem a great deal. Many people live sufficiently above the level of high water not to be directly affected. However, half of humanity inhabits the coastal zones around the world. Within these, the lowest lying are some of the most fertile and densely populated. To people living in these areas, even a fraction of a metre increase in sea level can add enormously to their problems. Some of the areas that are especially vulnerable are, firstly, large river delta areas, for instance Bangladesh; secondly, areas very close to sea level where sea defences are already in place, for instance the Netherlands; and, thirdly, small low-lying islands in the Pacific and other oceans. We shall look at these in turn.

Bangladesh is a densely populated country of about 120 million people located in the complex delta region of the Ganges, Brahmaputra and Meghna Rivers.[6] About ten per cent of the country's habitable land (with about six million population) would be lost with half a metre of sea level rise and about twenty per cent (with about 15 million population) would be lost with a 1-m rise[7] (Figure 7.4). Estimates of the sea level rise are of about 1 m by 2050 (compounded of 70 cm due to subsidence because of land movements and removal of groundwater and 30 cm from the effects of global warming) and nearly 2 m by 2100 (1.2 m due to subsidence and 70 cm from global warming)[8] – although there is a large uncertainty in these estimates.

It is quite impractical to consider full protection of the long and complicated coastline of Bangladesh from sea level rise. Its most obvious effect, therefore, is that substantial amounts of good agricultural land will be lost. This is serious: half the country's economy comes from

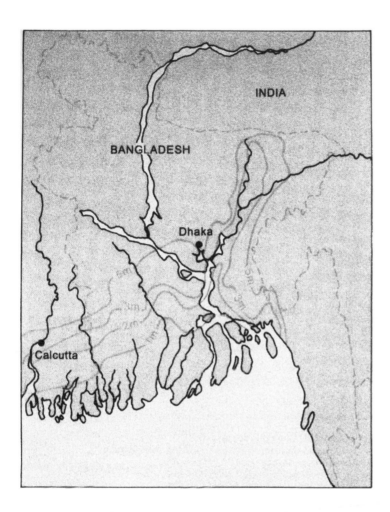

Figure 7.4 Land affected in Bangladesh by various amounts of sea level rise.

agriculture and eighty-five per cent of the nation's population depends on agriculture for its livelihood. Many of these people are at the very edge of subsistence.

But the loss of land is not the only effect of sea level rise. Bangladesh is extremely prone to damage from storm surges. Every year, on average, at least one major cyclone attacks Bangladesh. During the past twenty-five years there have been two very large disasters with extensive flooding and loss of life. The storm surge in November 1970 is probably the largest of the world's natural disasters in recent times; it is estimated to have claimed the lives of over a quarter of a million people. Well over a 100 000 are thought to have lost their lives in a similar storm in April 1991. Even small rises in sea level add to the vulnerability of the region to such storms.

There is a further effect of sea level rise on the productivity of agricultural land; that is, the intrusion of saltwater into fresh groundwater resources. At the present time, it is estimated that in some parts of Bangladesh saltwater extends seasonally inland over 150 km. With a 1-m rise in sea level, the area affected by saline intrusion could increase substantially[9] although, since it is also likely that climate change will bring increased monsoon rainfall, some of the intrusion of saltwater could be alleviated.[10]

What possible responses can Bangladesh make to these likely future problems? Over the timescale of change that is currently envisaged it can be supposed that the fishing industry can relocate and respond with flexibility to changing fishing areas and changing conditions. It is less easy to see what the population of the affected agricultural areas can do to relocate or to adapt. No significant areas of agricultural land are available elsewhere in Bangladesh to replace that lost to the sea, nor is there anywhere else in Bangladesh where the population of the delta region can easily be located. It is clear that very careful study and management of all aspects of the problem is required. The sediment brought down by the rivers into the delta region is of particular importance. The amount of sediment and how it is used can have a large effect on the level of the land affected by sea level rise. Careful management is therefore required upstream as well as in the delta itself; groundwater and sea defences must also be managed carefully if some alleviation of the effects of sea level rise is to be achieved.

A similar situation exists in the Nile delta region of Egypt. The likely rise in sea level this century is made up from local subsidence and global warming in much the same way as for Bangladesh – approximately 1 m by 2050 and 2 m by 2100. About twelve per cent of the country's arable land with a population of over seven million people would be affected by a 1-m rise of sea level.[11] Some protection from the sea is afforded by the extensive sand dunes but only up to half a metre or so of sea level rise.[12]

Many other examples of vulnerable delta regions, especially in south-east Asia and Africa, can be given where the problems would be similar to those in Bangladesh and in Egypt. For instance, several large and low-lying alluvial plains are distributed along the eastern coastline of China. A sea level rise of just half a metre would inundate an area of about 40 000 km^2 (about the area of The Netherlands)[13] where over thirty million people currently live. A particular delta which has been extensively studied is that of the Mississippi in North America. These studies underline the point that human activities and industry are already exacerbating the potential problems of sea level rise due to global warming. Because of river management little sediment is delivered by the river to the delta to counter the subsidence occurring because of long-term movements of the Earth's crust. Also, the building of canals and dykes has inhibited the input of sediments from the ocean.[14] Studies of this kind emphasise the importance of careful management of all activities influencing such regions, and the necessity of making maximum use of natural processes in ensuring their continued viability.

We now turn to The Netherlands, a country more than half of which consists of coastal lowlands, mainly below present sea level. It is one of the most densely populated areas in the world; eight of the fourteen million inhabitants of the region live in large cities such as Rotterdam, The Hague and Amsterdam. An elaborate system of about 400 km of dykes and coastal dunes, built up over many years, protects it from the sea. Recent methods of protection, rather than creating solid bulwarks, make use of the effects of various forces (tides, currents, waves, wind and gravity) on the sands and sediments so as to create a stable barrier against the sea – similar policies are advocated for the protection of the Norfolk coast in eastern England.[15] Protection against sea level rise next century will require no new technology. Dykes and sand dunes will need to be raised; additional pumping will also be necessary to combat the incursion of saltwater into freshwater aquifers. It is estimated[16] that an expenditure of about twelve thousand million dollars (US) would be required for protection against a sea level rise of 1 m.

The third type of area of especial vulnerability is the low-lying small island.[17] Half a million people live in archipelagos of small islands and coral atolls, such as the Maldives in the Indian Ocean, consisting of 1190 individual islands, and the Marshall Islands in the Pacific, which lie almost entirely within 3 m of sea level. Half a metre or more of sea level rise would reduce their areas substantially – some would have to be abandoned – and remove up to fifty per cent of their groundwater. The cost of protection from the sea is far beyond the resources of these islands' populations. For coral atolls, rise in sea level at a rate of up to about half a metre per century can be matched by coral growth, providing that growth is not disturbed by human interference and providing also that the growth is not inhibited by a rise in the maximum sea temperature exceeding about 1–2 °C.[18]

These are some examples of areas particularly vulnerable to sea level rise. Many other areas in the world will be affected in similar, although perhaps less dramatic, ways. Many of the world's cities are close to sea level and are being increasingly affected by subsidence because of the withdrawal of groundwater. The rise of sea level due to global warming will add to this problem. There is no technical difficulty for most cities in taking care of these problems, but the cost of doing so must be included when calculating the overall impact of global warming.

So far, in considering the impact of sea level rise, places of dense population where there is a large effect on people have been considered. There are also areas of importance where few people live. The world's wetlands and mangrove swamps currently occupy an area of about a million square kilometres (the figure is not known very precisely), equal approximately to twice the area of France. They contain much biodiversity and their biological productivity equals or exceeds that of any other natural or agricultural system. Over two-thirds of the fish caught for human consumption, as well as many birds and animals, depend on coastal marshes and swamps for part of their life cycles, so they are vital to the total world ecology. Such areas can adjust to slow levels of sea level rise, but there is no evidence that they could keep pace with a rate of rise of greater than about 2 mm per year – 20 cm per century. What will tend to occur, therefore, is that the area of wetlands will extend inland, sometimes with a loss of good agricultural land. However, because in many places such extension will be inhibited by the presence of flood embankments and other human constructions, erosion of the seaward boundaries of the wetlands will lead more usually to a loss of wetland area. Because of a variety of human activities (such as shoreline protection, blocking of sediment sources, land reclamation, aquaculture development and oil, gas and water extraction), coastal wetlands are currently being lost at a rate of 0.5–1.5% per year. Sea level rise because of climate change would further exacerbate this loss.[19]

To summarise the impact of the half-metre or so of sea level rise due to global warming which could occur during the twenty-first century: global warming is not the only reason for sea level rise but it is likely to exacerbate the impacts of other environmental problems. Careful management of human activities in the affected areas can do a lot to alleviate the likely effects, but substantial adverse impacts will remain. In delta regions, which are particularly vulnerable, sea level rise will lead to substantial loss of agricultural land and salt intrusion into freshwater resources. In Bangladesh, for instance, over ten million people are likely to be affected by such loss. A further problem in Bangladesh and other low-lying tropical areas will be the increased intensity and frequency of disasters because of storm surges. Each year, the number of people worldwide experiencing flooding because of storm surges is estimated now at about forty million. With a 40-cm sea level rise by the 2080s this number is estimated to quadruple – a number that might be reduced by half if coastal protection is enhanced in proportion to gross domestic product (GDP) growth.[20] Low-lying small islands will also suffer loss of land and freshwater supplies. Countries like The Netherlands and many cities in coastal regions will have to spend substantial sums on protection against the sea. Significant amounts of land will also be lost near the important wetland areas of the world. Attempts to put costs against these impacts, in both money and human terms, will be considered later in the chapter.

In this section we have considered the impacts of sea level rise for the twenty-first century. Because, as we have seen, the ocean takes centuries to adjust to an increase in surface temperature, the longer-term impacts of sea level rise also need to be emphasised. Even if the concentrations of greenhouse gases in the atmosphere were stabilised so that anthropogenic climate change is halted, the sea level will continue to rise for many centuries as the whole ocean adjusts to the new climate.

Increasing human use of fresh water resources

The global water cycle is a fundamental component of the climate system. Water is cycled between the oceans, the atmosphere and the land surface (Figure 7.5). Through evaporation and condensation it provides the main means whereby energy is transferred to the atmosphere and within it. Water is essential to all forms of life; the main reason for the wide range of life forms, both plant and animal, on the Earth is the extremely wide range of variation in the availability of water. In wet

Figure 7.5 The global water cycle (in thousands of cubic kilometres per year), showing the key processes of evaporation, precipitation, transport as vapour by atmospheric movements and transport from the land to the oceans by runoff or groundwater flow.

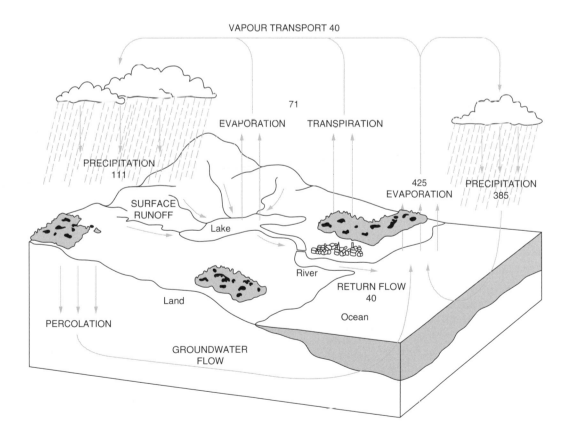

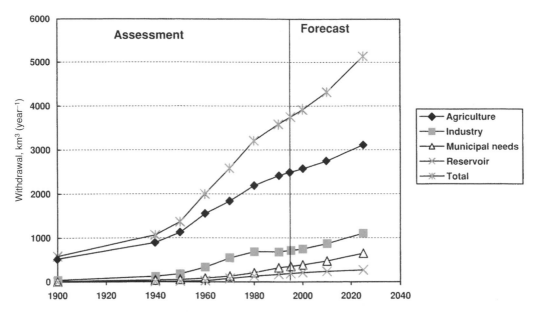

Figure 7.6 Global water withdrawal for different purposes, 1900–1995, and projected to the year 2025 in cubic kilometres per year. Losses from reservoirs are also included. As some water withdrawn is reused, the total water consumption amounts to about sixty per cent of the total water withdrawal.

tropical forests, the jungle teems with life of enormous variety. In drier regions sparse vegetation exists, of a kind which can survive for long periods with the minimum of water; animals there are also well adapted to dry conditions.

Water is also a key substance for humankind; we need to drink it, we need it for the production of food, for health and hygiene, for industry and transport. Humans have learnt that the ways of providing for livelihood can be adapted to a wide variety of circumstances regarding water supply except, perhaps, for the completely dry desert. Water availability for domestic, industrial and agricultural use averaged per capita in different countries varies from less than 100 m³ (22 000 imperial gallons) per year to over 100 000 m³ (22 million imperial gallons)[21] – although quoting average numbers of that kind hides the enormous disparity between those in very poor areas who may walk many hours each day to fetch a few gallons and many in the developed world who have access to virtually unlimited supplies at the turn of a tap.

The demands of increased populations and the desire for higher standards of living have brought with them much greater requirements for fresh water. During the last fifty years water use worldwide has grown over threefold (Figure 7.6); it now amounts to about ten per cent of the estimated global total of the river and groundwater flow from land to sea

(Figure 7.5). Two-thirds of human water use is currently for agriculture, much of it for irrigation; about a quarter is used by industry; only ten per cent or so is used domestically. Increasingly, water stored over hundreds or thousands of years in underground aquifers is being tapped for current use. With this rapid growth of demand comes greatly increased vulnerability regarding water supplies.

The extent to which a country is *water stressed* is related to the proportion of the available freshwater supply that is withdrawn for use. Withdrawal exceeding twenty per cent of renewable water supply has been used as a threshold of water stress. Under this definition approximately 1.7 billion people, one-third of the world's population, live in water-stressed countries. This number is projected to rise to around five billion by 2025, dependent on the rate of population growth, this without taking into account any effect on water supplies due to climate change. For instance, in India about seventy-five per cent of available water is currently used for irrigation. There is therefore very little surplus for future growth, the only river in north India with a surplus is the Brahmaputra. Similar situations where nearly all the available water is currently used prevail in much of central and western Asia.[22] Many other developing countries especially in Africa face similar situations.

A further vulnerability arises because many of the world's major sources of water are shared. About half the land area of the world is within water basins which fall between two or more countries. There are forty-four countries for which at least eighty per cent of their land areas falls within such international basins. The Danube, for instance, passes through twelve countries which use its water, the Nile water through nine, the Ganges-Bramaputra through five. Other countries where water is scarce are critically dependent on sharing the resources of rivers such as the Euphrates and the Jordan. The achievements of agreements to share water often bring with them demands for more effective use of the water and better management. Failure to agree brings increased possibility of tension and conflict. The former United Nations Secretary General, Boutros Boutros-Ghali, has said that 'The next war in the Middle East will be fought over water, not politics'.[23]

The impact of climate change on fresh water resources

The availability of fresh water will be substantially changed in a world affected by global warming. We saw in Chapter 6 (Figure 6.5(b)) that, although there remains substantial uncertainty in model predictions of precipitation change, it is possible to identify some areas where it is

Figure 7.7 Simulations of average monthly runoff in the Sacramento basin of California comparing (a) current climate with (b) changed climates with a 4 °C temperature increase and a twenty per cent increase in rainfall and (c) with the same temperature increase but with a twenty per cent decrease of rainfall.

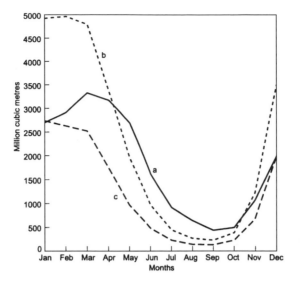

likely that there will be significant increases or decreases. For instance, precipitation is expected to increase in northern high latitudes in winter and the monsoon regions of south-east Asia in summer, while other regions (e.g. southern Europe, Central America, southern Africa and Australia) can expect significantly drier summers. Further, increase of temperature will mean that a higher proportion of the water falling on the Earth's surface will evaporate. In regions with increased precipitation, some or all of the loss due to evaporation may be made up. However, in regions with unchanged or less precipitation, there will be substantially less water available at the surface. The combined effect of less rainfall and more evaporation means less soil moisture available for crop growth and also less runoff – in regions with marginal rainfall this loss of soil moisture can be critical.

The runoff in rivers and streams is what is left from the precipitation that falls on the land after some has been taken by evaporation and by transpiration from plants; it is the major part of what is available for human use. The amount of runoff is highly sensitive to changes in climate; even small changes in the amount of precipitation or in the temperature (affecting the amount of evaporation) can have a big influence on it. To illustrate this, Figure 7.7 shows simulations, carried out for the Sacramento Basin in California, USA (a region where water is stored for some of the year in mountain snow), of changes in runoff with changed climate conditions. With a 4 °C regional temperature rise and twenty per cent decreased rainfall, the runoff in the summer months falls to between twenty per cent and fifty per cent of its normal value. Even with twenty per cent increased rainfall and the same temperature increase, summer runoff still remains well below normal. Watersheds in arid or semi-arid regions are especially sensitive because the annual runoff is in any case highly variable.

Some watersheds in mid latitudes in the Northern Hemisphere, where snowmelt is an important source of runoff, can also be severely affected. For these places, as temperatures rise, winter runoff will increase substantially and spring high water will be much reduced. Further, as we saw earlier in the chapter, up to one-half of the mass of mountain glaciers and small ice caps may melt away over the next hundred years which could substantially change the seasonal distribution of river flow and water supply for hydroelectric generation and agriculture.

A detailed hydrological study of Asia under climate change illustrates the very different impacts on different parts of the continent.[24] Projections from a climate model for 2050 under a scenario similar to A1B were introduced into a hydrological model of different river catchments and the changes in runoff in the river basins estimated. In arid or semi-arid regions of Asia, surface runoff is expected to decrease drastically so affecting the volume of water available for irrigation and other purposes. Average annual runoff in the basins of the Tigris, Euphrates, Indus and Brahmaputra rivers was estimated to decline by twenty-two, twenty-five, twenty-seven and fourteen per cent respectively. Other areas will experience substantial increases in runoff, for instance by thirty-seven and twenty-six per cent, respectively, in the Yangtze (Changjiang) and Huang He rivers. Substantial increases were also projected for Siberian rivers.

Watersheds that are particularly vulnerable to climate change can be identified by asking certain questions about them.[25]

- How much water storage is there in the watershed relative to the annual flow? In Colorado in the United States, for instance, the storage is four times the annual flow, whereas in the Atlantic States it is only one-tenth of the annual flow.
- How large is the demand as a percentage of the potential supply? This varies a great deal. For instance, in North America, for the Rio Grande and for the lower Colorado demand approximately equals supply and very little of either of these rivers reaches the sea. Therefore, though the Colorado has substantial storage and is therefore not very sensitive to annual variations, the amount of use in its lower reaches means that, over a number of years, any reduction of its flow is bound to imply lower water availability.
- How much groundwater is being used? There are many places in the world where groundwater is being used faster than it is being replenished. To give two examples, for more than half the land area of the

United States over a quarter of the groundwater withdrawn is not being replenished, so every year the water has to be extracted from deeper levels; and in Beijing in China the water table is falling by 2 m a year as its groundwater is pumped out.

• How variable are the stream and river flows? This question is particularly relevant to arid and semi-arid areas. Detailed studies taking these criteria into account have been carried out for a number of areas; one example for the MINK (Missouri, Iowa, Nebraska and Kansas) region of the United States is shown in the box.

Study of the 'MINK' region in the United States

The United States Department of Energy has carried out a detailed study[26] of the likely effects of climate change on a region (known as the MINK region) in the centre of the United States comprising the states of Missouri, Iowa, Nebraska and Kansas. Included within the region are parts of four major river basins – the Missouri, the Arkansas, the Upper and the Lower Mississippi. Water is already a scarce resource within the MINK states; much of the area's irrigation relies on non-renewable groundwater supplies. These will diminish with time, so that even in the absence of climate change less water will be available, especially for irrigation.

To provide an analogue of the climate which might be expected with increased carbon dioxide, the period of the 1930s was chosen, when the average temperature in the region was about 1 °C warmer than in the period 1950–1980 (the 'control' period) and the average precipitation about ten per cent lower than in the control period.

Water would become scarcer under the analogue climate compared with the control.[27] The hotter and drier conditions would increase evaporation and reduce runoff. Streamflow would drop by about thirty per cent in the Missouri and the Upper Mississippi basins and by about ten per cent in the Arkansas. Most streams would fall well short of supplying both the desired instream flows and the current levels of consumption use.

Under the analogue climate, irrigated agriculture would be bound to decline substantially because of the increased constraints on groundwater use coupled with less water availability from other sources. This would also result in a drive to increased efficiency, albeit at greater cost. Maintaining the high priority currently given to navigation on the main stem of the Missouri would become very costly.

So far when mentioning changes in temperature or rainfall it is changes in the average with which we have been concerned; for instance, the simulations in Figure 7.7 are for average conditions. But, as has been constantly emphasised, the severity of climate impacts depends to a great degree on extreme conditions. This is well illustrated by looking at the scale of natural disasters involving water – either too much water in floods or too little in droughts. Some of the most damaging floods of recent years were mentioned in Chapter 1 (page 5) – see also Table 7.3 (page 183). Droughts do not appear high up on the table of natural disasters, not because they are unimportant, but because, unlike most other disasters, their effects tend to be felt over a long period of time. The 'dust bowl' years in the 1930s in the United States are still within living memory, as are the droughts and famines in India in 1965–7 which, it is estimated, claimed one and a half million lives. Recent decades have seen a series of damaging droughts in the Sahel region and in other parts of Africa[28] – which are still recurring only too frequently on that continent.

Any temperature or rainfall record shows a large variability. The inevitable result of variability added to higher average temperatures (meaning higher evaporation) and higher average rainfall will be a greater number and greater intensity of both droughts and floods.[29] For instance, associated with the substantial changes in average runoff expected by 2050 in parts of Asia mentioned above will be increases in the number and intensity of floods and droughts. Some of the areas likely to be affected are just those areas that are particularly vulnerable at the moment – although, as was also implied in Chapter 6, droughts and floods are increasingly likely to occur in some locations where, at present, such disasters are rare. Very few quantitative estimates have been made of the likely increase in floods or droughts as a result of the increase of greenhouse gases. One estimate quoted in Chapter 6 (page 131) projected an increase of a factor of five in intense precipitation events in parts of Europe under a doubled atmospheric carbon dioxide concentration.

The monsoon regions of southeast Asia are an example of an area that may be particularly vulnerable to both floods and droughts. Figure 7.8 shows the predicted change in summer precipitation over the Indian subcontinent as simulated by a regional climate model (RCM) for 2050 under a scenario similar to SRES A1B. Note the improvement in detail of the precipitation pattern that results from the use of the increased resolution of the regional model compared with the global model (GCM), for instance over the Western Ghats (the mountains that rise steeply from India's west coast) there are large increases not present in the global model simulation. The most serious reductions in water availability simulated by the regional model are in the arid regions of northwest India and Pakistan where average precipitation is reduced to less than 1 mm day^{-1} – that coupled with higher temperatures leads to a sixty per cent decline in soil moisture. Substantial increases in average precipitation are projected for areas in eastern India and in flood-prone

Figure 7.8 Predicted changes in monsoon rainfall (mm/day) over India between the present day and the middle of the twenty-first century from a 300-km resolution GCM and from a 50-km resolution RCM. The RCM pattern is very different in some respects from the coarser resolution pattern of the GCM.

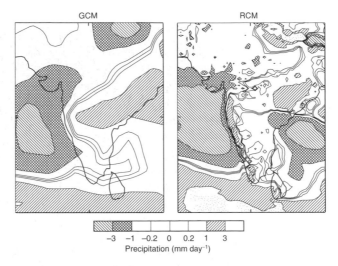

Bangladesh where the projected increase is about twenty per cent. What is urgently required for this part of the world and elsewhere is much better information linking changes in average parameters with likely changes in frequency, intensity and location of extreme events.

There is another reason, not unconnected with global warming, for the vulnerability of water supplies: the link between rainfall and changes in land use. Extensive deforestation can lead to large changes in rainfall (see box on page 173). A similar tendency to reduced rainfall can be expected if there is a reduction in vegetation over large areas of semi-arid regions. Such changes can have a devastating and widespread effect and assist in the process of desertification. This is a potential threat to the drylands covering about one-quarter of the land area of the world (see box on page 163).

What sort of action can be taken to lessen the vulnerability of human communities to changes in water availability or supply? Irrigation accounts for about two-thirds of world water use, and is of great importance to world agriculture. Irrigation is applied to about one-sixth of the world's farmland, which produces about one-third of the world's crops. In some areas the ratio is much higher; for instance, over eighty per cent of the agricultural land in California is irrigated. Most irrigation is through open ditches, which is very wasteful of water; over sixty per cent is lost through evaporation and seepage. Microirrigation techniques, in which perforated pipes deliver water directly to the plants, provide large opportunities for water conservation, making it possible to expand irrigated fields without building new dams.[30] Management of the existing infrastructure can be improved, for instance by arranging for the integration of different supplies, and conservation in the domestic and industrial sectors can be encouraged. Most of these actions will cost

Desertification

Drylands (defined as those areas where precipitation is low and where rainfall typically consists of small, erratic, short, high-intensity storms) cover about forty per cent of the total land area of the world and support over one-fifth of the world's population. Figure 7.9 shows how these arid areas are distributed over the continents.

Desertification in these drylands is the degradation of land brought about by climate variations or human activities that have led to decreased vegetation, reduction of available water, reduction of crop yields and erosion of soil. The United Nations Convention to Combat Desertification (UNCCD) set up in 1996 estimates that over seventy per cent of these drylands, covering over twenty-five per cent of the world's land area, are degraded[33] and therefore affected by desertification. The degradation can be exacerbated by excessive land use or increased human needs (generally because of increased population), or political or economic pressures (for instance, the need to grow cash crops to raise foreign currency). It is often triggered or intensified by a naturally occurring drought.

The progress of desertification in some of the drylands will be increased by the more frequent or more intense droughts that are likely to result from climate change during the twenty-first century.

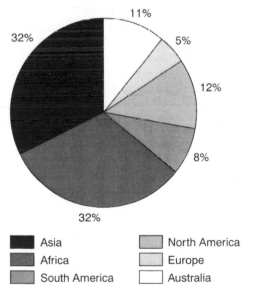

Figure 7.9 The world's drylands, by continent. The total area of drylands is about sixty million square kilometres (about forty per cent of the total land area), of which ten million are hyper-arid deserts.

money, although they may be much more cost-effective ways of coping with future change in water resources than attempting to develop major new facilities.[31]

In summary, what are the likely effects of global warming on water supplies? Firstly, the current vulnerability of many communities to water shortage should be noted. This is especially true of arid and semi-arid regions where the increasing demands of human communities mean that droughts, even for short periods, are more disastrous than before. Vulnerability is well demonstrated in many areas of the world where the amounts of groundwater extraction greatly exceed its replenishment – a situation that cannot continue for very long into the future. Because of population growth these vulnerabilities will increase and will exacerbate the negative effects of global warming.

Secondly, climate change because of global warming will result in large changes in water supplies in many places. Although the present state of knowledge regarding regional and local climate change does not allow scientists to identify precisely the most vulnerable areas, they are able to indicate the sort of area which will be most affected. Such areas are those arid and semi-arid areas with reduced rainfall leading to greater aridity and even desertification; continental areas where decreased summer rainfall and increased temperature result in a substantial loss in soil moisture and much increased vulnerability to drought; and areas where increased rainfall could lead to a greater incidence of floods. The changing pattern of climate extremes, especially droughts and floods, will be the cause of most of the problems. It is also the case that regions such as southeast Asia that are dependent on unregulated river systems are more sensitive to change than regions such as western Russia and the western United States that have large, regulated water resource systems. Thirdly, some of the adverse impact of climate change on water supplies can be reduced by taking appropriate alleviating action, by introducing more careful and integrated water management[32] and by introducing more effective disaster preparedness in the most vulnerable areas.

Impact on agriculture and food supply

Every farmer understands the need to grow crops or rear animals that are suited to the local climate. The distribution of temperature and rainfall during the year are key factors in making decisions regarding what crops to grow. These will change in the world influenced by global warming. The patterns of what crops are grown where will therefore also change. But these changes will be complex; economic and other factors will take their place alongside climate change in the decision-making process.

There is enormous capacity for adaptation in the growth of crops for food – as is illustrated by what was called the Green Revolution of the 1960s, when the development of new strains of many species of crops resulted in large increases in productivity. Between the mid 1960s and the mid 1980s global food production rose by an average annual rate of 2.4% – faster than global population – more than doubling over that thirty-year period. Grain production grew even faster, at an annual rate of 2.9%.[34] There are concerns that factors such as the degradation of many of the world's soils largely through erosion and the slowed rate of expansion of irrigation because less fresh water is available will tend to reduce the potential for increased agricultural production in the future. However, with declining rates of population growth, there remains optimism that, in the absence of major climate change, the growth in world food supply is likely to continue to match the growth in demand at least during the early decades of the twenty-first century.[35]

What will be the effect of climate change on agriculture and food supply? With the detailed knowledge of the conditions required by different species and the expertise in breeding techniques and genetic manipulation available today, there should be little difficulty in matching crops to new climatic conditions over large parts of the world. At least, that is the case for crops that mature over a year or two. Forests reach maturity over much longer periods, from decades up to a century or even more. The projected rate of climate change is such that, during this time, trees may find themselves in a climate to which they are far from suited. The temperature regime or the rainfall may be substantially changed, resulting in stunted growth or a greater susceptibility to disease and pests. The impact of climate change on forests is considered in more detail in the next section.

An example of adaptation to changing climate is the way in which farmers in Peru adjust the crops they grow depending on the climate forecast for the year.[36] Peru is a country whose climate is strongly influenced by the cycle of El Niño events described in Chapters 1 and 5. Two of the primary crops grown in Peru, rice and cotton, are very sensitive to the amount and the timing of rainfall. Rice requires large amounts of water; cotton has deeper roots and is capable of yielding greater production during years of low rainfall. In 1983, following the 1982–3 El Niño event, agricultural production dropped by fourteen per cent. By 1987 forecasts of the onset of El Niño events had become sufficiently good for Peruvian farmers to take them into account in their planning. In 1987, following the 1986–7 El Niño, production actually increased by three per cent, thanks to a useful forecast.

Three factors are particularly important in considering the effect of climate change on agriculture and food production. The availability of water is the most important of the factors. The vulnerability of water supplies to climate change carries over into a vulnerability in the growing of crops and the production of food. Thus the arid or semi-arid areas, mostly in developing countries, are most at risk. A second factor, which tends to lead to increased production as a result of climate change, is the boost to growth that is given, particularly to some crops, by increased atmospheric carbon dioxide (see box below). A third factor is the effect of temperature changes; in particular, under very high temperatures, yields of some crops are substantially reduced.

The carbon dioxide 'fertilisation' effect

An important positive effect of increased carbon dioxide (CO_2) concentrations in the atmosphere is the boost to growth in plants given by the additional CO_2. Higher CO_2 concentrations stimulate photosynthesis, enabling the plants to fix carbon at a higher rate. This is why in glasshouses additional CO_2 may be introduced artificially to increase productivity. The effect is particularly applicable to what are called C3 plants (such as wheat, rice and soya bean), but less so to C4 plants (for example, maize, sorghum, sugar-cane, millet and many pasture and forage grasses). Under ideal conditions it can be a large effect; for C3 crops under doubled CO_2, an average of +30%.[37] However, under real conditions on the large scale where water and nutrient availability are also important factors influencing plant growth, experiments show that the increases, although difficult to measure accurately, tend to be substantially less than the ideal.[38] In experimental work, grain and forage quality declines with CO_2 enrichment and higher temperatures. More research is required especially for many tropical crop species and for crops grown under suboptimal conditions (low nutrients, weeds, pests and diseases).

Detailed studies have been carried out of the sensitivity to climate change during the twenty-first century of the major crops which make up a large proportion of the world's food supply (see box below). They have used the results of climate models to estimate changes in temperature and precipitation. Many of them study the effect of CO_2 fertilisation and some also model the effects of climate variability as well as changes in the means. Some also include the possible effects of economic factors and of modest levels of adaptation. These studies in general indicate that the benefit of increased CO_2 concentration on crop growth and yield does not always overcome the effects of excessive heat and drought. For cereal crops in mid latitudes, potential yields are projected to increase for small increases in temperature (2–3 °C) but decrease for larger temperature rises.[39] In most tropical and subtropical regions, potential yields are projected to decrease for most increases in temperature; this is because such crops are near their maximum temperature tolerance. Where there is a large decrease in rainfall, tropical crop yields would be even more adversely affected.

Taking the supply of food for the world as a whole, studies tend to show that, with appropriate adaptation, the effect of climate change on total global food supply is not likely to be large. However, none of them have adequately taken into account the likely effect on food production of climate extremes (especially of the incidence of drought), of increasingly limited water availability or of other factors such as the integrity of the world's soils, which are currently being degraded at an alarming rate.[40] A serious issue exposed by the studies is that climate change is likely to affect countries very differently. Production in developed countries with relatively stable populations may increase, whereas that in many developing countries (where large increases in population are occurring) is likely to decline as a result of climate change. The disparity between developed and developing nations will tend to become much larger, as will the number of those at risk from hunger. The surplus of food in developed countries is likely to increase, while developing countries will face increasing deprivation as their declining food availability becomes much less able to provide for the needs of their increasing populations. Such a situation will raise enormous problems, one of which will be that of employment. Agriculture is the main source of employment in developing countries; people need employment to be able to buy food. With changing climate, as some agricultural regions shift, people will tend to attempt to migrate to places where they might be employed in agriculture. With the pressures of rising populations, such movement is likely to be increasingly difficult and we can expect large numbers of environmental refugees.

In looking to future needs, two activities that can be pursued now are particularly important. Firstly, there is large need for technical advances in agriculture in developing countries requiring investment and widespread local training. In particular, there needs to be continued development of programmes for crop breeding and management, especially in conditions of heat and drought. These can be immediately useful in the improvement of productivity in marginal environments today. Secondly, as was seen earlier when considering fresh water supplies, improvements need to be made in the availability and management of water for irrigation, especially in arid or semi-arid areas of the world.

Modelling the impact of climate change on world food supply

An example illustrating the key elements of a detailed study of the impact of climate change on world food supply is shown in Figure 7.10.[41]

A climate change scenario is first set up with a climate model of the kind described in Chapter 5. Models of different crops that include the effects of temperature, precipitation and CO_2 are applied to 124 different locations in 18 countries to produce projected crop yields that can be compared with projected yields in the absence of climate change. Included also are farm-level adaptations, e.g. planting date shifts, more climatically adapted varieties, irrigation and fertiliser application. These estimates of yield are then aggregated to provide yield-change estimates by crop and country or region.

These yield changes are then employed as inputs to a world food trade model that includes assumptions about global parameters such as population growth and economic change and links together national and regional economic models for the agricultural sector through trade, world market prices and financial flows. The world food trade model can explore the effects of adjustments such as increased agricultural investment, reallocation of agricultural resources according to economic returns (including crop switching) and reclamation of additional arable land as a response to higher cereal prices. The outputs from the total process provide information projected up to the 2080s on food production, food prices and the number of people at risk of hunger (defined as the population with an income insufficient either to produce or to procure their food requirements).

The main results with this model for the 2080s regarding the impact of climate change following the IS 92a scenario (Figure 6.1) are that yields at mid to high latitudes are expected to increase, and at low latitudes (especially the arid and sub-humid tropics) to decrease. This pattern becomes more pronounced as time progresses. The African continent is particularly likely to experience marked reductions in yield, decreases in production and an estimated sixty million or more additional people at risk of hunger as a result of climate change.

The authors emphasise that, although the models and the methods they have employed are comparatively complex, there are many factors that have not been taken into account. For instance, they have not adequately considered the impact of changes in climate extremes, the availability of water supplies for irrigation or the effects of future technological change on agricultural productivity. Further (see Chapter 6), scientists are not yet very confident in the regional detail of climate change. The results, therefore, although giving a general indication of the changes that could occur, should not be treated as a detailed prediction. They highlight the importance of studies of this kind as a guide to future action.

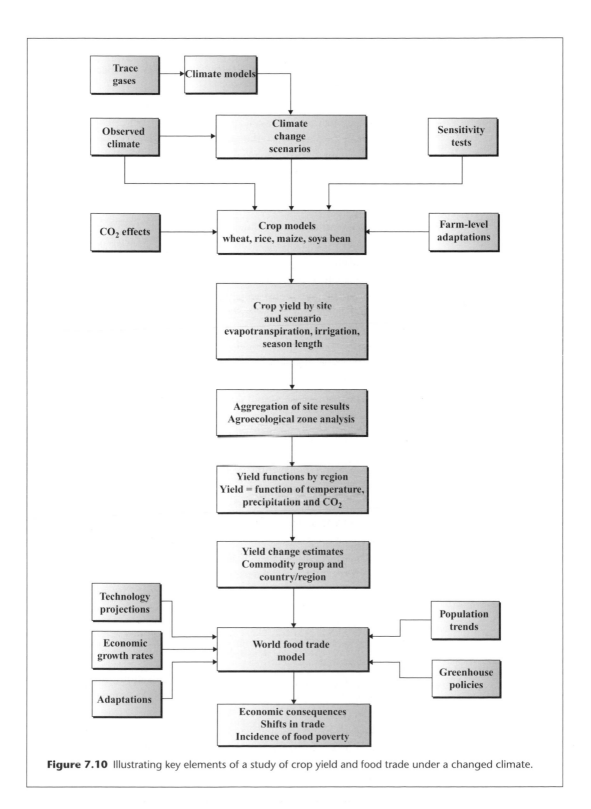

Figure 7.10 Illustrating key elements of a study of crop yield and food trade under a changed climate.

The impact on ecosystems

A little over ten per cent of the world's land area is under cultivation – that was the area addressed in the last section. The rest is to a greater or lesser extent unmanaged by humans. Of this about thirty per cent is natural forest and between one and two per cent plantation forest. The variety of plants and animals that constitute a local ecosystem is sensitive to the climate, the type of soil and the availability of water. Ecologists divide the world into biomes – regions characterised by their distinctive vegetation. This is well illustrated by information about the distribution of vegetation over the world during past climates (e.g. for the part of North America shown in Figure 7.11), which indicates what species and what ecosystems are most likely to flourish under different climatic regimes.

Changes in climate alter the suitability of a region for different species, and change their competitiveness within an ecosystem, so that even relatively small changes in climate will lead, over time, to large changes in the composition of an ecosystem. Since climate is the dominant factor determining the distribution of biomes (Figure 7.12), information gleaned from paleo sources could be used to produce maps of the optimum distribution of natural vegetation under the climate scenarios expected to occur with global warming.

However, changes of the kind illustrated in Figure 7.11 took place over thousands of years. With global warming similar changes in climate occur over a few decades. Most ecosystems cannot respond or migrate that fast. Fossil records indicate that the maximum rate at which most plant species have migrated in the past is about 1 km per year. Known constraints imposed by the dispersal process (e.g. the mean period between germination and the production of seeds and the mean distance that an individual seed can travel) suggest that, without human intervention, many species would not be able to keep up with the rate of movement of their preferred climate niche projected for the twenty-first century, even if there were no barriers to their movement imposed by land use.[42] Natural ecosystems will therefore become increasingly unmatched to their environment. How much this matters will vary from species to species: some are more vulnerable to changes in average climate or climate extremes than others. But all will become more prone to disease and attack by pests. Any positive effect from added 'fertilisation' due to increased carbon dioxide is likely to be more than outweighed by negative effects from other factors.

Trees are long-lived and take a long time to reproduce, so they cannot respond quickly to climate change. Further, many trees are surprisingly sensitive to the average climate in which they develop. The environmental

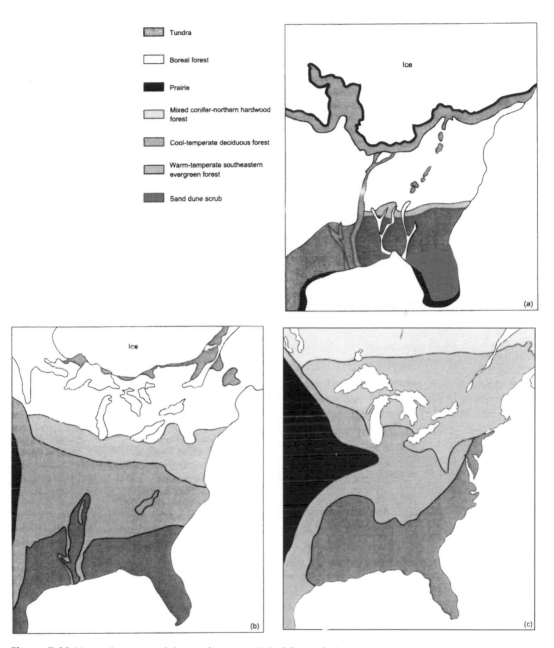

Figure 7.11 Vegetation maps of the south-eastern United States during past climate regimes: (a) for 18 000 years ago at the maximum extent of the last ice age, (b) for 10 000 years ago, (c) for 5000 years ago when conditions were similar to present. A vegetation map for 200 years ago is similar to that in (c).

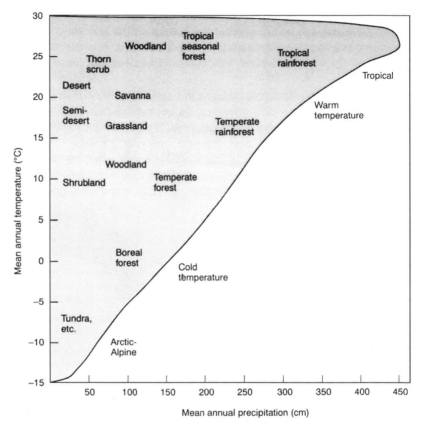

Figure 7.12 The pattern of world biome types related to mean annual temperature and precipitation. Other factors, especially the seasonal variations of these quantities, affect the detailed distribution patterns (after Gates).

conditions (e.g. temperature and precipitation) under which a species can exist and reproduce are known as its niche. Climate niches for some typical tree species are illustrated in Figure 7.13; under some conditions a change as small as 1 °C in annual average temperature can make a substantial difference to a tree's productivity. For the likely changes in climate in the twenty-first century, a substantial proportion of existing trees will be subject to unsuitable climate conditions. This will be particularly the case in the boreal forests of the Northern Hemisphere where, as trees become less healthy, they will be more prone to pests, die-back and forest fires. One estimate projects that, under a doubled CO_2 scenario, up to sixty-five per cent of the current boreal forested area could be affected.[43]

A decline in the health of many forests in recent years has received considerable attention, especially in Europe and North America where much of it has been attributed to acid rain and other pollution originating

Forests, deforestation and climate change

Extensive changes in the area of forests due to deforestation can seriously affect the climate in the region of change. Also, changes in temperature or rainfall that occur because of long-term changes in climate can also have a major impact on forests. We look at these effects in turn.

Changes in land use such as those brought about by deforestation can affect the amount of rainfall, for three main reasons. Over a forest there is a lot more evaporation of water (through the leaves of the trees) than there is over grassland or bare soil, hence the air will contain more water vapour. Also, a forest reflects twelve to fifteen per cent of the sunlight that falls on it, whereas grassland will reflect about twenty per cent and desert sand up to forty per cent. A third reason arises from the roughness of the surface where vegetation is present.

An American meteorologist, Professor Jules Charney, suggested in 1975 that, in the context of the drought in the Sahel, there could be an important link between changes of vegetation (and hence changes of reflectivity) and rainfall. The increased energy absorbed at the surface when vegetation is present and the increased surface roughness both tend to stimulate convection and other dynamic activity in the atmosphere so leading to the production of rainfall.

Early experiments with numerical models that included these physical processes demonstrated the effect and indicated, for instance, a reduction of about fifteen per cent in rainfall if the forest north of 30 °S in South America were removed and replaced by grassland.[44] Similar model experiments for Zaire over a smaller region showed an average reduction in rainfall of over thirty per cent.[45] A much more drastic experiment in which the Amazonian forest was removed and replaced with a desert surface showed a reduction in rainfall by seventy per cent to levels similar to those of the semi-arid regions of the Sahel part of Africa.[46] Such a model experiment does not represent a realistic situation, but it illustrates the significant impact that widespread deforestation could have on the local climate.

More recent work has been with interactive models that include not just the effect of changes in land use or forestation on the climate but also, in a dynamic way, the effect on forests and other vegetation of changes in climate. In an experiment with such a model that assumes carbon dioxide emissions following the IS 92a scenario,[47] substantial reductions in precipitation are projected for areas of Amazonia, that lead to die-back of the Amazonian forest and significant release of carbon to the atmosphere (one of the positive climate feedbacks mentioned in the box on page 40). As the forest dies back, the rainfall is further reduced because of the change in properties of the land surface, leading, by the end of the twenty-first century, to the replacement of much of the forest cover by semi-arid conditions. Such results are still subject to considerable uncertainties (for instance, those associated with the model simulations of El Niño events under climate change conditions and the connections between these events and the climate over Amazonia), but they illustrate the type of impacts that might occur and emphasise the importance of understanding the interactions between climate and vegetation.

from heavy industry, power stations and motor cars. Not all damage to trees, however, is thought to have this origin. Studies in several regions of Canada, for instance, indicate that the die-back of trees there is related to changes in climatic conditions, especially to successions of warmer winters and drier summers.[48] In some cases it may be the double effect of pollution and climate stress causing the problem; trees already weakened

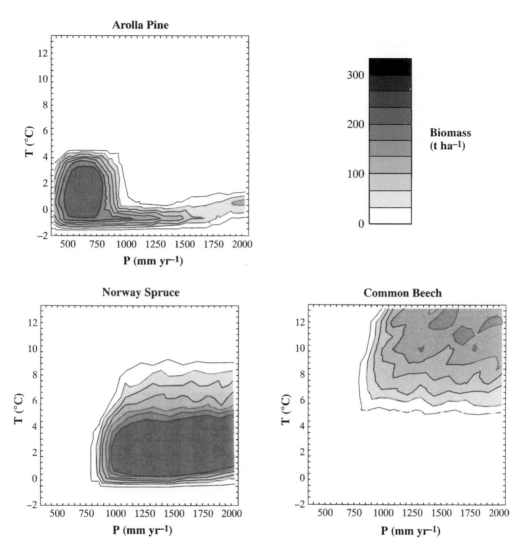

Figure 7.13 Simulated environmental realised niches (the realised niche describes the conditions under which the species is actually found) for three tree species, Arolla Pine, Norway Spruce and Common Beech. Plots are of biomass generated per year against annual means of temperature (T) and precipitation (P). Arolla Pine is a species with a particularly narrow niche. The narrower the niche, the greater the potential sensitivity to climate change.

by the effects of pollution fail to cope with climate stress when it comes. The assessment of the impact of climate change carried out for the MINK region of the United States (see box on page 160) concluded that, under the warmer, drier conditions of the analogue climate they studied, decline and die-back of the forested part of the region would reduce the mass of timber in the forest by ten per cent over twenty years.[49] The results of these studies are indicative of the more serious levels of forest die-back that are likely to occur with the rapid rate of climate change expected with global warming (see box on page 173).These stresses on the world's forests due to climate change will be concurrent with other problems associated with forests, in particular those of continuing tropical defor- estation and of increasing demand for wood and wood products resulting from rapidly increasing populations especially in developing countries.

If a stable climate is eventually re-established, given adequate time (which could be centuries), different trees will be able to find again at some location their particular climatic niche. It is during the period of rapid change that most trees will find themselves unsatisfactorily located from the climate point of view.

It was mentioned in Chapter 3 that forests represent a large store of carbon; eighty per cent of above-ground and forty per cent of below- ground terrestrial carbon is in forests. We also saw in Chapter 3 that tropical deforestation due to human activities is probably releasing be- tween 1 and 2 Gt of carbon into the atmosphere each year. If, because of the rate of climate change, substantial stress and die-back occurs in boreal and tropical forests (see box on page 173) a release of carbon will occur. This positive feedback was mentioned in Chapter 3 (see the box on page 40). Just how large this will be is uncertain but estimates as high as 240 Gt over the twenty-first century for the above-ground component alone have been quoted.[50]

The above discussion has largely related to the impact of climate change on natural forests where the likely impacts are largely negative. Studies of the impacts on managed forests are more positive.[51] They sug- gest that with appropriate adaptation and land and product management, even without forestry projects that increase the capture and storage of carbon (see Chapter 10), a small amount of climate change could in- crease global timber supply and enhance existing market trends towards rising market share in developing countries.

A further concern about natural ecosystems relates to the diversity of species that they contain and the loss of species and hence of biodiversity due to the impact of climate change. Significant disruptions of ecosys- tems from disturbances such as fire, drought, pest infestation, invasion of species, storms and coral bleaching events are expected to increase. The stresses caused by climate change, added to other stresses on ecological systems (e.g. land conversion, land degradation, deforestation, harvest- ing and pollution) threaten substantial damage to or complete loss of some unique ecosystems, and the extinction of some endangered species. Coral reefs and atolls, mangroves, boreal and tropical forests, polar and alpine ecosystems, prairie wetlands and remnant native grasslands are examples of systems threatened by climate change. In some cases the threatened ecosystems are those that could mitigate against some climate change impacts (e.g. coastal systems that buffer the impact of storms). Possible adaptation methods to reduce the loss of biodiversity include the establishment of refuges, parks and reserves with corridors to allow migration of species, and the use of captive breeding and translocation of species.[52]

So far we have been considering ecosystems on land. What about those in the oceans; how will they be affected by climate change? Although we know much less about ocean ecosystems, there is considerable evidence that biological activity in the oceans has varied during the cycle of ice ages. Chapter 3 noted (see box on page 35) the likelihood that it was these variations in marine biological activity which provided the main control on atmospheric carbon dioxide concentrations during the past million years (see Figure 4.4). The changes in ocean water temperature and the possible changes in some of the patterns of ocean circulation are likely to result in changes in the regions where upwelling occurs and where fish congregate. Some fisheries could collapse and others expand. At the moment the fishing industry is not well adapted to address major change.[53]

Some of the most important marine ecosystems are found within coral reefs that occur in many locations throughout the tropical and subtropical world. They are especially rich in biodiversity and are particularly threatened by global warming. Within them the species diversity contains more phyla than rainforests and they harbour more than twenty-five per cent of all known marine fish.[54] They represent a significant source of food for many coastal communities. Corals are particularly sensitive to sea surface temperature and even one degree Celsius of persistent warming can cause bleaching (paling in colour) and extensive mortality accompanies persistent temperature anomalies of 3 °C or more. Much recent bleaching, for instance that in 1998, have been associated with El Niño events.[55]

The impact on human health

Human health is dependent on a good environment. Many of the factors that lead to a deteriorated environment also lead to poor health. Pollution of the atmosphere, polluted or inadequate water supplies, and poor soil (leading to poor crops and inadequate nutrition) all present dangers to human health and wellbeing and assist the spread of disease. As has been seen so far in considering the impacts of global warming, many of these factors will tend to be exacerbated through the climate change that will occur in the warmer world. The greater likelihood of extremes of climate, such as droughts and floods, will also bring greater risks to health from increased malnutrition and from a prevalence of conditions more likely to lead to the spread of diseases from a variety of causes.

How about direct effects of the climate change itself on human health?[56] Humans can adapt themselves and their buildings so as to live satisfactorily in very varying conditions and have great ability to adapt to a wide range of climates. The main difficulty in assessing the impact of climate change on health is that of unravelling the influences of climate from the large number of other factors (including other environmental factors) that affect health.

The main direct effect on humans will be that of heat stress in the extreme high temperatures that will become more frequent and more widespread especially in urban populations (see box and Figure 6.6). In large cities where heat waves commonly occur death rates can be doubled or tripled during days of unusually high temperatures.[57] Although such episodes may be followed by periods with fewer deaths showing that some of the deaths would in any case have occurred about that time, most of the increased mortality seems to be directly associated with the excessive temperatures with which old people in particular find it hard to cope. On the positive side, mortality due to periods of severe cold in winter will be reduced. The results of studies are equivocal regarding

Heat waves in Europe, 2003

Record extreme temperatures were experienced in Europe during June, July and August 2003. At many locations temperature rose over 40 °C. In France, Italy, the Netherlands, Portugal and Spain, over 21 000 additional deaths were attributed to the unrelenting heat. Spain, Portugal, France and countries in Central and Eastern Europe suffered from intense forest fires.[58] Figure 7.14 illustrates the extreme rareness of this event.

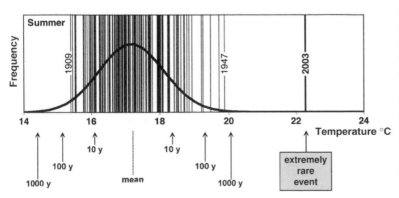

Figure 7.14 Distribution of average summer temperatures (June, July, August) in Switzerland from 1864–2003 showing a fitted gaussian probability distribution – standard deviation 0.94 °C. Average temperatures with 10,100 and 1000 year return periods are also indicated. The 2003 value is 5.4 standard deviations from the mean showing it as an extremely rare event (from Schar et al. 2004, *Nature* 427, 332–6).

whether the reduction in winter mortality will be greater or less than the increase in summer mortality. These studies have largely been confined to populations in developed countries, precluding a more general comparison between changes in summer and winter mortality.

A further likely impact of climate change on health is the increased spreading of diseases in a warmer world. Many insect carriers of disease thrive better in warmer and wetter conditions. For instance, epidemics of diseases such as viral encephalitides carried by mosquitoes are known to be associated with the unusually wet conditions that occur in the Australian, American and African continents associated with different phases of the El Niño cycle.[59] Some diseases, currently largely confined to tropical regions, with warmer conditions could spread into mid latitudes. Malaria is an example of such a disease that is spread by mosquitoes under conditions which are optimum in the temperature range of 15–32 °C with humidities of fifty to sixty per cent. It currently represents a huge global public health problem, causing annually around 300 million new infections and over one million deaths. Under climate change scenarios, most predictive model studies indicate a net increase in the geographic range (and in the populations at risk) of potential transmission of malaria and dengue infections, each of which currently impinge on forty to fifty per cent of the world's population. Other diseases that are likely to spread for the same reason are yellow fever and some viral encephalitis. In all cases, however, actual disease occurrence will be strongly influenced by local environmental conditions, socio-economic circumstances and public health infrastructure.

The potential impact of climate change on human health could be large. However, the factors involved are highly complex; any quantitative conclusions will require careful study of the direct effects of climate on humans and of the epidemiology of the diseases that are likely to be particularly affected. Some remarks about how the health impacts of extremes and disasters might be reduced are given in the next section.

Adaptation to climate change

As we have seen, some of the impacts of climate change are already becoming apparent. A degree of adaptation therefore has already become a necessity. Numerous possible adaptation options for responding to climate change have already been identified. These can reduce adverse impacts and enhance beneficial effects of climate change and can also produce immediate ancillary benefits, but they cannot prevent all damages. Examples of adaptation options are listed in Table 7.1. Many of the options listed are presently employed to cope with current climate variability and extremes; their expanded use can enhance both current and future capacity to cope. But such actions may not be as effective in the future as the amount and rate of climate change increase. To make a list of possible adaptation options is relatively easy. If they are to be applied effectively, much more information is urgently required regarding the detail and the cost of their application over the wide range of circumstances where they will be required.

Of particular importance is the requirement for adaptation to extreme events and disasters such as floods, droughts and severe storms.[60] Vulnerability to such events can be substantially reduced by much more adequate preparation.[61] For instance, following hurricanes George and Mitch, the Pan American Health Organisation (PAHO) identified a range of policies to reduce the impact of such events[62]:

- Undertaking vulnerability studies of existing water supply and sanitation systems and ensuring that new systems are built to reduce vulnerability.
- Developing and improving training programmes and information systems for national programmes and international cooperation on emergency management.
- Developing and testing early warning systems that should be coordinated by a single national agency and involve vulnerable communities. Provision is also required for providing and evaluating mental care, particularly for those who may be especially vulnerable to the adverse psychosocial effects of disasters (e.g. children, the elderly and the bereaved).

Costing the impacts: extreme events

In the previous paragraphs the impacts of climate change have been described in terms of a variety of measures; for instance, the number of people affected (e.g. by mortality, disease or by being displaced), the gain or loss of agricultural or forest productivity, the loss of biodiversity, the increase in desertification, etc. However, the most widespread measure, looked for by many policymakers, is monetary cost or benefit. But before describing what has been done so far to estimate the overall costs of impacts, we need to consider what is known about the cost of damage due to extreme events (such as floods, droughts or windstorms). As has been constantly emphasised in this chapter these probably constitute the most important element in climate change impacts.

Table 7.1	Examples of adaptation options for selected sectors.
Sector/System	**Adaptation Options**
Water	Increase water-use efficiency with 'demand-side' management (e.g. pricing incentives, regulations, technology standards). Increase water supply, or reliability of water supply, with 'supply-side' management (e.g. construct new water storage and diversion infrastructure). Change institutional and legal framework to facilitate transfer of water among users (e.g. establish water markets). Reduce nutrient loadings of rivers and protect/augment streamside vegetation to offset eutrophying effects of higher water temperatures. Reform flood management plans to reduce downstream flood peaks; reduce paved surfaces and use vegetation to reduce storm runoff and increase water infiltration. Reevaluate design criteria of dams, levees, and other infrastructure for flood protection.
Food and fibre	Change timing of planting, harvesting, and other management activities. Use minimum tillage and other practices to improve nutrient and moisture retention in soils and to prevent soil erosion. Alter animal stocking rates on rangelands. Switch to crops or crop cultivars that are less water-demanding and more tolerant of heat, drought, and pests. Conduct research to develop new cultivars. Promote agroforestry in dryland areas, including establishment of village woodlots and use of shrubs and trees for fodder. Replant with mix of tree species to increase diversity and flexibility. Promote revegetation and reforestation initiatives. Assist natural migration of tree species with connected protected areas and transplanting. Improve training and education of rural work forces. Establish or expand programs to provide secure food supplies as insurance against local supply disruptions. Reform policies that encourage inefficient, non-sustainable, or risky farming, grazing, and forestry practices (e.g. subsidies for crops, crop insurance, water).
Coastal areas and marine fisheries	Prevent or phase-out development in coastal areas vulnerable to erosion, inundation, and storm-surge flooding. Use 'hard' (dikes, levees, seawalls) or 'soft' (beach nourishment, dune and wetland restoration, afforestation) structures to protect coasts. Implement storm warning systems and evacuation plans. Protect and restore wetlands, estuaries, and floodplains to preserve essential habitat for fisheries. Modify and strengthen fisheries management institutions and policies to promote conservation of fisheries. Conduct research and monitoring to better support integrated management of fisheries.
Human health	Rebuild and improve public health infrastructure. Improve epidemic preparedness and develop capacities for epidemic forecasting and early warning. Monitor environmental, biological, and health status. Improve housing, sanitation, and water quality. Integrate urban designs to reduce heat island effect (e.g. use of vegetation and light coloured surfaces). Conduct public education to promote behaviors that reduce health risks.
Financial services	Risk spreading through private and public insurance and reinsurance. Risk reduction through building codes and other standards set or influenced by financial sector as requirements for insurance or credit.

See Table 3.6 on p. 76 of Watson, R. *et al.* (eds.) 2001, *Climate Change 2001: Synthesis Report.*

Because the incidence of such extreme events has increased significantly in recent decades, information about the cost of the damage due to them has been tracked by insurance companies. They have catalogued both the insured losses and, so far as they have been able to estimate, the total economic losses – these latter have shown an approximately tenfold increase from the 1950s to the 1990s (see Figure 1.2 and box below). Although factors other than climate change have contributed to this increase, climate change is probably the factor of most significance. The estimates for the 1990s of annual economic losses from weather-related disasters amount to approximately 0.2% of global world product (GWP) and vary from about 0.3% of aggregate GDPs for the North and Central American and the Asian regions to less than 0.1% for Africa (Table 7.2). These average figures hide big regional and temporal variations. For instance, the annual loss in China from natural disasters from 1989 to 1996 is estimated to range from three to six per cent of GDP, averaging nearly four per cent[63] – over ten times the world average. The reason why the percentage for Africa is so low is not because there are no disasters there – Africa on the whole has more than its fair share – but because most of the damage in African disasters is not realised in economic terms, nor does it appear in economic statistics. Further such averaged numbers hide the severe impact of disasters on individual countries or regions which, as we mention below with the example of Hurricane Mitch, can prove to be very large indeed.

The percentages we have quoted are conservative in that they do not represent all relevant costs. They relate to direct economic costs only and do not include associated or knock-on costs of disasters. This means, for instance, that the damage due to droughts is seriously underestimated. Droughts tend to happen slowly and many of the losses may not be recorded or borne by those not directly affected. Another reason for treating the information in the box with caution is because of the large disparities between different parts of the world and countries regarding per capita wealth, standard of living and degree of insurance cover. For instance, probably the most damaging hurricane ever, Hurricane Mitch, that hit Central America in 1998 does not appear in Table 7.3 as the total insured losses were less than one billion dollars. In that storm, 600 mm of rainfall fell in forty-eight hours, there were 9000 deaths and economic losses estimated at over six billion dollars. The losses in Honduras and Nicaragua amounted to about seventy and forty-five per cent respectively of their annual gross national product (GNP). Another example that does not appear in Table 7.3 for the same reason is the floods in central Europe in 1997 that caused the evacuation of 162 000 people and over five billion dollars of economic damage.

How about the likely costs of extreme events in the future? To estimate those we need much more quantitative information about their likely future incidence and intensity. Very few such estimates exist. One was mentioned in Chapter 6 (Figure 6.8) – a possible factor of five in extreme precipitation events in Europe under doubled pre-industrial carbon dioxide concentration. A speculative but probably conservative calculation of a global average figure for the future might be obtained as follows. Beginning with the 0.2% or 0.3% of GDP from the insurance companies' estimate of the current average costs due to weather-related

Table 7.2 *Fatalities, economic losses and insured losses (both in 1999 US dollars) for disasters in different regions as estimated by the insurance industry for the period 1985–99. The percentage from weather-related disasters (including windstorms, floods, droughts, wildfire, landslides, land subsidence, avalanches, extreme temperature events, lightning, frost and ice/snow damages) is indicated in each case. Total losses are higher than those summarised in Figure 1.2 because of the restriction of Figure 1.2 to losses from large catastrophic events*

	Africa	America: South	America: North, Central, Caribbean	Asia	Australia	Europe	World
Number of events	810	610	2260	2730	600	1810	8820
Weather-related	91%	79%	87%	78%	87%	90%	85%
Fatalities	22990	56080	37910	429920	4400	8210	559510
Weather-related	88%	50%	72%	70%	95%	96%	70%
Economic losses							
(current $US billion)	7	16	345	433	16	130	947
Weather-related	81%	73%	84%	63%	84%	89%	75%
Insured losses							
(current US$ billion)	0.8	0.8	119	22	5	40	187
Weather-related	100%	69%	86%	78%	74%	98%	87%

Data from Munich Re, presented in Figure 8.6 in Vellinga, P., Mills, E. *et al.* 2001. In McCarthy, *Climate Change 2001: Impacts*, Chapter 8.

The insurance industry and climate change

The impact of climate on the insurance industry is mainly through extreme weather events. In developing countries there may be very high mortality from extreme weather but relatively small costs to the industry because of low insurance penetration. In developed countries the loss of life may be much less but the costs to the insurance industry can be very large. Figure 1.2 illustrates the large growth in weather-related disasters and the associated economic and insured losses since the 1950s and Table 7.2 the distribution of the disasters, fatalities and economic losses from 1985 to 1999 around the continents. Some idea of the types of disaster that cause the largest economic loss can be gleaned from Table 7.3.

Part of the observed upward trend in historical disaster losses is linked to socio-economic factors such as population growth, increased wealth and urbanisation in vulnerable areas; part is linked to climatic factors such as changes in precipitation, flooding and drought events. There are differences in balance between the causes by region and type of event. Because of the complexities involved in delineating both the socio-economic and the climatic factors, the proportion of the contribution from human-induced climate change cannot be defined with any certainty – although it is interesting to note that the growth rate in damage cost of weather-related events was three times that of non-weather-related events for the period 1960–99.

Recent history has shown that weather-related losses can stress insurers to the point of bankruptcy. Hurricane Andrew in 1992 broke the twenty billion dollar barrier for insured loss and served as a wake-up call to the industry. The insurance industry, therefore, is very concerned to make estimates, as accurately as possible, for future trends as the rate of climate change increases.

Table 7.3 *Individual events included in the aggregates in Table 7.2 that incurred over five billion dollars of economic loss and over one billion dollars of insured loss*

Year	Event	Area	Economic losses (bn $US)	Ratio: insured/ economic losses
1995	Earthquake	Japan	112.1	0.03
1994	Northridge Earthquake	USA	50.6	0.35
1992	Hurricane Andrew	USA	36.6	0.57
1998	Floods	China	30.9	0.03
1993	Floods	USA	18.6	0.06
1991	Typhoon Mireille	Japan	12.7	0.54
1989	Hurricane Hugo	Caribbean, USA	12.7	0.50
1999	Winterstorm Lothar	Europe	11.1	0.53
1998	Hurricane Georges	Caribbean, USA	10.3	0.34
1990	Winterstorm Daria	Europe	9.1	0.75
1993	Blizzard	USA	5.8	0.34
1996	Hurricane Fran	USA	5.7	0.32
1987	Winterstorm	W. Europe	5.6	0.84
1999	Typhoon Bart	Japan	5.0	0.60

Data from Munich Re, presented in Table 8.3 in Vellinga, P., Mills, E. *et al.* 2001. In McCarthy, *Climate Change 2001: Impacts*, Chapter 8.

extreme events, then multiplying by two to allow for the factors mentioned above (e.g. associated or knock-on costs) and further multiplying by three to allow for the possible increase in extreme events, say by the middle of the twenty-first century, we end up with a figure of between one and two per cent of GDP. Further, this again is a 'money' estimate. The real total costs of extreme events taking into account all damages (including those that cannot be expressed in money terms) are likely to be very significantly larger especially in many developing countries.

Costing the total impacts

We now turn to consider all the impacts of anthropogenic climate change, attempts that have been made to express their cost in monetary terms and the validity of the methods employed. The IPCC 1995 Report contained a review of four cost studies[64] of the impacts of climate change in a world where the atmospheric carbon dioxide concentration had doubled from its pre-industrial level. The most detailed studies had been carried out for the United States. For those impacts against which some value of damage can be placed, estimates fell in the range of fifty-five to seventy-five thousand million dollars per annum or between 1.0% and 1.5% of the US GDP in 1990. For other countries in the developed world, estimates of the cost of impacts in terms of percentage of GDP were similar. For the developing world, estimates of annual cost were typically around five per cent of GDP (with a range of from two to nine per cent of GDP). Aggregated over the world the estimates are between 1.5% and 2% of globally aggregated GDP (sometimes called global world product or GWP). These studies provided the first indication of the scale of the problem in economic terms. However, as the authors of these economic studies explain, their estimates were crude, were based on very broad assumptions, were mostly calculated in terms of the impact on today's economies rather than future ones and should not be considered as precise values.

More recent studies have given more consideration to the possibilities for adaptation (not forgetting the cost of adaptation) through which there is large potential to reduce the damage cost of climate change. This is especially the case in the agricultural sector.[65] In that sector, again for doubling of atmospheric carbon dioxide concentration, studies of global aggregate economic impact vary from the slightly negative to the moderately positive depending on underlying assumptions (see also Impact on agriculture and food supply, pages 164ff.). But the aggregate hides large regional differences. Beneficial effects are expected predominantly in the developed world; strongly negative effects are expected for populations that are poorly connected to regional and global trading systems. Regions that will get drier or are already quite hot for agriculture also will suffer, as will countries that are less well prepared to adapt (e.g. because of lack of infrastructure, capital or education). Overall, climate change is likely to tip agriculture production in favour of well-to-do and well-fed regions at the expense of less well-to-do and less well-fed regions. However, these studies have largely ignored the increasing influence of climate extremes and as yet inadequately considered important factors such as water availability – largely because of the lack of detailed information regarding these.

A further factor to which more consideration has recently been given is the possibility of what are often called 'singular events' or irreversible events of large or unknown impact. Some of these have been mentioned earlier in this chapter or in previous chapters. Some examples are given in Table 7.4. It is clearly extremely difficult to provide quantitative estimates of the probability of such events. Nevertheless it is important that they are not ignored. One recent study[66] has allocated a potential damage cost to these of about one per cent of GWP for a warming of 2.5 °C and about seven per cent of GWP for a warming of 6 °C. Such calculations are necessarily based on highly speculative assumptions, but in that particular study these singular events represent the largest single contributor to the total overall cost.

These further factors, such as more adaptation and singular events, have worked in both directions in the estimates of cost, some reducing and some increasing them. There has tended therefore to be a greater spread in the overall results, thus emphasising the large uncertainties hidden in the calculations.[67] Further, little or no allowance has been made in recent studies for the important influence of extreme events.

If some allowance is made for the impact of extreme events (as described in the last section), the studies so far suggest, in very general terms, that the cost of damage due to climate change induced by a doubling of the pre-industrial carbon dioxide concentration that can be expressed in monetary terms is typically around one or two per cent of GDP for developed countries and perhaps five per cent or more of GDP in developing countries. However, what is also clear is that any estimates at the moment must be considered as preliminary and uncertain, because of inadequacies in both the assumptions that have to be made and the available data necessary for the calculations.

An initial inspection of the sort of cost figures we have been presenting in the last few paragraphs might suggest that the costs of global

Table 7.4 Examples of singular non-linear events and their impacts[a]

Singularity	Causal process	Impacts
Non-linear response of thermohaline circulation (THC)	Changes in thermal and freshwater forcing could result in complete shutdown of North Atlantic THC or regional shutdown in the Labrador and Greenland Seas. In the Southern Ocean, formation of Antarctic bottom water could shut down. Such events are simulated by models and also found in the paleoclimatic record.	Consequences for marine ecosystems and fisheries could be severe. Complete shutdown would lead to a stagnant deep ocean, with reducing deepwater oxygen levels and carbon uptake, affecting marine ecosystems. It would also represent a major change in heat budget and climate of northwest Europe.
Disintegration of West Antarctic Ice Sheet (WAIS)	WAIS may be vulnerable to climate change because it is grounded below sea level. Its disintegration could raise global sea level by four to six metres. Significant sea level rise from this cause is very unlikely during the twenty-first century, but a contribution of up to three metres over the next 1000 years is considered possible.	Considerable and rapid sea level rise would widely exceed adaptive capacity for most coastal structures and ecosystems.
Positive feedbacks in the carbon cycle	Climate change could reduce the efficiency of current oceanic and biospheric carbon sinks. Under some conditions the biosphere could become a source.[b] Gas hydrate reservoirs also may be destabilised, releasing large amounts of methane to the atmosphere.	Rapid, largely uncontrollable increases in atmospheric carbon concentrations and subsequent climate change would increase all impact levels and strongly limit adaptation possibilities.
Destabilisation of international order by environmental refugees and emergence of conflicts as a result of multiple climate change impacts	Climate change – alone or in combination with other environmental pressures – may exacerbate resource scarcities in developing countries. These effects are thought to be highly non-linear, with potential to exceed critical thresholds along each branch of the causal chain.	This could have severe social effects, which, in turn, may cause several types of conflict, including scarcity disputes between countries, clashes between ethnic groups and civil strife and insurgency, each with potentially serious repercussions for the security interests of the developed world.

[a] Table 19.6 from Smith, J. B. *et al.* 2001. Vulnerability to climate change and reasons for concern: a synthesis. In McCarthy, *Climate Change 2001: Impacts*, Chapter 19.

[b] See box on climate carbon cycle feedbacks in Chapter 3, page 40.

warming, although large, at a few per cent of GWP are not imposs-
ibly large, and that we might be able to buy our way out of the problems
of its impact. There are, however, two factors which have been omitted
from the studies cited above. The first is that the studies have only been
concerned with the impacts of global warming up to about the middle
of the next century when, for a business-as-usual scenario, the equivalent carbon dioxide concentration will have doubled from its pre-
industrial value. The longer-term impacts, if the growth of greenhouse
gases continues under business-as-usual, are likely to be much greater.[68]
The second factor is that impacts cannot be quantified in terms of monetary cost alone. For instance, the loss of life (see question 7), human
amenity, natural amenity or the loss of species cannot be easily expressed
in money terms. This can be illustrated by focusing on those who are
likely to be particularly disadvantaged by global warming. Most of them
will be in the developing world at around the subsistence level. They
will find their land is no longer able to sustain them because it has been
lost either to sea level rise or to extended drought. They will therefore
wish to migrate and will become environmental refugees.

It has been estimated that, under a business-as-usual scenario, the to
tal number of persons displaced by the impacts of global warming could
total of the order of 150 million by the year 2050 (or about three million
per year on average) – about 100 million due to sea level rise and coastal
flooding and about fifty million due to the dislocation of agricultural production mainly due to the incidence and location of areas of drought.[69]
The cost of resettling three million displaced persons per year (assuming
that is possible) has been estimated at between $US1000 and $US5000
per person, giving a total of about ten thousand million US dollars per
year.[70] What the estimated cost for resettlement does not include, however (as the authors of the study themselves emphasise), is the human
cost associated with displacement. Nor does it include the social and political instabilities that ensue when substantial populations are seriously
disrupted because their means of livelihood has disappeared. The effects
of these could be very large.

The overall impact of global warming

The incidence of various impacts of global warming is complex and far
from uniform over the world. The following is a brief summary of what
we have learnt in this chapter.

- There are many ways in which our current environment is being degraded due to human activities; global warming will tend to exacerbate
  these degradations. Sea level rise will make the situation worse for

low-lying land that is subsiding because of the withdrawal of ground-water and because the amount of sediment required to maintain the level of the land has been reduced. The loss of soil due to overuse of land or deforestation will be accelerated, with increasing droughts or floods in some areas. In other places, extensive deforestation will lead to drier climates and less sustainable agriculture.

• To respond to the impacts from the changes brought about by global warming, it will be necessary to adapt. In many cases this will involve changes in infrastructure, for instance new sea defences or water supplies. Many of the impacts of climate change will be adverse, but even when the impacts in the long term turn out to be beneficial, in the short term the process of adaptation will mostly have a negative impact and involve cost.

• Many of the important impacts of climate change are likely to arise because of changes in the frequency and/or intensity of extreme events (see Table 7.5 for a summary). For example, some parts of the world are expected to become warmer and drier, especially in summer, with a greater likelihood of droughts and heat waves; in other parts a greater incidence of floods is expected.

• Through adaptation to different crops and practices, first indications are that the total of world food production will not be seriously affected by climate change – although studies have not yet taken into account the likely occurrence of climate extremes. However, the disparity in per capita food supplies between the developed and the developing world will almost certainly become larger.

• Because of the likely rate of climate change, there will also be a serious impact on natural ecosystems, especially at mid to high latitudes. Forests especially will be affected by increased climate stress causing substantial die-back and loss of production, associated with which there is the positive feedback of additional carbon dioxide emissions. In a warmer world longer periods of heat stress will have an effect on human health; warmer temperatures will also encourage the spread of certain tropical diseases, such as malaria, to new areas.

• Economists have attempted to estimate the average annual cost in monetary terms of the impacts that would arise under the climate change due to a doubling of pre-industrial atmospheric carbon dioxide concentration. If some allowance is added for the impact of extreme events, the estimates are typically around one or two per cent of GDP for developed countries and around five per cent or more for developing countries. Later chapters will compare them with the cost of taking action to slow the onset of global warming or reduce its overall magnitude. But it is important to realise that these attempts at monetary

Table 7.5 *Examples of climate variability and extreme climate events and their impacts*[*]

Projected changes during the twenty-first century in extreme climate phenomena and their likelihood	Representative examples of projected impacts[a] (all high confidence of occurrence in some areas)
Higher maximum temperatures, more hot days and heat waves over nearly all land areas (*very likely*)	Increased incidence of death and serious illness in older age groups and urban poor. Increased heat stress in livestock and wildlife. Shift in tourist destinations. Increased risk of damage to a number of crops. Increased electric cooling demand and reduced energy supply reliability.
Higher (increasing) minimum temperatures, fewer cold days, frost days and cold waves over nearly all land areas (*very likely*)	Decreased cold-related human morbidity and mortality. Decreased risk of damage to a number of crops, and increased risk to others. Extended range and activity of some pest and disease vectors. Reduced heating energy demand.
More intense precipitation events (*very likely*, over many areas)	Increased flood, landslide, avalanche and mudslide damage. Increased soil erosion. Increased flood runoff could increase recharge of some floodplain aquifers. Increased pressure on government and private flood insurance systems and disaster relief.
Increased summer drying over most mid-latitude continental interiors and associated risk of drought (*likely*)	Decreased crop yields. Increased damage to building foundations caused by ground shrinkage. Decreased water resource quantity and quality. Increased risk of forest fire.
Increase in tropical cyclone peak wind intensities, mean and peak precipitation intensities (*likely*, over some areas)[b]	Increased risks to human life, risk of infectious disease epidemics and many other risks. Increased coastal erosion and damage to coastal buildings and infrastructure. Increased damage to coastal ecosystems such as coral reefs and mangroves.
Intensified droughts and floods associated with El Niño events in many different regions (*likely*) (see also under droughts and intense precipitation events)	Decreased agricultural and rangeland productivity in drought- and flood-prone regions. Decreased hydro-power potential in drought-prone regions.
Increased Asian summer monsoon precipitation variability (*likely*)	Increase in flood and drought magnitude and damages in temperate and tropical Asia.
Increased intensity of mid-latitude storms (little agreement between current models)	Increased risks to human life and health. Increased property and infrastructure losses. Increased damage to coastal ecosystems.

[a] These impacts can be lessened by appropriate response measures.

[b] Changes in regional distribution of tropical cyclones are possible but have not been established.

[*] Table SPM-2 from Watson, R. *et al.* (eds.) 2001. *Climate Change 2001: Synthesis Report. Contributions of Working Groups I, II and III to the Third Assessment Report of the Intergovernmental Panel on Climate Change.* Cambridge: Cambridge University Press. For definition of *likely*, *very likely,* etc. see Note 1 in Chapter 4.

costing can only represent a part of the overall impact story. Any assessment of impacts has to take into account the cost in human terms and the large social and political disruption some of the impacts will bring. In particular, it is estimated that there could be up to three million new environmental refugees each year or over 150 million by the middle of the twenty-first century.

It is important to bear in mind that these estimates of overall impact have concentrated on the doubled carbon dioxide scenario (in other words, the next fifty or sixty years). Soon after the end of the twenty-first century, under the scenarios with higher carbon dioxide emissions (in other words, if strong action is not taken to curb emissions), a further doubling of the equivalent carbon dioxide concentration will have occurred and it will be continuing to rise. The impacts of the additional climate change which would occur with a second effective doubling of carbon dioxide will be substantially more severe than those of the first doubling.

That, of course, is a lot further away in time; perhaps for that reason it has not been given much attention. However, because of the long life-time of some greenhouse gases, because of the long memory of the climate system, because some of the impacts may turn out to be irreversible and also because of the time taken for human activities and ecosystems to respond and change course, it is important to have an eye on the longer term. The much more severe impacts that can be expected at longer time horizons (see Table 7.4) increase the imperative now to take the necessary action.

However, many will ask why we should be concerned about the state of the Earth so far ahead in the future. Can we not leave it to be looked after by future generations? The next chapter will give something of my personal motivation for caring about what happens to the Earth in the future as well as now.

Questions

1   For your local region, find out about its water supply and how the water is used (e.g. by domestic users, agriculture, industry, etc.). What are likely to be the trends in its use over the next fifty years due, for instance, to population changes or changes in agriculture or industry? What are the possibilities for increased supply and how might these be affected by climate change?

2   For your local area, find out about current environmental problems such as sea level rise due to subsidence, over-use of ground water, air pollution affecting forests. Which of these are likely to be exacerbated by climate change? Try to estimate by how much.

3   For your local region, identify the possible impacts of climate change over the next one hundred years and quantify them as far as you can. Attempt to make an estimate of the cost of the damage for each impact. How far could adaptation reduce each type of damage?

4   From the information in Chapter 6, make estimates of possible climate change by the middle of next century for typical regions of boreal forest. Then estimate from Figure 7.13 for the each of the three tree species what loss of productivity might occur in each case.

5   Make an estimate of the total volume of ice in the Greenland and Antarctic ice-caps. What proportion would have to melt to increase the sea level by the six metres or so which occurred during the last interglacial period?

6   In the past, human communities have adapted to changes of many kinds including some changes in climate. It is sometimes argued that, because the adaptability of human beings is not fully allowed for, the likely damage from the impacts of climate change in the future tends to be overestimated. Do you agree?

7   In economic cost–benefit analyses, it is often necessary to attach a value to a 'statistical life'.[71] It is not human life itself that is being valued but a change in the risk of death averaged over a population of human beings. One way of attempting this valuation is to consider a person as an economic agent capable of producing economic output. However, the preferred approach is to value a statistical life on the basis of what individuals are willing to pay or accept for changes in the risk of death. This approach tends to produce very different money values between developed countries and developing countries. Do you think this is defensible? Give up to five examples of the analysis of particular environmental problems for which you think it would be useful to include the valuation of a statistical human life. Look for values that have been attributed in different circumstances. Do questions of equity have any relevance in your examples?

Notes for Chapter 7

1   For comprehensive detail about climate change impacts, see McCarthy, J. J., Canziani, O., Leary, N. A., Dokken, D. J., White, K. S. (eds.) 2001. *Climate Change 2001: Impacts, Adaptation and Vulnerability. Contribution of Working Group II to the Third Assessment Report of the Intergovernmental Panel on Climate Change.* Cambridge: Cambridge University Press.

2   From Summary for policymakers. McCarthy, *Climate Change 2001: Impacts*, p. 6.

3   See, for instance, *Global Environmental Outlook 3 (UNEP Report).* 2002. London: Earthscan. See also Goudie, A. 2000. *The Human Impact on the Natural Environment*, fifth edition. Massachusetts: MIT Press.

4   For further details see Church, J. A., Gregory, J. M. *et al.* 2001. Changes in sea level. In Houghton, J. T., Ding, Y., Griggs, D. J., Noguer, M., van der Linden, P. J., Dai, X., Maskell, K., Johnson, C. A. (eds.) *Climate Change 2001: The Scientific Basis. Contribution of Working Group I to the Third Assessment Report of the Intergovernmental Panel on Climate Change.* Cambridge: Cambridge University Press, Chapter 11.

5   See, for instance, van der Veen, C. J. 1991. State of balance of the cryosphere. *Reviews of Geophysics*, **29**, pp. 433–55.

6   For a comprehensive account of the impact of climate change on Bangladesh see Warrick, R. A., Ahmad, Q. K. (eds.) 1996. *The Implications of Climate and Sea Level Change for Bangladesh*. Dordrecht: Kluwer.

7   Nicholls, R. J., Mimura, N. 1998. Regional issues raised by sea level rise and their policy implications. *Climate Research*, **11**, pp. 5–18.

8   Broadus, J. M. 1993. Possible impacts of, and adjustments to, sea-level rise: the case of Bangladesh and Egypt. In Warrick, R. A., Barrow, E. M., Wigley, T. M. L. (eds.) 1993 *Climate and Sea-level Change: Observations, Projections and Implications*. Cambridge: Cambridge University Press, pp. 263–75. Note that, because of variations in the ocean structure, sea-level rise would not be the same everywhere. In Bangladesh it would be somewhat above average (see Gregory, J. M. 1993. Sea-level changes under increasing CO_2 in a transient coupled ocean-atmosphere experiment. *Journal of Climate*, **6**, pp. 2247–62).

9   From Muhtab, F. U. 1989. *Effect of Climate Change and Sea Level Rise on Bangladesh*. London: Commonwealth Secretariat.

10 See Chapter 4 in Warrick, R. A., Ahmad, Q. K. (eds.) 1996. *The Implications of Climate and Sea Level Change for Bangladesh*. Dordrecht: Kluwer.

11 Broadus, 1993. In Warrick, *Climate and Sea-level Change*, pp. 263–75.

12 Milliman, J. D. 1989. Environmental and economic implications of rising sea level and subsiding deltas: the Nile and Bangladeshi examples. *Ambio*, **18**, pp. 340–5.

13 From a report entitled *Climate Change due to the Greenhouse Effect and its Implications for China*. 1992. Gland, Switzerland: Worldwide Fund for Nature.

14 Day, J. W. *et al*. 1993. Impacts of sea-level rise on coastal systems with special emphasis on the Mississippi river deltaic plain. In Warrick, *Climate and Sea-level Change*, pp. 276–96.

15 Clayton, K. M. 1993. Adjustment to greenhouse gas induced sea-level rise on the Norfolk coast: a case study. In Warrick, *Climate and Sea-level Change*, pp. 310–21.

16 Nicholls, R. J., Mimura, N. 1998. Regional issues raised by sea-level rise and their policy implications. *Climate Research*, **11**, 5–18. See also de Ronde, J. G. 1993. What will happen to the Netherlands if sea-level rise accelerates? In Warrick, *Climate and Sea-level Change*, pp. 322–35.

17 See Nurse, L., Sem, G. *et al*. 2001. Small island states. In McCarthy, *Climate Change 2001: Impacts*, Chapter 17.

18 Bijlsma, L. 1996. Coastal zones and small islands. In Watson, R. T., Zinyowera, M. C., Moss, R. H. (eds.) 1996. *Climate Change 1995: Impacts, Adaptations and Mitigation of Climate Change: Scientific-Technical Analyses. Contribution of Working Group II to the Second Assessment Report of the Intergovernmental Panel on Climate Change*. Cambridge: Cambridge University Press, Chapter 9.

19 McLean, R. F., Tsyban, A. *et al*. 2001. Coastal zones and marine ecosystems. In McCarthy, *Climate Change 2001: Impacts*, Chapter 6.

20   From Figure 3.6 in Watson, R. *et al.* (eds.) 2001. *Climate Change 2001: Synthesis Report. Contribution of Working Groups I, II and III to the Third Assessment Report of the Intergovernmental Panel on Climate Change.* Cambridge: Cambridge University Press.

21   See Table 11.8 from Shiklomanov, I. A., Rodda, J. C. (eds.) 2003. *World Water Resources at the Beginning of the 21st Century.* Cambridge: Cambridge University Press.

22   Lal, M. *et al.* 2001. In McCarthy, *Climate Change 2001: Impacts,* Chapter 11.

23   Quoted by Geoffrey Lean in 'Troubled waters', in the colour supplement to the *Observer* newspaper, 4 July 1993.

24   Lal, M. *et al.* 2001. In McCarthy, *Climate Change 2001: Impacts,* Chapter 11. See also Arnell, N. W. 1999. Climate change and global water resources. *Global Environmental Change,* **9**, S51–S67.

25   Gleick, P. H. 1990. Vulnerability of water systems. In Waggoner, P. E. 1990. *Climate Change and US Water Resources.* New York: Wiley, pp. 223–40.

26   Rosenberg, N. J. (ed.) 1993. Towards an integrated impact assessment of climate change: the MINK study. *Climatic Change,* **24**, nos. 1–2. Dordrecht: Kluwer Academic Publishers.

27   The MINK study on water resources is described in Frederick, K. D. 1993. *Climatic Change,* **24**, pp. 83–115.

28   Over the 1980s and the 1990s ten major droughts each affecting five to ten million people were recorded in the Sahel – see the OFDA/CRED International Data Base, www.cred.be/emdat, and *Global Environmental Outlook 3 (UNEP Report).* 2002. London: Earthscan, pp. 276–8.

29   Dracup, J. A., Kendall, D. R. 1990. Floods and droughts. In Waggoner, *Climate Change and US Water Resources,* pp. 243–67.

30   See, for instance, Maurits la Riviere, J. W. 1989. Threats to the world's water. *Scientific American,* **261**, pp. 48–55. See also Bullock, P. 1996. Land degradation and desertification. In Watson, R. T., Zinyowera, M. C., Moss, R. H. (eds.) *Climate Change 1995: Impacts, Adaptations and Mitigation of Climate Change: Scientific-Technical Analyses. Contribution of Working Group II to the Second Assessment Report of the Intergovernmental Panel on Climate Change.* Cambridge: Cambridge University Press, Chapter 4.

31   Waggoner, P. E. 1990. *Climate Change and US Water Resources.* New York: Wiley.

32   See Arnell, N., Liu, C. *et al.* 2001. Hydrology and water resources. In McCarthy, *Climate Change 2001: Impacts,* Chapter 4.

33   See UNCCD website, www.unccd.int/.

34   Crosson, P. R., Rosenberg, N. J. 1989. Strategies for agriculture. *Scientific American,* **261**, September, pp. 78–85.

35   Gitay, H. *et al.* 2001. Ecosystems and their goods and services. In McCarthy, *Climate Change 2001: Impacts,* Chapter 5, Section 3.

36   Information in proposal for an International Research Institute for Climate Prediction. Report by Moura, A. D. (ed.) 1992. Prepared for the International Board for the TOGA project. Geneva: World Meteorological Organisation.

37   Reilly, J. *et al*. 1996. Agriculture in a changing climate. In Watson, *Climate Change 1995: Impacts*, Chapter 13.

38   More detail in Gitay, H. *et al*. 2001. Ecosystems and their goods and services. In McCarthy, *Climate Change 2001: Impacts*, Chapter 5, Section 3; also for an estimate of the global effect of CO_2 fertilisation on ecosystems see Melillo, J. M. *et al*. 1993. Global climate change and terrestrial net primary production. *Nature*, **363**, pp. 234–40.

39   The statement in this sentence and the sentence following are from The summary for policymakers in Watson, R. *et al*. (eds.) 2001. *Climate Change 2001: Synthesis Report. Contribution of Working Groups I, II and III to the Third Assessment Report of the Intergovernmental Panel on Climate Change*. Cambridge: Cambridge University Press. However, because of the many factors involved in the studies, regarding many of which there are large uncertainties, the IPCC only ascribed medium confidence (see Note 1. in Chapter 4 for explanations of IPCC confidence statements) to these statements.

40   See *Global Environmental Outlook 3 (UNEP Report)*. 2002. London: Earthscan, pp. 63–5.

41   Parry, M. *et al*. 1999. Climate change and world food security: a new assessment. *Global Environmental Change*, **9**, S51–S67.

42   From Watson, *Climate Change 2001: Synthesis Report*, paragraph 5.17.

43   Miko, U. F. *et al*. 1996. Climate change impacts on forests. In Watson, *Climate Change 1995: Impacts*, Chapter 1. See also Gitay, H. *et al*. 2001. Ecosystems and their goods and services. In McCarthy, *Climate Change 2001: Impacts*, Chapter 5, Section 5.6.3.

44   Lean, J., Rowntree, P. R. 1993. A simulation of the impact of Amazonian deforestation on climate using an improved canopy representation. *Quarterly Journal of the Royal Meteorological Society*, **119**, pp. 509–30. Similar results with somewhat larger reductions in rainfall have been reported by Henderson-Sellers, A. *et al*. 1993. Tropical deforestation: modelling local to regional scale climate change. *Journal of Geophysics Research*, **98**, pp. 7289–315.

45   Mylne, M. F., Rowntree, P. R. 1992. Modelling the effects of albedo change associated with tropical deforestation. *Climatic Change*, **21**, pp. 317–43.

46   Quoted in Mitchell, J. F. B. *et al*. 1990. Equilibrium climate change and its implications for the future. In Houghton, J. T., Jenkins, G. J., Ephraums, J. J. (eds.) 1990. *Climate Change: the IPCC Scientific Assessments*. Cambridge: Cambridge University Press, pp.131–72.

47   Cox, P. M., Betts, R. A., Collins, M., Harris, P., Huntingford, C., Jones, C. D. 2004. Amazon dieback under climate-carbon cycle projections for the 21st century. *Theoretical and Applied Climatology*, in press.

48   Gates, D. M. 1993. *Climate Change and its Biological Consequences*. Sunderland, Mass.: Sinauer Associates Inc., p. 77.

49   Bowes, M. D., Sedjo, R. A. 1993. Impacts and responses to climate change in forests of the MINK region. *Climatic Change*, **24**, pp. 63–82.

50 Melillo, J. M. *et al.* 1996. Terrestrial biotic responses to environmental change and feedbacks to climate. In Houghton, J. T., Meira Filho, L. G., Callander, B. A., Harris, N., Kattenberg, A., Maskell, K. (eds.) *Climate Change 1995: the Science of Climate Change.* Cambridge: Cambridge University Press, Chapter 9. See also Miko, U. F. *et al.* 1996. Climate change impacts on forests. In Watson, *Climate Change 1995: Impacts,* Chapter 1.

51 Gitay, H. *et al.* 2001. Ecosystems and their goods and services. In McCarthy, *Climate Change 2001: Impacts,* Technical Summary Section 4.3.

52 Detail in The summary for policymakers in Watson, R. *et al.* (eds.) 2001. *Climate Change 2001: Synthesis Report. Contribution of Working Groups I, II and III to the Third Assessment Report of the Intergovernmental Panel on Climate Change.* Cambridge: Cambridge University Press, pp. 68–69, paragraph 3.18. Myers, N. *et al.* 2000. *Nature,* **403**, pp. 853–8 has proposed concentrating conservation effort in selected places with exceptional concentrations of biodiversity.

53 From McG. Tegert, W. J., Sheldon, G. W., Griffiths, D. C. (eds.) 1990. *Climate Change: the IPCC Impacts Assessment.* Canberra: Australian Government Publishing Service, pp. 6–20. Although made in 1990 this statement remains true in 2003.

54 Sale, P. F. 1999. *Nature,* **397**, pp. 25–7. More information regarding diversity in corals available on World Resources Institute Website www.wri.org/wri/marine

55 More information about impact on corals in McLean, R. F. A. *et al.* 2001. Coastal zones and marine ecosystems. In McCarthy, *Climate Change 2001: Impacts,* Chapter 6, Section 6.4.5 .

56 For more detail see, McMichael, A. J., Githeko, A. *et al.* 2001. Human health. In McCarthy, *Climate Change 2001: Impacts,* Chapter 9. See also McMichael, A. J. 1993. *Planetary Overload.* Cambridge: Cambridge University Press.

57 Kalkstein, L. S. 1993. Direct impact in cities. *Lancet,* **342**, pp. 1397–9.

58 Information on India from Dr. Rajendra K. Pachauri, Tata Energy Research Institute. Information regarding Europe from World Meterological Organization, Geneva.

59 Nicholls, N. 1993. El Niño-Southern Oscillation and vector-borne disease. *Lancet,* **342**, pp. 1284–5. The El Niño cycle is described in Chapter 5.

60 See *Global Environmental Outlook 3 (UNEP Report).* 2002. London: Earthscan, pp. 274–5.

61 As an example of progress with respect to disaster preparedness, the International Red Cross has recently formed a Climate Change Unit based in the Netherlands.

62 PAHO Report 1999 Conclusions and Recommendations: Meeting on Revaluation of Preparedness and Response to Hurricanes George and Mitch, quoted in McMichael, A. *et al.* 2001. Human health. In McCarthy, *Climate Change 2001: Impacts,* Chapter 9.

63 See *Global Environmental Outlook 3 (UNEP Report)*. 2002. London: Earthscan, pp. 272.

64 Studies by Cline, Fankhauser, Nordhaus and Tol presented in Pearce, D. W. *et al.* 1996. The social costs of climate change. In Bruce, J., Hoesung Lee, Haites, E. (eds.) 1996. *Climate Change 1995: Economic and Social Dimensions of Climate Change.* Cambridge: Cambridge University Press, Chapter 6.

65 Smith, J. B. *et al.* 2001. Vulnerability to climate change and reasons for concern: a synthesis. In McCarthy, *Climate Change 2001: Impacts*, Chapter 19, Box 19.

66 Nordhaus, W. D., Boyer, J. 2000. *Warming the World: Economic Models of Global Warming.* Massachusetts: MIT Press, pp. 87–91.

67 Mendelsohn, R., Neumann, J. E. (eds.) 1999. *The Impact of Climate Change on the United States Economy.* Cambridge: Cambridge University Press. As an example, this recent set of studies for the United States finds that assuming more realistic adaptation and the application of damage estimates to a 2060 economy rather than to a 1990 economy tends to reduce the impact estimates in the early 1990 studies and for some cases produces net benefits for the US economy. However, no account has been taken of the impact of extreme events in these studies.

68 Cline, W. R. 1992. *The Economics of Global Warming.* Washington DC: Institute for International Economics, Chapter 2.

69 Myers, N., Kent, J. 1995. *Environmental Exodus: an Emergent Crisis in the Global Arena.* Washington DC: Climate Institute; also Adger, N., Fankhauser, S. 1993. Economic analysis of the greenhouse effect: optimal abatement level and strategies for mitigation. *International Journal of Environment and Pollution*, **3**, pp. 104–19.

70 Adger, N., Fankhauser, S. 1993. Economic analysis of the greenhouse effect: optimal abatement level and strategies for mitigation. *International Journal of Environment and Pollution*, **3**, pp. 104–19.

71 See, for instance, Pearce, D. W. *et al.* 1996. The social costs of climate change. In Bruce, J., Hoesung Lee, Haites, E. (eds.) 1996. *Climate Change 1995: Economic and Social Dimensions of Climate Change.* Cambridge: Cambridge University Press, Chapter 6, pp. 196–7.

Chapter 8
Why should we be concerned?

I have been describing the likely changes in climate that may occur as a result of human activities, and the impact these may have in different parts of the world. But large and potentially devastating changes are likely to be a generation or more away. So why should we be concerned? What responsibility, if any, do we have for the planet as a whole and the great variety of other forms of life that inhabit it, or for future generations of human beings? And does our scientific knowledge in any way match up with other insights, for instance ethical and religious ones, regarding our relationship with our environment? In this chapter I want to digress from the detailed consideration of global warming (to which I shall return) in order briefly to explore these fundamental questions and to present something of my personal viewpoint on them.

Earth in the balance

Al Gore, Vice-President of the United States in the Clinton Administration, entitled his book on the environment *Earth in the Balance*,[1] implying that there are balances in the environment that need to be maintained. A small area of a tropical forest possesses an ecosystem that contains some thousands of plant and animal species, each thriving in its own ecological niche in close balance with the others. Balances are also important for larger regions and for the Earth as a whole. These balances can be highly precarious, especially where humans are concerned.

One of the first to point this out was Rachel Carson in her book *Silent Spring*,[2] first published in 1962, which described the damaging effects of pesticides on the environment. Humans are an important part of the global ecosystem; as the size and scale of human activities continue to escalate, so can the seriousness of the disturbances caused to the overall balances of nature. Some examples of this were given in the last chapter.

It is important that we recognise these balances, in particular the careful relationship between humans and the world around us. It needs to be a balanced and harmonious relationship in which each generation of humans should leave the Earth in a better state, or at least in as good a state as they found it. The word that is often used for this is sustainability – politicians talk of sustainable development (see box in Chapter 9, page 226). This principle, and its link with the harmonious relationship between humans and nature, was given prominent place by the United Nations Conference on Environment and Development held at Rio de Janeiro in Brazil in June 1992. The first principle in a list of twenty-seven at the Rio Declaration adopted by the Conference is 'Human beings are at the centre of concerns for sustainable development. They are entitled to a healthy and productive life in harmony with nature.'[3]

However, despite such statements of principle from a body such as the United Nations, many of the attitudes that we commonly have towards the Earth are not balanced, harmonious or sustainable. Some of these are briefly outlined in the following paragraphs.

Exploitation

Humankind has over many centuries been exploiting the Earth and its resources. It was at the beginning of the Industrial Revolution some two hundred years ago that the potential of the Earth's minerals began to be realised. Coal, the result of the decay of primaeval forests and laid down over many millions of years, was the main source of energy for the new industrial developments. Iron ore to make steel was mined in vastly increased quantities. The search for other metals such as zinc, copper and lead was intensified until today many millions of tonnes are mined each year. Around 1960, oil took over from coal as the dominant world source of energy; oil and gas between them now supply over twice the energy supplied by coal.

We have not only been exploiting the Earth's mineral resources. The Earth's biological resources have also been attacked. Forests have been cut down on a large scale to make room for agriculture and for human habitation. Tropical forests are a particularly valuable resource, important for maintaining the climate of tropical regions. They have also been estimated to contain perhaps half of all the Earth's biological species. Yet only about half of the mature tropical forests that existed a few hundred years ago still stand.[4] At the present rate of destruction virtually all will be gone by the end of the twenty-first century.

Great benefits have come to humankind through the use of fossil fuels, minerals and other resources. Yet, much of this exploitation has been carried out with little or no thought as to whether this use of natural resources has been a responsible one. Early in the Industrial Revolution it seemed that resources were essentially limitless. Later on, as one source ran out others became available to more than take its place. Even now, for most minerals new sources are being found faster than present sources are being used. But the growth of use is such that this situation cannot continue. In many cases known reserves or even likely reserves will begin to run out during the next hundred or few hundred years. These resources have been laid down over many millions if not billions of years. Nature took about a million years to lay down the amount of fossil fuel that we now burn worldwide every year – and in doing so it seems that we are causing rapid change of the Earth's climate. Such a level of exploitation is clearly not in balance, not harmonious and not sustainable.

'Back to nature'

Almost the reverse of this attitude is the suggestion that we all adopt a much more primitive lifestyle and give up a large part of industry and intensive farming – that we effectively put the clock back two or three hundred years to before the Industrial Revolution. That sounds very seductive and some individuals can clearly begin to live that way. But there are two main problems.

The first is that it is just not practical. The world population is now some six times what it was two hundred years ago and about three times that of fifty years ago. The world cannot be adequately fed without farming on a reasonably intensive scale and without modern methods of food distribution. Further, most people that have them would not be prepared to be without the technical aids – electricity, central heating, refrigerator, washing machine, television and so on – which give the freedom, the interest and the entertainment which is so much taken for granted. Moreover, increasing numbers of people in the developing world are also taking advantage of and enjoy these aids to a life of less drudgery and more freedom.

The second problem is that it fails to take account of human creativity. Human scientific and technical development cannot be frozen at a given point in history, insisting that no further ideas can be developed. A proper balance between humans and the environment must leave room for humans to exercise their creative skills.

Again, therefore, a 'back to nature' viewpoint is neither balanced nor sustainable.

The technical fix

A third common attitude to the Earth is to invoke the 'technical fix'. As a senior environmental official from the United States said to me some years ago, 'We cannot change our lifestyle because of the possibility of climate change, we just need to fix the biosphere.' It was not clear just what he supposed the technical fixes would turn out to be. The point that he was making is that, in the past, humans have been so effective at developing new technology to meet the problems as they arise, can it not be assumed that this will continue? Concern about the future then turns into finding the 'fixes' as they are required.

On the surface the 'technical fix' route may sound a good way to proceed; it demands little effort and no foresight. It implies that damage can be corrected when it has been created rather than avoided in the first place. But damage already done to the environment by human activities is causing problems now. It is as if, in looking after my home, I decided not to carry out any routine maintenance but 'fixed' the failures as they occurred. For my home that would be a high-risk route to follow: failure to rewire when necessary could easily lead to a disastrous fire. A similar attitude to the Earth is both arrogant and irresponsible. It fails to recognise the vulnerability of nature to the large changes that human activities are now able to generate.

Science and technology possess enormous potential to assist in caring for the Earth, but they must be employed in a careful, balanced and responsible way. The 'technical fix' approach is neither balanced nor sustainable.

Future generations

Having described attitudes that are not balanced or harmonious in their relationship to the Earth and that fail to contribute to sustainability, I now turn to describe attitudes to the environment that are more acceptable in terms of the criteria I have set.

Firstly, there is our responsibility to future generations. It is a basic instinct that we wish to see our children and our grandchildren well set up in the world and wish to pass on to them some of our most treasured possessions. A similar desire would be that they inherit from us an Earth which has been well looked after and which does not pose to them more difficult problems than those we have had to face. But such an attitude is not universally held. I remember well, after a presentation I made on global warming to the British Cabinet at Number Ten, Downing Street in London, a senior politician commented that the problem would not become serious in his lifetime and could be left for its solution to the next generation. I do not think he had appreciated that the longer we delay in taking action, the larger the problem becomes and the more difficult to solve. There is a need to face up to the problem now for the sake of the next and subsequent generations. We have no right to act as if there is no tomorrow. We also have a responsibility to give to those who follow us a pattern for their future based on the principle of sustainable development.

The unity of the Earth

A second point of view sees us as having some responsibility, not just for all generations of humanity, but also for the larger world of all living things. We are, after all, part of that larger world. There is good scientific justification for this. We are becoming increasingly aware of our dependence on the rest of nature and of the interdependencies that exist between different forms of life, between living systems and the physical and chemical environment that surrounds life on the Earth – and indeed between ourselves and the rest of the universe.

The scientific theory named Gaia after the Greek Earth goddess and publicised particularly by James Lovelock emphasises these interdependencies. Lovelock[5] points out that the chemical composition of the Earth's atmosphere is very different from that of our nearest planetary neighbours, Mars and Venus. Their atmospheres, apart from some water vapour, are almost pure carbon dioxide. The Earth's atmosphere, by contrast, is seventy-eight per cent nitrogen, twenty-one per cent oxygen and only 0.03% carbon dioxide. So far as the major constituents are concerned, this composition has remained substantially unchanged over many millions of years – a fact that is very surprising when it is realised that it is a composition that is very far from chemical equilibrium.

This very different atmosphere on the Earth has come about because of the emergence of life. Early in the history of life, plants appeared which photosynthesise, taking in carbon dioxide and giving out oxygen. There followed other living systems which 'breathe', taking in oxygen and giving out carbon dioxide. The presence of life therefore influences and effectively controls the environment to which living systems in turn adapt. It is the close match of the environment to the needs of life and its development which seems so remarkable and which Lovelock has emphasised. He gives many examples; I will quote one concerned with oxygen in the atmosphere. There is a critical connection between the oxygen concentration and the frequency of forest fires.[6] Below an oxygen concentration of fifteen per cent, fires cannot be started even in dry twigs. At concentrations above twenty-five per cent fires burn extremely fiercely even in the damp wood of a tropical rain forest. Some species are dependent on fires for their survival; for instance, some conifers require the heat of fire to release their seeds from the seed pods. Above twenty-five per cent concentration of oxygen there would be no forests; below fifteen per cent, the regeneration that fires provide in the world's forests would be absent. The oxygen concentration of twenty-one per cent is ideal.

It is this sort of connection that has driven Lovelock to propose that there is tight coupling between the organisms that make up the world of living systems and their environment. He has suggested a simple model of an imaginary world called Daisyworld (see box below), which illustrates the type of feedback mechanisms that can lead to tight coupling and exert control. This model is similar to the one he proposed for the biological and chemical history of the Earth during the first 1000 million years after primitive life first appeared on the Earth some 3500 million years ago.

The real world is, of course, enormously more complex than Daisy-world, which is why the Gaia hypothesis has led to so much debate. Lovelock's first statement in 1972 of the hypothesis[7] was that 'Life, or the biosphere, regulates or maintains the climate and the atmospheric composition at an optimum for itself.' In his later writings he introduced the analogy between the Earth and a living organism, introducing a new science which he calls geophysiology[8] – a more recent book is entitled *Gaia, the Practical Science of Planetary Medicine*.

An advanced organism such as a human being has many built-in mechanisms for controlling the interactions between different parts of the organism and for self-regulation. In a similar way, Lovelock argues, the ecosystems on the Earth are so tightly coupled to their physical and chemical environments that the ecosystems and their environment could be considered as one organism with an integrated 'physiology'. In this sense he believes that the Earth is 'alive'.

That elaborate feedback mechanisms exist in nature for control and for adaptation to the environment is not in dispute. But many scientists feel that Lovelock has gone too far in suggesting that ecosystems and their environment can be considered as a single organism. Although Gaia has stimulated much scientific comment and research, it remains a hypothesis.[9] What the debate has done, however, is to emphasise the interdependencies that connect all living systems to their environment – the biosphere is a system in which is incorporated a large measure of self-control.

There is the hint of a suggestion in the Gaia hypothesis that the Earth's feedbacks and self-regulation are so strong that we humans need not be concerned about the pollution we produce – Gaia has enough control to take care of anything we might do. Such a view fails to recognise

Daisyworld and life on the early Earth

Daisyworld is an imaginary planet spinning on its axis and orbiting a sun rather like our own. Only daisies live in Daisyworld; they are of two hues, black and white. The daisies are sensitive to temperature. They grow best at 20 °C, below 5 °C they will not grow and above 40 °C they wilt and die. The daisies influence their own temperature by the way they absorb and emit radiation: black ones absorb more sunlight and therefore keep warmer than white ones.

In the early period of Daisyworld's history (Figure 8.1), the sun is relatively cool and the black daisies are favoured because, by absorbing sunlight, they can keep their temperature closest to 20 °C. Most of their white cousins die because they reflect sunlight and fail to keep above the critical 5 °C. However, later in the planet's history, the sun becomes hotter. Now the white daisies can also flourish; both sorts of daisies are present in abundance. Later still as the sun becomes even hotter the white daisies become dominant as conditions become too warm for the black ones. Eventually, if the sun continues to increase its temperature even the white ones cannot keep below the critical 40 °C and all the daisies die.

Daisyworld is a simple model employed by Lovelock[11] to illustrate the sort of feedbacks and self-regulation that occur in very much more complex forms within the living systems on the Earth.

Lovelock proposes a similar simple model as a possible description of the early history of life on the Earth (Figure 8.2). The dashed line shows the temperature which would be expected on a planet possessing no life but with an atmosphere consisting, like our present atmosphere, mostly of nitrogen with about ten per cent carbon dioxide. The rise in temperature occurs because the sun gradually became hotter during this period. About 3500 million years ago primitive life appeared. Lovelock, in this model, assumes just two forms of life, bacteria that are anaerobic photosynthesisers – using carbon dioxide to build up their bodies but not giving out oxygen – and bacteria that are decomposers, converting organic matter back to carbon dioxide and methane. As life appears the temperature decreases as the concentration of the greenhouse gas, carbon dioxide, decreases. At the end of the period about 2300 million years ago, more complicated life appears; there is an excess of free oxygen and the methane abundance falls to low values, leading to another fall in temperature, methane also being a greenhouse gas. The overall influence of these biological processes has been to maintain a stable and favourable temperature for life on the Earth.

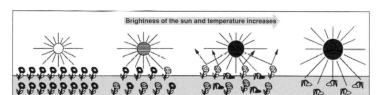

Figure 8.1 Daisyworld.

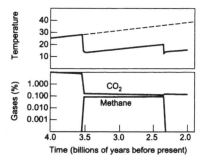

Figure 8.2 Model of the Earth's early history, as proposed by Lovelock.

the the effect on the Earth's system of substantial disturbances, in particular vulnerability of the environment with respect to its suitability for humans. To quote Lovelock,[10] 'Gaia, as I see her, is no doting mother tolerant of misdemeanours, nor is she some fragile and delicate damsel in danger from brutal mankind. She is stern and tough, always keeping the world warm and comfortable for those who obey the rules, but ruthless in her destruction of those who transgress. Her unconscious goal is a planet fit for life. If humans stand in the way of this, we shall be eliminated with as little pity as would be shown by the micro-brain of an intercontinental ballistic nuclear missile in full flight to its target.'

The Gaia scientific hypothesis can help to bring us back to recognise two things: firstly, the inherent value of all parts of nature, and secondly our dependence, as human beings, on the Earth and on our environment. Gaia remains a scientific theory. But some have been quick to see it as a religious idea, supporting ancient religious beliefs. Many of the world's religions have drawn attention to the close relationship between humans and the Earth.

The Native American tribes of North America lived close to the Earth. One of their chiefs when asked to sell his land expressed his dismay at the idea and said,[12] 'The Earth does not belong to man, man belongs to the Earth. All things are connected like the blood that unites us all.' An ancient Hindu saying,[13] 'The Earth is our mother, and we are all her children' also emphasises a feeling of closeness to the Earth. Those who have worked closely with indigenous peoples have given many examples of the care with which, in a balanced way, they look after the trees, plants and animals in their local ecosystem.[14]

The Islamic religion teaches the value of the whole environment, for instance in a saying of the prophet Mohammed: 'He who revives a dead land will be rewarded accordingly, and that which is eaten by birds, insects and animals out of that land will be charity provided by God' – so emphasising both our duty to care for the natural environment and our obligation to allow all living creatures their rightful place within it.[15]

Judaism and Christianity share the stories of creation in the early chapters of the Bible that emphasise the responsibility of humans to care for the Earth – we shall refer to these stories again later on in the chapter. Further on in the Old Testament detailed instructions are given regarding care for the land and the environment.[16] Christianity was described by William Temple, Archbishop of Canterbury sixty years ago, as 'the most materialistic of the great religions'. Because of its central belief that God became human in Jesus (an event Christians call the incarnation), Temple goes on to say 'by the very nature of its central doctrine Christianity is committed to a belief . . . in the reality of matter and its place in the divine scheme'.[17] For the Christian, the twin doctrines of creation and incarnation demonstrate God's interest in and concern for the Earth and the life it contains.

In looking for themes that emphasise the unity between humans and their environment, we need not confine ourselves to the Earth. There is a very much larger sphere in which a similar perspective of unity is becoming apparent. Some astronomers and cosmologists, overwhelmed by the size, scale, complexity, intricacy and precision of the universe, have begun to realise that their quest for an understanding of the evolution of the universe right from the 'Big Bang' some fifteen thousand million years ago is not just a scientific project but a search for meaning.[18] Why else has Stephen Hawking's book *A Brief History of Time*,[19] in selling over six million copies, become one of the bestsellers of our time?

In this new search for meaning, the perspective has arisen that the universe was made with humans in mind – an idea expressed in some formulations of the 'anthropic principle'.[20] Two particular pointers emphasise this. Firstly, we have already seen that the Earth itself is fitted in a remarkable way for advanced forms of life. Cosmology is telling us that, in order for life on our planet to be possible, the universe itself at the time of the Big Bang and in its early history needed to be 'fine-tuned' to an incredible degree.[21] Secondly, there is the remarkable fact that human minds, themselves dependent on the whole universe for their existence, are able to appreciate and understand to some extent the fundamental mathematical structure of the universe's design.[22] As Albert Einstein commented, 'The most incomprehensible thing about the universe is that it is comprehensible.' In the theory of Gaia, the Earth itself is central and humans are just one part of life on Earth; the insights of cosmology suggest that humans have a particular place in the whole scheme of things.

This section has recognised the intrinsic unity and interdependencies that exist not only on our Earth but also within the whole universe, and the particular place that we humans have in the universe. Being aware of these has large implications for our attitude to our environment.

Environmental values

What do we value in the environment and how do we decide what we need to preserve, to foster or improve? At the basis of our discussion so far have been several assumptions regarding the value or importance of different fundamental attitudes or actions, some of which I have associated with ideas that come from the underlying environmental science. Is it legitimate, however, to make connections of this kind between science and values? It is often argued that science itself is value free. But science is not an activity in isolation. As Michael Polanyi[23] has pointed out, the facts of science cannot sensibly be considered apart from the participation and the commitment of those who discover those facts or incorporate them into wider knowledge.

In the methodology and the practice of science are many assumptions of value. For instance, that there is an objective world of value out there to discover, that there is value in the qualities of elegance and economy in scientific theory, that complete honesty and cooperation between scientists are essential to the scientific enterprise.

Values can also be suggested from the perspective of the underlying science as we have shown earlier in the chapter.[24] For instance, we have described the Earth in terms of balance, interdependency and unity. Since all of these are critical to the Earth as we know it, we can argue that they are of fundamental value and worth preserving. We have also provided some scientific evidence that humans have a particular place in the overall scheme of the natural world, that they possess special knowledge – which suggests that they also possess special responsibility.

Moving away from science, we have already referred to values related to the environment that come from our basic experiences as human beings. These are often called 'shared values' because they are common to different members of a human community – which may be a local community, a nation or ultimately the global community taking in the whole human race. An outstanding example is the conservation of the Earth and its resources, not just for our generation but for future generations. Other examples may involve how resources are used now for the benefit of the present generation of humans and how they are shared between different communities or nations. Holmes Rolston shows that in these areas of shared values, *natural* values (valuing the natural world) and *cultural* values (interpersonal, social and community values) belong together. He writes of 'a domain of hybrid values ... the resultant of integrated influences from nature and culture.'[25]

When shared values are applied to real situations, however, conflicts often arise. For instance, how much should we forego now in order to make provision for future generations, or how should resources be shared between different countries, for instance between those in the relatively rich 'North' and those in the relatively poor 'South'? How do we exercise our responsibility as humans to share the Earth with other parts of the creation? How much resource should be deployed to maintain particular ecosystems or to prevent loss of species? How do we apply principles of justice and equity in the real world? Discussion within and between human communities can assist in the definition and application of such shared values.

Many of these shared values have their origins in the cultural and religious backgrounds of human communities. Discussions about values need therefore to recognise fully the cultural and religious traditions, beliefs and assumptions that underlie many of our attitudes and reasoning about ethical concerns.

An obstacle to the recognition of religious assumptions in the attempt to establish environmental values is the view that religious belief is not consistent with a scientific outlook. Some scientists maintain that only science can provide real explanations based on provable evidence whereas the assertions of religion cannot be tested in an objective way.[26] Other scientists, however, have suggested that the seeming inconsistency between science and religion arises because of misunderstandings about the questions being addressed by the two disciplines and that there is more in common between the methodologies of science and religion than is commonly thought.[27]

Scientists are looking for descriptions of the world that fit into an overall scientific picture. They are working towards making this picture as complete as possible. For instance, scientists are looking for mechanisms to describe the 'fine-tuning' of the universe (these are known as 'Theories of Everything'!) mentioned earlier. They are also looking for mechanisms to describe the interdependencies between living systems and the environment.

But the scientific picture can only depict part of what concerns us as human beings. Science deals with questions of 'how' not questions of 'why'. Most questions about values are 'why' questions. Nevertheless, scientists do not always draw clear distinctions between the two. Their motivations have often been associated with the 'why' questions. That was certainly true of the early scientists in the sixteenth and seventeenth centuries, many of whom were deeply religious and whose main driving force in pursuit of the new science was that they might 'explore the works of God'.[28]

That science and religion should be seen as complementary ways of looking at truth is a point made strongly by Al Gore in *Earth in the Balance*[29] which lucidly discusses current environmental issues such as global warming. He blames much of our lack of understanding of the environment on the modern approach, which tends to separate scientific study from religious and ethical issues. Science and technology are often pursued with a clinical detachment and without thinking about the ethical consequences. 'The new power derived from scientific knowledge could be used to dominate nature with moral impunity,' he writes.[30] He goes on to describe the modern technocrat as 'this barren spirit, precinct of the disembodied intellect, which knows the way things work but not the way they are'.[31] However, he also points out[32] that 'there is now a powerful impulse in some parts of the scientific community to heal the breach' between science and religion. In particular, as we pursue an understanding of the Earth's environment, it is essential that scientific studies and technological inventions are not divorced from their ethical and religious context.

Stewards of the Earth

The relationship between humans and the Earth that I have been advocating is often described as one of stewardship. We are on the Earth as its stewards. The word implies that we are carrying out our duty as stewards on behalf of someone else – but whom? Some environmentalists see no need to answer the question specifically, others might say we are stewards on behalf of future generations or on behalf of a generalised humanity. A religious person would want to be more specific and say that we are stewards on behalf of God. The religious person would also argue that to associate the relationship of humans to God with the relationship of humans to the environment is to place the latter relationship in a wider, more integrated, context – providing additional insights and a more complete basis for environmental stewardship.[33]

In the Judaeo-Christian tradition in the story of creation in the early chapters of the Bible is a helpful 'model' of stewardship – that of humans being 'gardeners' of the Earth. It is not only appropriate for those from those particular traditions – it is a model that can be widely applied. That story tells that humans were created to care for the rest of creation – the idea of human stewardship of creation is a very old one – and were placed in a garden, the Garden of Eden, 'to work it and take care of it'.[34] The animals, birds and other living creatures were brought to Adam in the garden for him to name them.[35] We are left with a picture of the first humans as 'gardeners' of the Earth – what does our work as 'gardeners' imply? I want to suggest four things:

- A garden provides food and water and other materials to sustain life and human industry. Part of the garden in the Genesis story contained mineral resources – 'the gold of that land is good; aromatic resin and onyx are also there'.[36] The Earth provides resources of many kinds for humans to use as they are needed.
- A garden is to be maintained as a place of beauty. The trees in the Garden of Eden were 'pleasing to the eye'.[37] Humans are to live in harmony with the rest of creation and to appreciate the value of all parts of creation. Indeed, a garden is a place where care is taken to

preserve the multiplicity of species, in particular those that are most vulnerable. Millions of people each year visit gardens that have been especially designed to show off the incredible variety and beauty of nature. Gardens are meant to be enjoyed.

- A garden is a place where humans, created as described in the Genesis story in the image of God,[38] can themselves be creative. Its resources provide for great potential. The variety of species and landscape can be employed to increase the garden's beauty and its productivity. Humans have learnt to generate new plant varieties in abundance and to use their scientific and technological knowledge coupled with the enormous variety of the Earth's resources to create new possibilities for life and its enjoyment. However, the potential of this creativity is such that increasingly we need to be aware of where it can take us; it has potential for evil as well as for good. Further, good gardeners intervene in natural processes with a good deal of restraint.

- A garden is to be kept so as to be of benefit to future generations. In this context, I shall always remember Gordon Dobson, a distinguished scientist, who in the 1920s developed new means for the measurement of ozone in the atmosphere. His home outside Oxford in England possessed a large garden with many fruit trees. When he was 85, a year or so before he died, I remember finding him hard at work in his garden replacing a number of apple trees; in doing so he clearly had future generations in mind.

How well do we humans match up to the description of ourselves as gardeners caring for the Earth? Not very well, it must be said; we are more often exploiters and spoilers than cultivators. Some blame science and technology for the problems, although the fault must lie with the craftsman rather than with the tools! Others have tried to place part of the blame on attitudes[39] that they believe originate in the early chapters of Genesis, which talk of human beings having rule over creation and subduing it.[40] Those words, however, should not be taken out of context – they are not a mandate for unrestrained exploitation. The Genesis chapters also insist that human rule over creation is to be exercised under God, the ultimate ruler of creation, and with the sort of care exemplified by the picture of humans as 'gardeners'. Why, therefore do humans so often fail to get their act together?

The will to act

Many of the principles I have been enunciating are included at least implicitly in the declarations, conventions and resolutions which came out of the United Nations Conference on Environment and Development held in Rio de Janeiro in June 1992; indeed, they form the background of many statements emanating from the United Nations or from official national sources. We are not short of statements of ideals. What tend to be lacking are the capability and resolve to carry them out. Sir Crispin Tickell, a British diplomat who has lectured widely on the policy implications of climate change, has commented 'Mostly we know what to do but we lack the will to do it'.[41]

Many recognise this lack of will to act as a 'spiritual' problem (using the word spiritual in a general sense), meaning that we are too obsessed with the 'material' and the immediate and fail to act according to generally accepted values and ideals particularly if it means some cost to ourselves or if it is concerned with the future rather than with the present. We are only too aware of the strong temptations we experience at both the personal and the national levels to use the world's resources to gratify our selfishness and greed. Because of this, it has been proposed that at the basis of stewardship should be a principle extending what has traditionally been considered wrong – or in religious parlance as sin – to include unwarranted pollution of the environment or lack of care for it.[42]

Those with religious belief tend to emphasise the importance of coupling together the relationship of humans to the environment to the relationship of humans to God.[43] It is here, religious believers would argue, that a solution for the problem of 'lack of will' can be found. That religious belief can provide an important driving force for action is often also recognised by those who look elsewhere than religion for a solution.

One of the main messages of this chapter is that action addressing environmental problems depends not only on knowledge about them but on the values we place on the environment and our attitudes towards it. In the chapter I have suggested that assessments of environmental value and appropriate attitudes can be developed from the following:

- The perspectives of balance, interdependence and unity in the natural world generated by the underlying science.
- A recognition – some would argue suggested by the science – that humans have a special place in the universe, which in turn implies that humans have special responsibilities with respect to the natural world.
- A recognition that to damage the environment or to fail to care for it is to do wrong.
- An interpretation of human responsibility in terms of stewardship of the Earth based on 'shared' values generally recognised by different

human communities and that strives for equity and justice as between different human communities and different generations.

- A recognition of the importance of the cultural and religious basis for the principles of stewardship – humans as 'gardeners' of the Earth is a possible 'model' of such stewardship.
- A recognition that, just as the totality of damage to the environment is the sum of the damage done by a large number of individuals, the totality of action to address environmental problems is the sum of a large number of individual actions to which we can all contribute.[44]

I shall return to the practical outworking of some of these issues in later chapters especially Chapter 12. Finally, let me recall some words of Thomas Huxley, an eminent biologist from last century, who emphasised the importance in the scientific enterprise of 'humility before the facts'. An attitude of humility is also one that lies at the heart of responsible stewardship of the Earth.

In the next chapter we shall reflect on the uncertainties associated with the science of global warming and consider how they can be taken into account in addressing the imperative for action. For instance, should action be taken now or should we wait until the uncertainties are less before deciding on the right action to take?

Questions

1 There is a debate regarding the relationship of humans to the environment. Should humans be at the centre of the environment with everything else and other life related to the human centre – in other words an anthropocentric view? Or should higher prominence be given to the non human part of nature in our scheme of things and in our consideration of values – a more ecocentric view? If so, what form should this higher prominence take?

2 How far can science be involved in the generation and application of environmental values?

3 How far do you think environmental values can be generated through debate and discussion in a human community without reference to the cultural or religious background of that community?

4 It has been suggested that religious belief (especially strongly held belief) is a hindrance in the debate about environmental values. Do you agree?

5 Should we strive for universally accepted values with respect to the environment? Or is it acceptable for different communities to possess different values?

6 Identify and list as many values as you can that belong to the categories *natural* and *cultural* (see page 206). In what ways do items in these categories 'belong together'?

7 An argument for religious belief which is sometimes put forward, irrespective of whether the belief is considered to have any foundation, is that such belief motivates people more strongly than other driving forces. Do you agree with this argument?

8 Explain how the cultural or religious traditions in which you have been brought up have influenced your view of environmental concern or action. How have these influences been modified because you now hold (or do not hold) definite religious beliefs?

9 Discuss the term 'stewardship', which is often used as a description of the relation of humans to the environment. Does it imply too anthropocentric a relationship?

10 Discuss the model of humans as 'gardeners' of the Earth. How adequate is the picture it presents of the relationship of humans to the environment?

11 Do you agree with Thomas Huxley when he spoke of the importance of humility before the scientific facts? How important do you think humility is in this context and in the wider context of the application of scientific knowledge to environmental concern?

12 Because of the formidability of the task of stewardship of the Earth, some have suggested that it is beyond the capability of the human race to tackle it adequately. Do you agree?

13 In Chapter 9 (see box on page 237) the concept of Integrated Assessment and Evaluation is introduced that involves all the natural and social science disciplines. In what ways could ethical or religious values be introduced into such evaluations? Is it appropriate and necessary that they be included?

Notes for Chapter 8

1 Gore, A. 1992. *Earth in the Balance*. New York: Houghton Mifflin Company.

2 Carson, R. 1962. *Silent Spring*. New York: Houghton Mifflin Company.

3 See box in Chapter 9 on page 231.

4 See Lean, G., Hinrichsen, D., Markham, A. 1990. *Atlas of the Environment*. London: Arrow Books.

5 Lovelock, J. E. 1979. *Gaia*. Oxford: Oxford University Press; Lovelock, J. E. 1988. *The Ages of Gaia*. Oxford: Oxford University Press.

6 Lovelock, *The Ages of Gaia*, pp. 131–3.

7 Lovelock, J. E., Margulis, L. 1974. *Tellus*, **26**, pp. 1–10.

8 Lovelock, J. E. 1990. Hands up for the Gaia hypothesis. *Nature*, **344**, pp. 100–12, also Lovelock, J. E. 1991. *Gaia: the Practical Science of Planetary Medicine*. London: Gaia Books.

9 Colin Russell discusses Gaia as a scientific hypothesis and also its possible religious connections in *The Earth, Humanity and God*. London: UCL Press, 1994.

10   Lovelock, *The Ages of Gaia*, p. 212.

11   For more details see Lovelock, *The Ages of Gaia*.

12   Quoted by Al Gore, *Earth in the Balance*, p. 259.

13   Quoted by Al Gore, *Earth in the Balance*, p. 261.

14   Ghillean Prance, Director of Kew Gardens in the UK, provides examples from his extensive work in countries of south America in his book *The Earth Under Threat*. Glasgow: Wild Goose Publications, 1996.

15   Khalil, M. H. 1993. Islam and the Ethic of Conservation. *Impact* (Newsletter of the Climate Network Africa), December, p. 8.

16   A number of injunctions were given to the Jews in the Old Testament regarding care for plants and animals and care for the land; for example, Leviticus 19:23–25, Leviticus 25:1–7, Deuteronomy 25:4.

17   Temple, W. 1964. *Nature, Man and God*. London: Macmillan (first edition 1934).

18   See for instance Davies, P. 1992. *The Mind of God*. London: Simon and Schuster. I have also addressed this theme in Houghton, J. T. 1995. *The Search for God: can Science Help?* London: Lion Publishing.

19   Hawking, S. 1989. *A Brief History of Time*. London: Bantam Press.

20   See for instance Davies, *The Mind of God*; also Barrow, J., Tipler, F. J. 1986. *The Anthropic Cosmological Principle*. Oxford: Oxford University Press

21   Barrow and Tipler, *The Anthropic Cosmological Principle*; and Gribbin, J., Rees, M. 1991. *Cosmic Coincidences*. London: Black Swan.

22   Davies, *The Mind of God*.

23   Polanyi, M. 1962. *Personal Knowledge*. London: Routledge and Kegan Paul.

24   The relation of science to value is explored in Rolston, H. III. 1999. *Genes, Genesis and God*. Cambridge: Cambridge University Press, Chapter 4.

25   Rolston, H. III. 1988. *Environmental Ethics*. Philadelphia: Temple University Press, p. 331.

26   See, for instance, Dawkins, R. 1986. *The Blind Watchmaker*. London: Longmans.

27   See, for instance, Polkinghorne, J. 1986. *One World*. London: SPCK; Polkinghorne, J. 1986. *Beyond Science*. Cambridge: Cambridge University Press; Houghton, *The Search for God*.

28   See, for instance, Russell, C. 1985. *Cross-currents: Interactions between Science and Faith*. Leicester, UK: Intervarsity Press.

29   Gore, *Earth in the Balance*.

30   Gore, *Earth in the Balance*, p. 252.

31   Gore, *Earth in the Balance*, p. 265.

32   Gore, *Earth in the Balance*, p. 254.

33   For expositions of a Christian view of the environment, see Elsdon, R. 1992. *Greenhouse Theology*. Tunbridge Wells: Monarch; Russell, *The Earth, Humanity and God*.

34   Genesis 2:15.

35   Genesis 2:19.

36 Genesis 2:12.
37 Genesis 2:9.
38 Genesis 1:27.
39 The best-known exposition of this position is, for instance, White, L. Jr. 1987. The historical roots of our ecological crisis. *Science*, **155**, pp. 1203–7; see Russell, *The Earth, Humanity and God*, for a commentary on this thesis.
40 Genesis 1:26–28.
41 *The Doomsday Letters*, broadcast on BBC Radio 4, UK, 1996.
42 This was the first of the principles that came out of a symposium (called the Patmos Principles since the climax of the symposium, held in celebration of the 1900th anniversary of the writing of the Book of Revelation, was on the island of Patmos) I attended in 1995 sponsored by the Ecumenical Patriarch Bartholomew I of the Greek Orthodox Church and Prince Philip in his capacity as President of the World Wildlife Fund. An extremely eclectic group, scientists, politicians, environmentalists and theologians attended from a wide range of religious backgrounds and beliefs. John, the Metropolitan of Pergamon, who was chairman of the symposium's scientific committee, kept emphasising that we should consider pollution of the environment, or lack of care for the environment, as a sin – not only against nature but a sin against God. His message struck a strong chord with the symposium. The principle goes on to explain that this new category of sin should include activities that lead to 'species extinction, reduction in genetic diversity, pollution of the water, land and air, habitat destruction and disruption of sustainable life styles'. The symposium's report is edited by Sarah Hobson and Jane Lubchenco and published under the title *Revelation and the Environment:– AD95–1995*. Singapore: World Scientific Publishing, 1997.
43 In Judaeo-Christian teaching the coupling of these two relationships begins with the Creation stories in Genesis. These stories go on to describe how humans disobeyed God (Chapter 3) and broke the partnership. But the Bible continually explains how God offers a way back to partnership. A few chapters on in Genesis (9:8–17), the basis of the relationship between God and Noah is a covenant agreement in which 'all life on the Earth' is included as well as humans. A relationship based on covenant is also the basis of the partnership between God and the Jewish nation in the Old Testament. But, after many times when that relationship was broken, the Old Testament prophets looked forward to a new covenant based not on law but on a real change of heart (Jeremiah 31:31–34). The New Testament writers (for example Hebrews 8:10–11) see this new covenant being worked out through the life and particularly through the death and resurrection of Jesus, the Son of God. Jesus promised his followers the Holy Spirit (John 15, 16), whose influence would enable the partnership between them and God to work. Paul, in his letters, is constantly referring to the dependent relationship which forms the basis of his own partnership with God

(Galatians 2:20, Philippians 4:13) and which has been the experience of millions of Christians down the centuries. Included in Paul's theology is the whole of creation (Romans 8:19–22).

44   Edmund Burke, a nineteenth century British politician said, 'no one made a greater mistake than he who did nothing because he could only do so little.' –quoted at the end of Chapter 12.

Chapter 9
Weighing the uncertainty

This book is intended to present clearly the current scientific position on global warming. A key part of this presentation concerns the uncertainty associated with all parts of the scientific description, especially with the prediction of future climate change, which forms an essential consideration when decisions regarding action are being taken. However, uncertainty is a relative term; utter certainty is not often demanded on everyday matters as a prerequisite for action. Here the issues are complex; we need to consider how uncertainty is weighed against the cost of possible action. First of all we address the scientific uncertainty.

The scientific uncertainty

Before considering the 'weighing' process and the cost of action, we begin by explaining the nature of the scientific uncertainty and how it has been addressed by the scientific community.

In earlier chapters I explained in some detail the science underlying the problem of global warming and the scientific methods that are employed for the prediction of climate change due to the increases in greenhouse gases. The basic physics of the greenhouse effect is well understood. If atmospheric carbon dioxide concentration doubles and nothing else changes apart from atmospheric temperature, then the average global temperature near the surface will increase by about 1.2 °C. That figure is not disputed among scientists.

However, the situation is complicated by feedbacks and regional variations. Numerical models run on computers are the best tools available for addressing these problems. Although they are highly complex, climate models are capable of giving useful information of a predictive kind. As was explained in Chapter 5, confidence in the models comes from the considerable skill with which they simulate present climate and its variations (including perturbations such as the Pinatubo volcanic eruption) and also from their success in simulating past climates; these latter are limited as much by the lack of data as by inadequacies in the models.

However, model limitations remain, which give rise to uncertainty (see box below). The predictions presented in Chapter 6 reflected these uncertainties, the largest of which are due to the models' failure to deal adequately with clouds and with the effects of the ocean circulation. These uncertainties loom largest when changes on the regional scale, for instance in regional patterns of rainfall, are being considered.

With uncertainty in the basic science of climate change and in the predictions of future climate, especially on the regional scale, there are bound also to be uncertainties in our assessment of the impacts of climate change. As Chapter 7 shows, however, some important general statements can be made with reasonable confidence. Under nearly all scenarios of increasing carbon dioxide emissions next century, the rate of climate change is likely to be large, probably greater than the Earth has seen for many millennia. Many ecosystems (including human beings) will not be able to adapt easily to such a rate of change. The most

---

The reasons for scientific uncertainty

The Intergovernmental Panel on Climate Change[1] has described the scientific uncertainty as follows.

There are many uncertainties in our predictions particularly with regard to the timing, magnitude and regional patterns of climate change, due to our incomplete understanding of:

- sources and sinks of greenhouse gases, which affect predictions of future concentrations,
- clouds, which strongly influence the magnitude of climate change,
- oceans, which influence the timing and patterns of climate change,
- polar ice-sheets, which affect predictions of sea level rise.

These processes are already partially understood, and we are confident that the uncertainties can be reduced by further research. However, the complexity of the system means that we cannot rule out surprises.

noticeable impacts are likely to be on the availability of water (especially on the frequency and severity of droughts and floods), on the distribution (though possibly not on the overall size) of global food production and on sea level in low-lying areas of the world. Further, although most of our predictions have been limited in range to the end of the twenty-first century, it is clear that by the century beyond 2100 the magnitude of the change in climate and the impacts resulting from that change are likely to be very large indeed.

The statement in the box regarding scientific uncertainty was formulated for the IPCC 1990 Report. Over ten years later it remains a good statement of the main factors that underly scientific uncertainty. That this is the case does not imply little progress in the past decade. On the contrary, as the subsequent IPCC Reports show, a great deal of progress has been made in scientific understanding and in the development of models. There is now much more confidence that the signal of anthropogenic climate change is apparent in the observed climate record. Models now include much more sophistication in their scientific formulations and possess increased skill in simulating the important climate parameters. For regional scale simulation and prediction, regional climate models (RCMs) with higher resolution have been developed that are nested within global models (see Chapters 5 and 6). These RCMs are beginning to bring more confidence to regional projections of climate change. Further, over the last decade, a lot of progress has been made with studies in various regions of the sensitivity to different climates of these regions' resources, such as water and food. Coupling such studies with regional scenarios of climate change produced by climate models enables more meaningful impact assessments to be carried out[2] and also enables appropriate measures to be assessed. Particularly in some regions large uncertainties remain; it will be seen for instance from Figure 6.5 that current models perform better for some regions than for others.

Summarised in Figure 9.1 are the various components that are included in the development of projections of climate change or its impacts. All of these possess uncertainties that need to be aggregated appropriately in arriving at estimates of uncertainties in different impacts.

The IPCC assessments

Because of the scientific uncertainty, it has been necessary to make a large effort to achieve the best assessment of present knowledge and to express it as clearly as possible. For these reasons the IPCC was set up jointly by two United Nations' bodies, the World Meteorological Organization (WMO) and the United Nations Environmental Programme

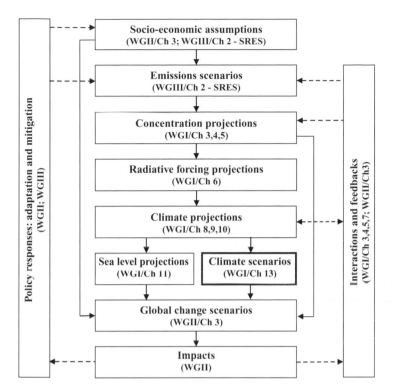

Figure 9.1 The cascade of uncertainties in projections to be considered in developing climate and related scenarios for climate change impact, adaptation and mitigation assessment. The chapters in the IPCC 2001 Report that deal with the various components are also identified.

(UNEP). The IPCC's first meeting in November 1988 was timely; it was held just as strong political interest in global climate change was beginning to develop. The Panel realised the urgency of the problem and, under the overall chairmanship of Professor Bert Bolin from Sweden, established three working groups, one to deal with the science of climate change, one with impacts and a third one to deal with policy responses. The IPCC has produced three main comprehensive Reports,[3] in 1990, 1995 and 2001, together with a number of special reports covering particular issues. Previous chapters have already referred widely to these reports.

I would like to say more about the Science Assessment Working Group (of which I was chairman from 1988 until 1992 and co-chairman from 1992 until 2002).[4] Its task has been to present in the clearest possible terms our knowledge of the science of climate change together with our best estimate of the climate change over the twenty-first century that is likely to occur as a result of human activities. In preparing its reports the Working Group realised from the start that if they were to be really authoritative and taken seriously, it would be necessary to involve as many as possible of the world scientific community in their production.

A small international organising team was set up at the Hadley Centre of the United Kingdom Meteorological Office at Bracknell and through meetings, workshops and a great deal of correspondence most of those scientists in the world (both in universities and government-supported laboratories) who are deeply engaged in research into the science of climate change were involved in the preparation and writing of the reports. For the first report, 170 scientists from 25 countries contributed and a further 200 scientists were involved in its peer review. For the third assessment report in 2001, these numbers had grown to 123 lead authors and 516 contributing authors involved with the various chapters, together with 21 review editors and 420 expert reviewers involved in the review process.

In addition to the comprehensive, thorough and intensively reviewed background chapters that form the basic material for each assessment, each report includes a Summary for Policymakers (SPM), the wording of which is approved in detail at a plenary meeting of the Working Group, the object being to reach agreement on the science and on the best way of presenting the science to policymakers with accuracy and clarity. The plenary meeting which agreed unanimously the 2001 SPM, held in Shanghai in January 2001,was attended by representatives of 99 countries and 45 scientists representing the lead authors of the scientific chapters together with a number of representatives from non-governmental organisations. There has been very lively discussion at these plenary meetings, most of which has been concerned with achieving the most informative and accurate wording rather than fundamental dispute over scientific content.

During the preparation of the reports, a considerable part of the debate amongst the scientists has centred on just how much can be said about the likely climate change in the twenty-first century. Particularly to begin with, some felt that the uncertainties were such that scientists should refrain from making any estimates or predictions for the future. However, it soon became clear that the responsibility of scientists to convey the best possible information could not be discharged without making estimates of the most likely magnitude of the change coupled with clear statements of our assumptions and the level of uncertainty in the estimates. Weather forecasters have a similar, although much more short-term responsibility. Even though they may feel uncertain about tomorrow's weather, they cannot refuse to make a forecast. If they do refuse, they withhold from the public most of the useful information they possess. Despite the uncertainty in a weather forecast, it provides useful guidance to a wide range of people. In a similar way the climate models, although subject to uncertainty, provide useful guidance for policy.

I have given these details of the work of the Science Assessment Group in order to demonstrate the degree of commitment of the scientific community to the understanding of global climate change and to the communication of the best scientific information to the world's politicians and policymakers. After all, the problem of global environmental change is one of the largest problems facing the world scientific community. No previous scientific assessments on this or any other subject have involved so many scientists so widely distributed both as regards their countries and their scientific disciplines. The IPCC Reports can therefore be considered as authoritative statements of the contemporary views of the international scientific community.

A further important strength of the IPCC is that, because it is an intergovernmental body, governments are involved in its work. In particular, government representatives assist in making sure that the presentation of the science is both clear and relevant from the point of view of the policymaker. Having been part of the process, the resulting assessments are in a real sense owned by governments as well as by scientists – an important factor when it comes to policy negotiations.

In the presentation of the IPCC assessments to politicians and policymakers, the degree of scientific consensus achieved has been of great importance in persuading them to take seriously the problem of global warming and its impact. In the run-up to the United Nations Conference on Environment and Development (UNCED) at Rio de Janeiro in June 1992, the fact that they accepted the reality of the problem led to the formulation of the Climate Convention. It has often been commented that without the clear message that came from the world's scientists, orchestrated by the IPCC, the world's leaders would never have agreed to sign the Climate Convention.

Since the publication of the reports, the debate concerning the scientific findings has continued in the world's press. Many have commented favourably on their clarity and accuracy. A few scientists have criticised because they feel the reports have insufficiently emphasised the uncertainties; others have expressed their disappointment that they have not spelt out the potential dangers to the world more forcefully. The scientific debate continues as indeed it must; argument and debate are intrinsic to the scientific process.

I have illustrated the work of the IPCC by describing in some detail the activity of the Science Assessment Working Group. The IPCC has two other Working Groups that have followed similar procedures and have dealt with the Impacts of Climate Change, with Adaptation and Mitigation strategies and with the Economics and Social Dimensions of Climate Change. Contributions to their work have not only come from natural scientists; increasingly social scientists, especially economists, have become involved. In these social science areas much fresh ground has been broken as consideration has been given to questions of what, in the global context, might form the basis of appropriate political and economic response to climate change. The rest of this chapter and the following chapters will draw heavily on their work.

Narrowing the uncertainty

A key question constantly asked by policymakers is, 'How long will it be before scientists are more certain about the projections of likely climate change, in particular concerning the regional and local detail?' They were asking that question over a decade ago and then I generally replied that in ten to fifteen years we would know a lot more. As we saw in the first section of this chapter, over the past decade a lot of progress in knowledge has been made. There is more confidence that anthropogenic climate change has been detected and more confidence too in climate change projections than was the case a decade ago. However, some of the key uncertainties remain and their reduction is urgently needed. Not surprisingly, policymakers are still asking for more certainty. What can be done to provide it?

For the science of change, the main tools of progress are observations and models. Both need further development and expansion. Observations are required to detect climate change in all its aspects as it occurs and also to validate models. That means that regular, accurate and consistent monitoring of the most important climate parameters is required with good coverage in both space and time. Monitoring may not sound very exciting work, often even less exciting is the rigorous quality control that goes with it, but it is absolutely essential if climate changes are to be observed and understood. Because of this, a major international programme, the Global Climate Observing System (GCOS) has been set up to orchestrate and oversee the provision of observations on a global basis. Models are needed to integrate all the scientific processes that are involved in climate change (most of which are non-linear, which means they cannot be added together in any simple manner) so that they can assist in the analysis of observations and provide a method of projecting climate change into the future.

Take, for instance, the example of cloud radiation feedback that remains the source of greatest single uncertainty associated with climate sensitivity.[5] It was mentioned in Chapter 5 that progress with understanding this feedback will be made by formulating better descriptions of cloud processes for incorporation into models and also by comparing model output, especially of radiation quantities, with observations especially those made by satellites. To be really useful such measurements need to be made with extremely high accuracy – to within the order of 0.1%

in the average radiation quantities – that is proving highly demanding. Associated with the better measurements of clouds is the need for all aspects of the hydrological (water) cycle to be better observed.

There is also inadequate monitoring at present of the major oceans of the world, which cover a large fraction of the Earth's surface. However, this is beginning to be remedied with the introduction of new methods of observing the ocean surface from space vehicles (see box below) and new means of observing the interior of the ocean. But not only are better physical measurements required: to be able to predict the detailed increases of greenhouse gases in the atmosphere, the problems of the carbon cycle must be unravelled; for this, much more comprehensive measurements of the biosphere in the ocean as well as that on land are needed.

Stimulated by internationally organised observing programmes such as the GCOS, space agencies around the world have been very active in the development of new instruments and the deployment of advanced space platforms that are beginning to provide many new observations relevant to the problems of climate change (see box below).

Space observations of the climate system

For forecasting the weather round the world – for airlines, for shipping, for many other applications and for the public – meteorologists rely extensively on observations from satellites. Under international agreements, five geostationary satellites are spaced around the equator for weather observation; moving pictures from them have become familiar to us on our television screens. Information from polar orbiting satellites is also available to the weather services of the world to provide input into computer models of the weather and to assist in forecasting (see for instance Figure 5.4).

These weather observations provide a basic input to climate models. But for climate prediction and research, comprehensive observations from other components of the climate system, the oceans, ice and land surface are required. ENVISAT, a satellite launched by the European Space Agency in 2002, is an example of the most recent generation of large satellites in which the latest techniques are directed to observing the Earth. The instruments are directed at the measurement of atmospheric temperature and composition (MIPAS, SCIAMACHY and GOMOS), sea surface temperature and topography, the latter for ocean current information (AATSR and RA-2), information about ocean biology and land surface vegetation (MERIS) and sea-ice coverage and ice-sheet topography (ASAR and RA-2).

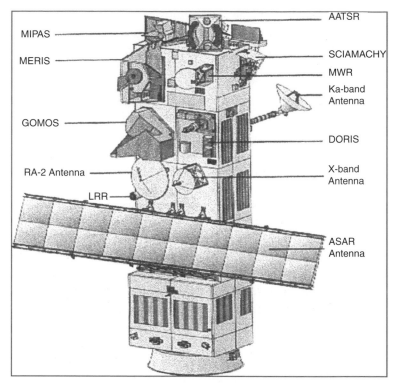

Figure 9.2 ENVISAT showing the instruments included in its payload: the Advanced Along-Track Scanning Radiometer (AATSR), the Michelson Interferometer for Passive Atmospheric Sounding (MIPAS), the MEdium Resolution Imaging Spectrometer (MERIS), the SCanning Image Absorption spectroMeter for Atmospheric CartograpHY (SCIAMACHY), the MicroWave Radiometer (MWR), the Global Ozone Monitoring by Observation of Stars (GOMOS), the Radar Altimeter – second generation (RA-2), the Advanced Synthetic Aperture Radar (ASAR), and other instruments for communication and exact tracking. DORIS stands for Doppler Orbitography and Radiopositioning Integrated by Satellite. In its 800-km Sun-synchronous orbit with the solar array deployed, it measures 26 m × 10 m × 5 m and weighs 8.1 tonnes.

Alongside the increased understanding and more accurate predictions of climate change coming from the community of natural scientists, much more effort is now going into studies of human behaviour and activities, how they will influence climate through changes in emissions of greenhouse gases and how they in turn might be affected by different degrees of climate change. Much better quantification of the impacts of climate change will result from these studies. Economists and other social scientists are pursuing detailed work on possible response strategies and the economic and political measures that will be necessary to achieve them. It is also becoming increasingly realised that there is an urgent need to interconnect more strongly research in the natural sciences with that in the social sciences. The integrated framework presented in Chapter 1 (Figure 1.5) illustrates the scope of interactions and of required integration between all the intellectual disciplines involved.

Sustainable development

So much for uncertainty in the science of global warming. But how does this uncertainty map on to the world of political decision making? A key idea is that of sustainable development.

One of the remarkable movements of the last few years is the way in which problems of the global environment have moved up the political agenda. In her speech at the opening in 1990 of the Hadley Centre at the United Kingdom Meteorological Office, Margaret Thatcher, the former British Prime Minister, explained our clear responsibility to the environment. 'We have a full repairing lease on the Earth. With the work of the IPCC, we can now say we have the surveyor's report; and it shows there are faults and that the repair work needs to start without delay. The problems do not lie in the future, they are here and now: and it is our children and grandchildren, who are already growing up, who will be affected.' Many other politicians have similarly expressed their feelings of responsibility for the global environment. Without this deeply felt and widely held concern, the UNCED conference at Rio, with environment as the number one item on its agenda, could never have taken place.

But, despite its importance, even when concentrating on the long term, the environment is only one of many considerations politicians must take into account. For developed countries, the maintenance of living standards, full employment (or something close to it) and economic growth have become dominant issues. Many developing countries are facing acute problems in the short term: basic survival and large debt repayment; others, under the pressure of large increases in population, are looking for rapid industrial development. However, an important characteristic of environmental problems, compared with many of the other issues faced by politicians, is that they are long-term and potentially irreversible – which is why Tim Wirth, the Under Secretary of State for Global Affairs in the United States Government during the Clinton Administration, said, 'The economy is a wholly owned subsidiary of the environment'.

A balance, therefore, has to be struck between the provision of necessary resources for development and the long-term need to preserve the environment. That is why the Rio Conference was about Environment and Development. The formula that links the two is called sustainable development (see box below) – development which does not carry with it the overuse of irreplaceable resources or irreversible environmental degradation.

The idea of sustainable development echoes what was said in Chapter 8, when addressing more generally the relationship of humans to their environment and especially the need for balance and harmony. The Climate Convention signed at the Rio Conference also recognised the need for this balance. In the statement of its objective (see box on page 243 in Chapter 10), it states the need for stabilisation of greenhouse gas concentrations in the atmosphere. It goes on to explain that this should be at a level and on a timescale such that ecosystems are allowed to adapt to climate change naturally, that food production is not threatened and that economic development can proceed in a sustainable manner.

Sustainable development: how is it defined?

A number of definitions of sustainable development have been produced. The following two well capture the idea.

According to the Bruntland Commission Report *Our Common Future* presented in 1987, sustainable development is 'meeting the needs of the present without compromising the ability of future generations to meet their own needs'.

A more detailed definition is contained in the White Paper *This Common Inheritance*, published by the United Kingdom Department of the Environment in 1990: 'sustainable development means living on the Earth's income rather than eroding its capital' and 'keeping the consumption of renewable natural resources within the limits of their replenishment'. It recognises the intrinsic value of the natural world explaining that sustainable development 'means handing down to successive generations not only man-made wealth (such as buildings, roads and railways) but also natural wealth, such as clean and adequate water supplies, good arable land, a wealth of wildlife and ample forests'.

The United Kingdom Government's first strategy report on sustainable development, issued in January 1994,[6] defined four principles that should govern necessary collective action:

- Decisions should be based on the best possible scientific information and analysis of risks.
- Where there is uncertainty and potentially serious risks exist, precautionary action may be necessary.
- Ecological impacts must be considered, particularly where resources are non-renewable or effects may be irreversible.
- Cost implications should be brought home directly to the people responsible – the 'polluter pays' principle.

It is also increasingly realised that the idea of sustainability not only applies to the environment but also to human communities. Sustainable development is often therefore assumed to include wider social factors as well as environmental and economic ones. The provision of social justice and equity are important components of a drive to sustainable communities. Considerations of equity include not just equity between nations but also equity between generations: we should not leave the world in a poorer state for the next generation.

Why not wait and see?

In taking action to satisfy the requirements of sustainable development a balance must be struck between many factors. The following sections address some of the arguments, issues and principles that are involved in this debate.

Firstly, in the light of scientific uncertainty, it is often argued that the case is not strong enough for any action to be taken now. What we should do is to obtain as quickly as possible, through appropriate research programmes, much more precise information about future climate change and its impact. We would then, so the argument goes, be in a much better position to decide on relevant action.

It is true that more accurate information is urgently needed so that decisions can be better informed. But in any sensible future planning, all information about likely future needs has to be taken properly into account. Decisions now should be informed by the best information available now, even if it is imperfect.

In the first place, quite a lot is already known – enough to scope the problem as a whole. There is general consensus amongst scientists about the most likely overall magnitude of climate change and there are good indications about its probable impact. Although we are not yet very confident regarding detailed predictions, enough is known to realise that the rate of climate change due to increasing greenhouse gases will almost certainly bring substantial deleterious effects and pose a large problem to the world. It will hit some countries much more than others. Those worst hit are likely to be those in the developing world that are least able to cope with it. Some countries may actually experience a more beneficial climate. But in a world where there is increasing interdependence between nations, no nation will be immune from the effects.

Secondly, the timescales of both atmospheric and human responses are long. Carbon dioxide emitted into the atmosphere today will contribute to the increased concentration of this gas and the associated climate change for over a hundred years. The more that is emitted now, the more difficult it will be to reduce atmospheric carbon dioxide concentration to the levels that will eventually be required. With regard to the human response, the major changes that are likely to be needed, for instance in large-scale infrastructure, will take many decades. Large power stations that will produce electricity in thirty or forty years' time are being planned and built today. The demands that are likely to be placed on all of us because of concerns about global warming need to be brought into the planning process now.

Thirdly, many of the required actions not only lead to substantial reductions in greenhouse gas emissions but they are good to do for other reasons which bring other direct benefits – such proposals for action are often described as 'no regrets' proposals. Many actions addressing increased efficiency lead also to net savings in cost (sometimes called 'win-win' measures). Other actions lead to improvements in performance or additional comfort.

Fourthly, there are more general beneficial reasons for some of the proposed actions. In Chapter 8 it was pointed out that humans are far too profligate in their use of the world's resources. Fossil fuels are burnt and minerals are used, forests are cut down and soil is eroded without any serious thought of the needs of future generations. The imperative of the global warming problem will help us to use the world's resources in a more sustainable way. Further, the technical innovation that will be required in the energy industry – in energy efficiency and conservation and in renewable energy development – will provide a challenge and opportunity to the world's industry to develop important new technologies – more of that in Chapter 11.

The Precautionary Principle

Some of these arguments for action are applications of what is often called the Precautionary Principle, one of the basic principles that was included in the Rio Declaration at the Earth Summit in June 1992 (see box below). A similar statement is contained in article 3 of the Framework Convention on Climate Change (see box on page 243 in Chapter 10).

We often apply the Precautionary Principle in our day-to-day living. We take out insurance policies to cover the possibility of accidents or losses; we carry out precautionary maintenance on housing or on vehicles, and we readily accept that in medicine prevention is better than cure. In all these actions we weigh up the cost of insurance or other precautions against the possible damage and conclude that the investment is worthwhile. The arguments are very similar as the Precautionary Principle is applied to the problem of global warming.

In taking out an insurance policy we often have in mind the possibility of the unexpected. In fact, when selling their policies, insurance companies often trade on our fear of the unlikely or the unknown,

especially of the more devastating possibilities. Although covering ourselves for the most unlikely happenings is not our main reason for taking
out the insurance, our peace of mind is considerably increased if the
policy includes these improbable events. In a similar way, in arguing for
action concerning global warming, some have strongly emphasised the
need to guard against the possibility of surprises (see examples in Table
7.4). They point out that, because of positive feedbacks that are not yet
well understood,[7] the increase of some greenhouse gases could be much
larger than is currently predicted. They also point to the evidence that
rapid changes of climate have occurred in the past (Figures 4.6 and 4.7)
possibly because of dramatic changes in ocean circulation; they could
presumably occur again.

The risk posed by such possibilities is impossible to assess. It is,
however, salutary to call attention to the discovery of the ozone 'hole'
over Antarctica in 1985. Scientific experts in the chemistry of the ozone
layer were completely taken by surprise by that discovery. In the years
since its discovery, the 'hole' has substantially increased in depth. Resulting from this knowledge, international action to ban ozone-depleting
chemicals has progressed much more rapidly. Ozone levels are beginning to recover   full recovery will take about a century. The lesson for
us here is that the climate system may be more vulnerable to disturbance
than we have often thought it to be. When it comes to future climate
change, it would not be prudent to rule out the possibility of surprises.

However, in weighing the action that needs to be taken with regard
to future climate change, although the possibility of surprises should be
kept in mind, that possibility must not be allowed to feature as the main
argument for action. Much stronger in the argument for precautionary
action is the realisation that significant anthropogenic climate change is
not an unlikely possibility but a near certainty; it is no change of climate
that is unlikely. The uncertainties that mainly have to be weighed lie in
the magnitude of the change and the details of its regional distribution.

An argument that is sometimes advanced for doing nothing now is
that by the time action is really necessary, more technical options will be
available. By acting now, we might foreclose their use. Any action taken
now must, of course, take into account the possibility of helpful technical
developments. But the argument also works the other way. The thinking
and the activity generated by considering appropriate actions now and
by planning for more action later will itself be likely to stimulate the sort
of technical innovation that will be required.

While speaking of technical options, I should briefly mention possible options to counteract global warming by the artificial modification
of the environment (sometimes referred to as geoengineering).[8] A number of proposals for 'technical fixes' of this kind have been put forward;

for instance, the installation of mirrors in space to cool the Earth by reflecting sunlight away from it; the addition of dust to the upper atmosphere to provide a similar cooling effect and the alteration of cloud amount and type by adding cloud condensation nuclei to the atmosphere. None of these has been demonstrated to be either feasible or effective. Further, they suffer from the very serious problem that none of them would exactly counterbalance the effect of increasing greenhouse gases. As has been shown, the climate system is far from simple. The results of any attempt at large-scale climate modification could not be perfectly predicted and might not be what is desired. With the present state of knowledge, artificial climate modification along any of these lines is not an option that needs to be considered.

The conclusion from this section – and the last one – is that to 'wait and see' would be an inadequate and irresponsible response to what we know. The Framework Convention on Climate Change (FCCC) signed in Rio (see box on page 243 in Chapter 10) recognised that some action needs to be taken now. Just what that action should be and how it fits in to a sensible scheme of sequential decision making will be the subject of the next chapter.

Principles for international action

From the three previous sections, four distinct principles can be identified to form the basis of international action. They are all contained in the Rio Declaration on Environment and Development (see box below) agreed by over 160 countries at the United Nations Conference on Environment and Development (the 'Earth Summit') held in Rio de Janeiro in 1992. They can also be identified in one form or another in the FCCC (see box on page 243 in Chapter 10). The Principles (with references to the Principles of the Rio Declaration and the Articles of the FCCC) are:

- The Precautionary Principle (Principle 15)
- The Principle of Sustainable Development (Principles 1 and 7)
- The Polluter-Pays Principle (Principle 16)
- The Principle of Equity – International and Intergenerational (Principles 3 and 5)

In the next chapter we shall consider how these principles can be applied.

Some global economics

So far in this chapter, our attempt to balance uncertainty against the need for action has been considered in terms of issues. Is it possible to carry

The Rio Declaration 1992

The Rio Declaration on Environment and Development was agreed by over 160 countries at the United Nations Conference on Environment and Development (the 'Earth Summit') held in Rio de Janeiro in 1992. Some examples of the twenty-seven principles enumerated in the Declaration are as follows:

Principle 1 Human beings are at the centre of concerns for sustainable development. They are entitled to a healthy and productive life in harmony with nature.

Principle 3 The right to development must be fulfilled so as to equitably meet developmental and environmental needs of present and future generations.

Principle 5 All States and all people shall cooperate in the essential task of eradicating poverty as an indispensable requirement for sustainable development, in order to decrease the disparities in standards of living and better meet the needs of the majority of the people of the world.

Principle 7 States shall cooperate in a spirit of global partnership to conserve, protect and restore the health and integrity of the Earth's ecosystem. In view of the different contributions to global environmental degradation, States have common but differentiated responsibilities. The developed countries acknowledge the responsibility they bear in the international pursuit of sustainable development in view of the pressures their societies place on the global environment and of the technologies and financial resources they command.

Principle 15 In order to protect the environment, the precautionary approach shall be widely applied by States according to their capabilities. Where there are threats of serious or irreversible damage, lack of full scientific certainty shall not be used as a reason for postponing cost-effective measures to prevent environmental degradation.

Principle 16 National authorities should endeavour to promote the internalisation of environmental costs and the use of economic instruments, taking into account the approach that the polluter should, in principle, bear the cost of pollution, with due regard to the public interest and without distorting international trade and investment.

out the weighing in terms of cost? In a world that tends to be dominated by economic arguments, quantification of the costs of action against the likely costs of the consequences of inaction must at least be attempted. It is also helpful to put these costs in context by comparing them with other items of global expenditure.

The costs of anthropogenic climate change fall into three parts. Firstly, there is the cost of the damage due to that change; for instance, the cost of flooding due to sea level rise or the cost of the increase in the number or intensity of disasters such as floods, droughts or windstorms, and so on. Secondly, there is the cost of adaptation that reduces the damage or the impact of the climate change. Thirdly, there is the cost of mitigating action to reduce the amount of climate change. The roles of adaptation and mitigation are illustrated in Figure 1.5. Because there is already a commitment to a significant degree of climate change, a need for significant adaptation is apparent. That need will continue to increase through the twenty-first century, an increase that will eventually be mollified as the effects of mitigation begin to bite. Mitigation is beginning now but the degree of mitigation that is eventually undertaken will depend on an assessment of the effectiveness and cost of adaptation. The costs, disadvantages and benefits of both adaptation and mitigation need therefore to be assessed and weighed against each other.

At the end of Chapter 7, estimates of the cost of damage from global warming were presented. Many of these estimates of damage cost also included some of the costs of adaptation; in general adaptation costs have not been separately identified. Many of these cost estimates assumed a situation for which, resulting from human activities, the increase in greenhouse gases in the atmosphere was equivalent to a doubling of the carbon dioxide concentration – under business-as-usual this is likely to occur around the middle of the twenty-first century. The estimates were typically around one to two per cent of gross domestic product (GDP) for developed countries. In developing countries, because of their greater vulnerability to climate change and because a greater proportion of their expenditure is dependent on activities such as agriculture and water, estimates of the cost of damage are greater, typically about five per cent of GDP or more. At the present stage of knowledge, these estimates are bound to be crude and subject to large uncertainties; nevertheless, they give a feel for the likely range of cost. It was also pointed out in Chapter 7 that the cost estimates only included those items that could be costed in money terms. Those items of damage or disturbance for which money is not an appropriate measure (e.g. the generation of large numbers of environmental refugees) also need to be to exposed and taken into account in any overall appraisal.

The longer-term damage, should greenhouse gases more than double in concentration, is likely to rise somewhat more steeply in relation to the concentration of carbon dioxide (Figure 9.3). For quadrupled equivalent carbon dioxide concentration, for instance, estimates of damage cost of the order of two to four times that for doubled carbon dioxide have been made – suggesting that the damage might follow something like a quadratic law relative to the expected temperature rise.[9] In addition the

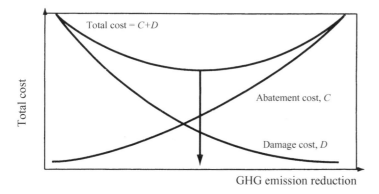

Figure 9.3 The application of classical cost benefit analysis to the mitigation of climate change, showing the shape of the curve of damage costs and mitigation costs as a function of atmospheric carbon dioxide emissions. On the assumption that economic costs are the only consideration, the arrow shows the 'optimal' reduction level.

much larger degree of climate change would considerably enhance the possibilities of singular events (see Table 7.4), irreversible change and of possible surprises.

Since the main contribution to global warming arises from carbon dioxide emissions, attempts have also been made to express these costs in terms of the cost per tonne of carbon as carbon dioxide emitted from human activities. A simple, but crude calculation can be carried out as follows. Consider the situation when carbon dioxide concentration in the atmosphere has doubled from its pre-industrial value, which will occur when an additional amount of carbon as carbon dioxide of about 800 Gt from anthropogenic sources has been emitted into the atmosphere (see Figure 3.1 and recall that about half the carbon dioxide emitted accumulates in the atmosphere). This carbon dioxide will remain in the atmosphere on average for about one hundred years. Assuming a figure of two per cent of global world product (GWP) – or $US 600 billion per annum – as the cost of the damage due to global warming in that situation, and assuming also that the damage remains over the one hundred years of the life-time of carbon dioxide in the atmosphere, the cost per tonne of carbon turns out to be about $US 70.

Calculations of the cost per tonne of carbon can be made with much more sophistication by considering that it is the *incremental* damage cost (that is, the cost of the damage due to one extra tonne of carbon emitted now) that is really required and also by allowing through a discount rate for the fact that it is damage some time in the future that is being costed now. Estimates made by different economists then range over $US 5–125 per tonne of carbon[10] – the very large range being due to the different assumptions that have been made. The estimates are particularly sensitive to the discount rate that is assumed; values at the top end of the range above about $US 50 have assumed a discount rate of less than two per cent; those at the bottom end have assumed a discount rate around five per cent.[11] The dominant effect of the discount rate will be clear when it is realised that over fifty years a two per cent discount rate devalues costs by a factor of about three while a five per cent rate discounts by a factor of thirteen. Over one hundred years the difference is even larger – a factor of seven for a two per cent rate and a factor of 170 for a five per cent rate. Amongst economists there has been much debate but no agreement about how to apply discount accounting to long-term problems of this sort or about what rate is most appropriate. However, as Partha Dasgupta points out,[12] 'the disagreement is not about economics nor about social cost-benefit analysis nor even about the numeracy of fellow scientists'. He explains, for instance, that the effects of carbon emissions could make substantial negative perturbations on future economies thus threatening the basis on which discount rates for future investment are set. Further, there are the likely damages that cannot easily be valued in money terms such as the large-scale loss of land – or even of whole countries – due to sea level rise or the large-scale loss of habitats or species. For these, even if valuation is attempted, discounting seems inappropriate. There seem cogent arguments that, if discount rates are applied to cost estimating for climate change, a smaller discount rate rather than a larger one should be employed. And in any case, for any cost estimate that is made the discount rate used should be adequately exposed. For our broad economic arguments in later chapters we shall therefore quote an estimate of damage cost in the range $US 50 to $US 100 per tonne of carbon emitted as carbon dioxide.

To slow the onset of climate change and to limit the longer-term damage, mitigating action can be taken by reducing greenhouse gas emissions, in particular the emissions of carbon dioxide. The cost of mitigation is very dependent on the amount of reduction required in greenhouse gas emissions; large reductions will cost proportionately more than small ones. It will also depend on the timescale of reduction. To reduce emissions drastically in the very near term would inevitably mean large reductions in energy availability with significant disruption to industry and large cost. However, more gradual reductions can be made with relatively small cost through actions of two kinds. Firstly, substantial efficiency gains in the use of energy can easily be achieved, many of which would lead to cost savings; these can be put into train now. Secondly, in the generation of energy, again proven technology exists for substantial efficiency improvements and also for the bringing into use of renewable sources of energy generation that are not dependent on fossil fuels. These can be planned for now and changes made as energy infrastructure, which has a typical life of thirty years or so, becomes ready for replacement. The next two chapters will present more detail about these possible actions and how they might be achieved.

Our purpose here is to look briefly at the likely overall cost of mitigation, much of which will arise in the energy or the transport sectors as

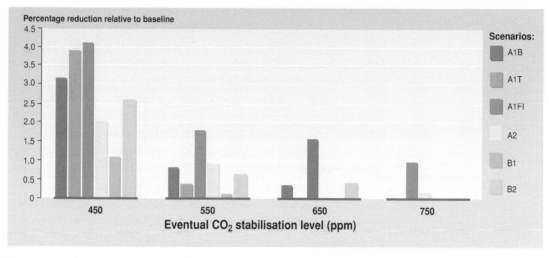

Figure 9.4 Global average GDP reduction in 2050 (relative to the baseline scenarios) for alternative stabilisation targets (see Chapter 10) and six SRES scenarios as baselines.

cheap fossil fuels are replaced by other energy sources that, at least in the short term, are likely to be more expensive. Some detail is provided in the next chapter of the profile of reductions in emissions that are required to stabilise carbon dioxide concentrations in the atmosphere at different levels over the next one or two hundred years. Figure 9.4 presents estimates from six economic models of the cost to the world's economy in the year 2050 of these reductions. As might be expected, the cost is substantially dependent on the target level of carbon dioxide concentration stabilisation. It is also dependent on the baseline scenario that is assumed. With typical levels of economic growth being between two and four per cent per annum, the cost of achieving reductions to meet any of the stabilisation levels in the figure, even that at 450 ppm, is less than one year's economic growth over fifty years.

However, it should be noted that, even if the carbon dioxide concentration is stabilised at 450 or 500 ppm, remembering that the effect of increases in the other greenhouse gases also has to be included (see Chapter 10, page 259), the world will have been committed to a significant degree of climate change (close to equivalent to doubling of atmospheric carbon dioxide concentration), bringing with it substantial costs and demands for adaptation. What is being mitigated is further and even more damaging climate change.

Although the economic studies I have mentioned have attempted to take into account many of the relevant factors, they are bound to be surrounded by substantial uncertainty. For instance, they have mostly not rigorously accounted for the economic effects of introducing new low-emission technologies, new revenue-raising instruments or adequate inter-regional financial and technology transfers, all elements which contribute to lower costs.[13] Further, one of the most difficult factors to take into account is that of likely future innovation. It is not easy to peer into the crystal ball of technical development; almost any attempt to do so is likely to underestimate its potential. For these reasons the estimates of mitigation cost are almost certainly on the high side.

The models that have been used to make the estimates of cost have all addressed limited parts of the whole problem. A complete assessment needs to address more completely the interactions between the factors that are driving climate change and its impacts both on humans and ecosystems, the human activities that are influencing those factors and the response to climate change both of humans and ecosystems – in fact all the elements illustrated in Figure 1.5. This is often called Integrated Assessment (see box below) and is supported by Integrated Assessment Models (IAMs) that are currently being built to address all the relevant elements in a more complete manner.

In considering the costs of the impacts of both global warming and adaptation or mitigation, figures of a small percentage of GDP have been mentioned. It is interesting to compare this with other items of expenditure in national or personal budgets. In a typical developed country, for example the United Kingdom, about five per cent of national income is spent on the supply of primary energy (basic fuel such as coal, oil and gas, fuel for electricity supply and fuel for transport), about nine per cent on health and three to four per cent on defence. It is, of course, clear that global warming is strongly linked to energy production – it is because of the way energy is provided that the problem exists – and this subject will be expanded in the next two chapters. But the impacts of global warming also have implications for health – such as the possible spread of disease – and for national security – for example, the possibility of wars fought over water, or the impact of large numbers of environmental refugees. Any thorough consideration of the economics of global warming needs therefore to assess the strength of these implications and to take them into account in the overall economic balance.

So far, on the global warming balance sheet we have estimates of costs and of benefits or drawbacks. What we do not have as yet is a capital account. Valuing human-made capital is commonplace, but in the overall accounting we are attempting, 'natural' capital must clearly be valued too. By 'natural' capital is meant, for instance, natural resources that may be renewable (such as a forest) or non-renewable (such as coal, oil or minerals).[16] Their value is clearly more than the cost of exploitation or extraction.

Integrated Assessment and Evaluation[14]

In the assessment and evaluation of the impacts of different aspects of global climate change with its large complexity, it is essential that all components are properly addressed. The major components are illustrated in Figure 1.5. They involve a very wide range of disciplines from natural sciences, technology, economics and the social sciences (including ethics). Take the example of sea level rise – probably the easiest impact to envisage and to quantify. From the natural sciences, estimates can be made of the amount and rate of rise and its characteristics. From various technologies, options for adaptation can be proposed. From economics and the social sciences, risks can be assessed and evaluated. The economic costs of sea level rise might be expressed, for instance, most simply as the capital cost of protection (where protection is possible) plus the economic value of the land or structures that may be lost plus the cost of rehabilitating those persons that could be displaced. But in practice the situation is more complex. For a costing to be at all realistic, especially when it is to apply to periods of decades into the future, it must account not only for direct damage and the cost of protection but also for a range of options and possibilities for adaptation other than direct protection. The likelihood of increased storm surges with the consequent damages and the possibility of substantial loss of life need also to be addressed. Further, there are other indirect consequences; for instance, the loss of fresh water because of salination, the loss of wetlands and associated ecosystems, wildlife or fisheries and the lives and jobs of people that would be affected in a variety of ways. In developed country situations rough estimates of the costs of some of these components can be made in money terms. For developing countries, however, the possible options can less easily be identified or weighed and even rough estimates of costs cannot be provided.

Integrated Assessment Models or IAMs are important tools for Integrated Assessment and Evaluation. They represent within one integrated numerical model the physical, chemical and biological processes that control the concentration of greenhouse gases in the atmosphere, the physical processes that determine the effect of changing greenhouse gas concentrations on climate and sea level, the biology and ecology of ecosystems (natural and managed), the physical and human impacts of climate change and the socio-economics of adaptation to and mitigation of climate change. Such models are highly sophisticated and complex although their components are bound to be very simplified. They provide an important means for studying the connections and interactions between the various elements of the climate change problem. Because of their complexity and because of the non-linear nature of many of the interactions, a great deal of care and skill is needed in interpreting the results from such models.

A number of the components of impact, even for the relatively simple situation of sea level rise, cannot be readily costed in money terms. For instance, the loss of ecosystems or wildlife as it impacts tourism can be expressed in money terms, but there is no agreed way of setting a money measure for the longer-term loss or the intrinsic value of unique systems. Or a further example is that, although the cost of rehabilitation for displaced people can be estimated, other social, security or political consequences of displacement (e.g. in extreme cases the loss of whole islands or even whole states) cannot be costed in terms of money. Any appraisal therefore of impacts of anthropogenic climate change will have to draw together components that are expressed in different ways or use different measures. Policy and decision makers need to find ways of considering alongside each other all the components that need to be aggregated in order to make appropriate judgements.[15]

Other items, some of which were mentioned at the end of Chapter 7, such as natural amenity and the value of species, can also be considered as 'natural' capital. I have argued (Chapter 8) that there is intrinsic value in the natural world – indeed, the value and importance of such 'natural' capital is increasingly recognised. The difficulty is that it is neither possible nor appropriate to express much of this value in money. Despite this difficulty, it is now widely recognised that national and global indicators of sustainable development should be prepared that include items of 'natural' capital and ways of including such items in national balance sheets are being actively pursued.

In summary, the items in the overall global warming balance sheet that have been identified are:

- Estimates of cost (for those items which can be quantified in terms of money) of the likely impacts of anthropogenic climate change. Most of the estimates to date have costed the impacts supposing the equivalent atmospheric carbon dioxide concentration were to double (which could occur during the second half of the twenty-first century). When some allowance is also included for the costs of extreme events (see Chapter 7, page 179) they are typically around one to two per cent of GDP in developed countries and typically five per cent or more in developing countries.
- Estimates of the cost of adaptation to anthropogenic climate change. Significant levels of adaptation will be required to the substantial degree of climate change to which the world is already committed. Even if the maximum possible mitigation takes place, because of the long time constants of change, the requirement for adaptation will continue for many centuries into the future, for instance to respond to the sea level rise that will continue for many centuries. Very few estimates have, as yet, been made of adaptation costs, although a few such costs are included in some of the impact studies.
- Estimates of the impacts of anthropogenic climate change that are difficult if not impossible to value in money terms; for instance, those with social consequences, those that affect human amenity and 'natural' capital or those that have implications for national security.
- Estimates of the cost of mitigation of anthropogenic climate change. For reductions in emissions leading to stabilisation of atmospheric carbon dioxide concentration (even at a level as low as 450 ppm – doubled pre-industrial carbon dioxide is 560 ppm) these are typically less than one year's economic growth by 2050.

There is already international acceptance that action to mitigate global warming is necessary. Such 'weighing' of the economics as has been possible so far brings two messages – that action must begin now to reduce emissions and slow the rate of change; that emissions of greenhouse gases up to the present have already produced a commitment to significant climate change over the next decades implying substantial requirements for adaptation and that more substantial emissions reductions will eventually be required for which planning must begin now.

The next chapter will consider some of the actions in more detail in the light of the principles we have enunciated in this chapter and in the context of the international Framework Convention on Climate Change.

Questions

1   It is sometimes argued that, in scientific enquiry, 'consensus' can never be achieved, because debate and controversy are fundamental to the search for scientific truth. Discuss what is meant by 'consensus' and whether you agree with this argument. Do you think the IPCC Reports have achieved 'consensus'?

2   How much do you think the value of IPCC Reports depends on (1) the peer review process to which they have been subjected, and (2) the involvement of governments in the presentation of scientific results?

3   Look out as many definitions of 'sustainable development' as you can find. Discuss which you think is the best.

4   Make a list of appropriate indicators that might be used to assess the degree to which a country is achieving sustainable development. Which do you think might be the most valuable?

5   Work out the value of a 'cost' today if it is twenty, fifty or one hundred years into the future and the assumed discount rate is one, two or five per cent. Look up and summarise the arguments for discounting future costs as presented for instance in various chapters of the IPCC 1995 and the IPCC 2001 Reports.[17] What do you think is the most appropriate discount rate to use?

6   Construct, as far as you are able, a set of environmental accounts for your country including items of 'natural' capital. Your accounts will not necessarily be all in terms of money.

7   Because of continuing economic growth, there is an expectation that the world will be very much richer by the middle of the twenty-first century and therefore, it is sometimes argued, in a better position than now to tackle the impacts or the mitigation of climate change. Do you agree with this argument?

Notes for Chapter 9

1   Houghton, J. T., Jenkins, G. J., Ephraums, J. J. (eds.) 1990. *Climate Change: the IPCC Scientific Assessments*. Cambridge: Cambridge University Press, p. 365; Executive Summary, p. xii. Similar but more elaborate statements are in the 1995 and the 2001 IPCC Reports.

2   For a detailed description of how the output from climate models can be com-
    bined with other information in climate studies see Mearns, L. O., Hulme,
    M. *et al.* 2001. Climate scenario development. In Houghton, J. T., Ding, Y.,
    Griggs, D. J., Noguer, M., van der Linden, P. J., Dai, X., Maskell, K., Johnson,
    C. A. (eds.) *Climate Change 2001: The Scientific Basis. Contribution of Work-
    ing Group I to the Third Assessment Report of the Intergovernmental Panel
    on Climate Change.* Cambridge: Cambridge University Press, Chapter 13.
3   Houghton, *Climate Change: the IPCC Scientific Assessment*, 1990.
    McG. Tegert, W. J., Sheldon, G. W., Griffiths, D. C. (eds.) 1990. *Climate
    Change: the IPCC Impacts Assessment.* Canberra: Australian Government
    Publishing Service.
    Houghton, J. T., Meira Filho, L. G., Callander, B. A., Harris, N., Kattenberg,
    A., Maskell, K. (eds.) 1996. *Climate Change 1995: the Science of Climate
    Change.* Cambridge: Cambridge University Press.
    Watson, R. T., Zinyowera, M. C., Moss, R. H. (eds.) 1996. *Climate Change
    1995: Impacts, Adaptations and Mitigation of Climate Change: Scientific-
    Technical Analyses. Contribution of Working Group II to the Second Assess-
    ment Report of the Intergovernmental Panel on Climate Change.* Cambridge:
    Cambridge University Press.
    Bruce, J., Hoesung Lee, Haites, E. (eds.) 1996. *Climate Change 1995: Eco-
    nomic and Social Dimensions of Climate Change.* Cambridge: Cambridge
    University Press.
    Houghton, J. T., Ding, Y., Griggs, D. J., Noguer, M., van der Linden, P. J., Dai,
    X., Maskell, K., Johnson, C. A. (eds.) *Climate Change 2001: The Scientific
    Basis. Contribution of Working Group I to the Third Assessment Report of the
    Intergovernmental Panel on Climate Change.* Cambridge: Cambridge Uni-
    versity Press.
    McCarthy, J. J., Canziani, O., Leary, N. A., Dokken, D. J., White, K. S. (eds.)
    2001. *Climate Change 2001: Impacts, Adaptation and Vulnerability. Con-
    tribution of Working Group II to the Third Assessment* Report of the *Inter-
    governmental Panel on Climate Change.* Cambridge: Cambridge University
    Press.
    Metz, B., Davidson, O., Swart, R., Pan, J. (eds.) 2001. *Climate Change 2001:
    Mitigation. Contribution of Working Group III to the Third Assessment Report
    of the Intergovernmental Panel on Climate Change.* Cambridge: Cambridge
    University Press.
4   See also Houghton, J. T. 2002. An overview of the IPCC and its process of
    science assessment. In Hester, R. E., Harrison, R. M. (eds.) *Global Environ-
    mental Change. Issues in Environmental Science and Technology*, No. 17.
    London: Royal Society of Chemistry.
5   Defined on page 120 in Chapter 6.
6   *Sustainable Development: the UK Strategy*. 1994. London: HMSO, Cm 2426,
    p. 7.
7   See Chapter 3.
8   Reviewed in *Policy Implications of Greenhouse Warming*. 1992. Washington
    DC: National Academy Press, pp. 433–64.

9   Pearce, D. W. *et al*. 1996. In Bruce, *Climate Change 1995: Economic and Social Dimensions*, Chapter 6.

10   Summary for policymakers. In Bruce, *Climate Change 1995: Economic and Social Dimensions*.

11   Cline argues for a low rate (Cline, W. R. 1992. *The Economics of Global Warming*. Washington DC: Institute for International Economics, Chapter 6). Nordhaus (Nordhaus, W. R. 1994. *Managing the Global Commons: the Economics of Climate Change*. Massachusetts: MIT Press) has used rates in the range of five to ten per cent; see also Tol, R. S. J. 1999. The marginal costs of greenhouse gas emissions. *The Energy Journal*, **20**, pp. 61–81.

12   Dasgupta, P. 2001. *Human Well-Being and the Natural Environment*. Oxford: Oxford University Press, p. 184, see also pp. 183–91; see also Markhndya, A., Halsnaes, K. *et al*. Costing methodologies. In Metz, *Climate Change 2001: Mitigation*, Chapter 7.

13   Further detail in Hourcade, J.-C., Shukla, P. *et al*. 2001. Global regional and national costs and ancillary benefits of mitigation. In Metz, *Climate Change 2001: Mitigation*, Chapter 8.

14   Weyant, J. *et al*. Integrated assessment of climate change. In Bruce, *Climate Change 1995: Economic and Social Dimensions*, Chapter 10.

15   A full discussion of such integrated appraisal can be found in *21st Report of the UK Royal Commission on Environmental Pollution*. London: Stationery Office.

16   For a discussion of this issue see Daly, H. E. 1993. From empty-world economics to full-world economics: a historical turning point in economic development. In Ramakrishna, K., Woodwell, G. M. (eds.) *World Forests for the Future*. Princeton: Yale University Press, pp. 79–91.

17   Bruce, *Climate Change 1995: Economic and Social Dimensions*; Metz, *Climate Change 2001: Mitigation*.

Chapter 10

A strategy for action to slow and stabilise climate change

Following the awareness of the problems of climate change aroused by the IPCC scientific assessments, the necessity of international action has been recognised. In this chapter I address the forms that action could take.

The climate convention

The United Nations Framework Convention on climate change signed by over 160 countries at the United Nations Conference on Environment and Development held in Rio de Janeiro in June 1992 came into force on 21 March 1994. It has set the agenda for action to slow and stabilise climate change. The signatories to the Convention (some of the detailed wording is presented in the box below) recognised the reality of global warming, recognised also the uncertainties associated with current predictions of climate change, agreed that action to mitigate the effects of climate change needs to be taken and pointed out that developed countries should take the lead in this action.

The Convention mentions one particular aim concerned with the relatively short-term and one far reaching objective. The particular aim is that developed countries (Annex I countries in Climate Convention parlance) should take action to return greenhouse gas emissions, in particular those of carbon dioxide, to their 1990 levels by the year 2000. The long-term objective of the Convention, expressed in Article 2, is that the concentrations of greenhouse gases in the atmosphere should be stabilised 'at a level which would prevent dangerous anthropogenic interference with the climate system', the stabilisation to be achieved within

242

Some extracts from the United Nations Framework Convention on climate change, signed by over 160 countries in Rio de Janeiro in June 1992

Firstly, some of the paragraphs in its preamble, where the parties to the Convention:

CONCERNED that human activities have been substantially increasing the atmospheric concentration of greenhouse gases, that these increases enhance the natural greenhouse effect, and that this will result on average in an additional warming of the Earth's surface and atmosphere and may adversely affect natural ecosystems and humankind.

NOTING that the largest share of historical and current global emissions of greenhouse gases has originated in developed countries, that per capita emissions in developing countries are still relatively low and that the share of global emissions originating in developing countries will grow to meet their social and development needs.

RECOGNISING that various actions to address climate change can be justified economically in their own right and can also help in solving other environmental problems.

RECOGNISING that low-lying and other small island countries, countries with low-lying coastal, arid and semi-arid areas or areas liable to floods, drought and desertification, and developing countries with fragile mountainous ecosystems are particularly vulnerable to the adverse effects of climate change.

AFFIRMING that responses to climate change should be coordinated with social and economic development in an integrated manner with a view to avoiding adverse impacts on the latter, taking into full account the legitimate priority needs of developing countries for the achievement of sustained economic growth and the eradication of poverty.

DETERMINED to protect the climate system for present and future generations, have AGREED as follows:
The Objective of the Convention is contained in Article 2 and reads as follows:

> The ultimate objective of this Convention and any related legal instruments that the Conference of the Parties may adopt is to achieve, in accordance with the relevant provisions of the Convention, stabilization of greenhouse gas concentrations in the atmosphere at a level that would prevent dangerous anthropogenic interference with the climate system. Such a level should be achieved within a time frame sufficient to allow ecosystems to adapt naturally to climate change, to ensure that food production is not threatened and to enable economic development to proceed in a sustainable manner.

Article 3 deals with principles and includes agreement that the Parties:

> take precautionary measures to anticipate, prevent or minimize the causes of climate change and mitigate its adverse effects. Where there are threats of serious or irreversible damage, lack of full scientific certainty should not be used as a reason for postponing such measures, taking into account that policies and measures to deal with climate change should be cost-effective so as to ensure global benefits at the lowest possible cost.

Article 4 is concerned with Commitments. In this article, each of the signatories to the Convention agreed:

> to adopt national policies and take corresponding measures on the mitigation of climate change, by limiting its anthropogenic emissions of greenhouse gases and protecting and enhancing its greenhouse sinks and reservoirs. These policies and measures will demonstrate that developed countries are taking the lead in modifying longer-term trends in anthropogenic emissions consistent with the objective of the Convention, recognizing that the return by the end of the present decade to earlier levels of anthropogenic emissions of carbon dioxide and other greenhouse gases not controlled by the Montreal Protocol would contribute to such modification . . .

Each signatory also agreed:

> in order to promote progress to this end . . . to communicate . . . detailed information on its policies and measures referred to above, as well as on its resulting projected anthropogenic emissions by sources and removals by sinks of greenhouse gases not covered by the Montreal Protocol . . . with the aim of returning individually or jointly to their 1990 levels these . . . emissions . . .

a time-frame sufficient to allow ecosystems to adapt naturally to climate change, to ensure that food production is not threatened and to enable economic development to proceed in a sustainable manner. In setting this objective, the Convention has recognised that it is only by stabilising the concentration of greenhouse gases (especially carbon dioxide) in the atmosphere that the rapid climate change which is expected to occur with global warming can be halted.

Up to the end of 2003, nine sessions of the Conference of the Parties to the Climate Convention have taken place. Those since November 1997 have largely been concerned with the Kyoto Protocol, the first formal binding legislation promulgated under the Convention. The following paragraphs will first outline the actions taken so far, then describe the Kyoto Protocol and address the further actions necessary to satisfy the Convention's objective to stabilise greenhouse gas concentrations. Scientific and technical details of the options available to achieve the reductions in emissions required will be described in Chapter 11.

Stabilisation of emissions

The target for short-term action proposed for developed countries by the Climate Convention was that, by the year 2000, greenhouse gas emissions should be brought back to no more than their 1990 levels. In the run-up to the Rio conference, before the Climate Convention was formulated, many developed countries had already announced their intention to meet such a target at least for carbon dioxide. They would do this mainly through energy-saving measures, through switching to fuels such as natural gas, which for the same energy production generates forty per cent less carbon dioxide than coal and twenty-five per cent less than oil. In addition those countries with traditional heavy industries (e.g. the iron and steel industry) were experiencing large changes which significantly reduce fossil fuel use. More detail of these energy-saving measures are given in the next chapter, which is devoted to a discussion of future energy needs and production.

By the year 2000, compared with 1990, global emissions from fossil fuel burning had risen by ten per cent. There was great variation between the emissions from different countries. In the USA they rose by seventeen per cent, in the rest of the OECD (Organization of Economic Cooperation and Development) they rose on average by five per cent. Emissions in countries in the former Soviet Union (FSU – also often called Economies in Transition) fell by around forty per cent because of the collapse of their economies, while the total of emissions from developing countries increased by around thirty-seven per cent (China and India by about nineteen and sixty-eight per cent respectively).

As we shall learn later in the chapter, stabilisation of carbon dioxide emissions would not lead in the foreseeable future to stabilisation of atmospheric concentrations. Stabilisation of emissions could only be a short-term aim. In the longer term much more substantial reductions of emissions are necessary.

The Montreal Protocol

The chlorofluorocarbons (CFCs) are greenhouse gases whose emissions into the atmosphere are already controlled under the Montreal Protocol on ozone-depleting substances. This control has not arisen because of their potential as greenhouse gases, but because they deplete atmospheric ozone (see Chapter 3). Emissions of CFCs have fallen sharply during the last few years and the growth in their concentrations has slowed; for some CFCs a slight decline in their concentration is now apparent. The phase-out of their manufacture in industrialised countries by 1996 and in developing countries by 2006 as required by the 1992 amendments to the Montreal Protocol will ensure that the profile of their atmospheric concentration will continue to decline. However, because of their long life in the atmosphere this decline will be slow; it will be a century or more before their contribution to global warming is reduced to a negligible amount.

The replacements for CFCs – the hydrochloro-fluorocarbons (HCFCs), which are also greenhouse gases though less potent than the CFCs – are required to be phased out by 2030. It will probably be close to that date before their atmospheric concentration stops rising and begins to decline.

Because of the international agreements which now exist for control of the production of the CFCs and many of the related species that contribute to the greenhouse effect, for these gases the stabilisation of atmospheric concentration required by the Climate Convention will in due course be achieved.

Other replacements for CFCs are the hydrofluorocarbons (HFCs), which are greenhouse gases but not ozone-depleting. The controls of the Montreal Protocol do not therefore apply and, as was mentioned in Chapter 3, any substantial growth in HFCs needs to be evaluated along with the other greenhouse gases. As we shall see in the next section, they are included in the 'basket' of greenhouse gases addressed by the Kyoto Protocol.

Table 10.1 *Emissions targets (1990*–2008/2012) for greenhouse gases under the Kyoto Protocol*

Country	Target (%)
EU-15**, Bulgaria, Czech Republic, Estonia, Latvia, Lithuania, Romania, Slovakia, Slovenia, Switzerland	−8
USA***	−7
Canada, Hungary, Japan, Poland	−6
Croatia	−5
New Zealand, Russian Federation, Ukraine	0
Norway	+1
Australia	+8
Iceland	+10

* Some economies in transition (EIT) countries have a baseline other than 1990.
** The fifteen countries of the European Union have agreed an average reduction; changes for individual countries vary from −28% for Luxembourg, −21% for Denmark and Germany to +25% for Greece and +27% for Portugal.
*** The USA has stated that it will not ratify the Protocol.

The Kyoto Protocol

At the first meeting after its entry into force held in Berlin in 1995, the Parties to the Climate Convention (i.e. all the countries that had ratified it) decided that they needed to negotiate a more specific and quantified agreement than the Convention on its own provided. Because of the principle in the Convention that industrialised countries should take the lead, a Protocol was formulated that required commitments from these countries (known as Annex I countries) for specific quantitative reductions in emissions (listed in Table 10.1) from their level in 1990 to their average from 2008–12, called the first commitment period. The Protocol also required that a second commitment period be defined for which negotiations must start no later than 2005. The Protocol carries inbuilt mechanisms that could lead to stronger action and be expanded over time to include developing countries.

The basic structure of the Protocol and the commitments required by different countries were agreed at a meeting of the Conference of the Parties in Kyoto in November 1997. But the Protocol is a highly complex agreement and over the next three years intense negotiations followed regarding the details – the range of gases covered, the basis for comparing them and the rules for monitoring, reporting and compliance. Further the Protocol incorporates a range of mechanisms (see box below) of a kind

Table 10.2 *Greenhouse gases covered by the Kyoto Protocol and their global warming potentials (GWPs) on a mass basis relative to carbon dioxide and for a time horizon of 100 years*

Greenhouse gas	Global warming potential (GWP)
Carbon dioxide (CO_2)	1
Methane (CH_4)	23
Nitrous oxide (N_2O)	296
Hydrofluorocarbons (HFCs)	from 12 to 12 000[a]
Perfluorocarbons (PFCs)	from 5000 to 12 000[a]
Sulphur hexafluoride (SF_6)	22200

Table 6.7 from Ramaswamy, V. *et al.* 2001. In Houghton, J. T., Ding, Y., Griggs, D. J., Noguer, M., van der Linden, P. J., Dai, X., Maskell, K., Johnson, C. A. (eds.) *Climate Change 2001: The Scientific Basis. Contribution of Working Group I to the Third Assessment Report of the Intergovernmental Panel on Climate Change.* Cambridge: Cambridge University Press.

[a] Range of values for different HFCs or PFCs – for more information about HFCs see Moomaw, W. R., Moreira, J. R. *et al.* 2001. Technological and economic potential of greenhouse gas emissions reduction. In Metz, B., Davidson, O., Swart, R., Pan, J. (eds.) 2001. *Climate Change 2001: Mitigation. Contribution of Working Group III to the Third Assessment Report of the Intergovernmental Panel on Climate Change.* Cambridge: Cambridge University Press, Chapter 3 and its appendix.

that are unprecedented in an international treaty and that enable countries to offset their domestic emission obligations against the absorption of emissions by 'sinks' (e.g. through forestation) or by investment in or trading with other countries where it might be cheaper to limit emissions.

The emissions controlled by the Protocol are from six greenhouse gases (Table 10.2) that can be converted into an amount of carbon-dioxide-equivalent through the use of their global warming potentials (GWPs) which were introduced in Chapter 3 page 52.

The details of the Protocol were finally agreed at a meeting of the Conference of the Parties in Marrakesh in October/November 2001. Much of the detailed discussion related to the inclusion of carbon sinks, especially from forests and from land-use change. Because of the large uncertainties regarding the magnitude of such sinks, considerable doubts were expressed regarding their inclusion in the Protocol arrangements. However, it was agreed that they should be included in a limited way and detailed regulations were agreed concerning the inclusion of afforestation, reforestation and deforestation activities and certain kinds of land-use change. Capping arrangements were also set up that limit the extent to which removals of carbon dioxide from these activities are allowed to offset emissions elsewhere.[1]

The Kyoto mechanisms

The Kyoto Protocol includes three special mechanisms to assist in emissions reductions.

Joint implementation (JI) allows industrialised countries to implement projects that reduce emissions or increase removals by sinks in the territories of other industrialised countries. Emissions reduction units generated by such projects can then be used by investing Annex I countries to help meet their emission targets. Examples of JI projects could be the replacement of a coal-fired power plant with a more efficient combined heat and power plant or the reforestation of an area of land. JI projects are expected to be mainly in EIT (economies in transition) countries where there is more scope for cutting emissions at low cost.

The Clean Development Mechanism (CDM) allows industrialised countries to implement projects that reduce emissions in developing countries. The certified emission reductions generated can be used by industrialised countries to help meet their emission targets, while the projects also help developing countries to achieve sustainable development and contribute to the objective of the Convention. Examples of CDM projects could be a rural electrification project using solar panels or the reforestation of degraded land.

Emissions Trading allows industrialised countries to purchase 'assigned amount units' of emissions from other industrialised countries that find it easier, relatively speaking, to meet their emissions targets. This enables countries to utilise lower cost opportunities to curb emissions or increase removals, irrespective of where those opportunities exist, in order to reduce the overall cost of mitigating climate change.

The detailed regulations concerning the implementation of these mechanisms state that projects will only be approved if they lead to real, measurable and long-term benefits related to the mitigation of climate change and that they are additional to any that would have occurred without the project.

Before the Marrakesh meeting in 2001 the United States had announced its withdrawal from the Protocol. Despite this by the end of 2003 120 countries had ratified the Protocol and the Annex I countries that had ratified represented 44% of Annex I country emissions. For the

Protocol to come into force fifty-five countries have to ratify together with sufficient Annex I countries to represent fifty-five per cent of Annex I country emissions. With ratification by Russia towards the end of 2004, the Protocol will come into force on 16 February 2005.

Concern has often been expressed about the likely cost of implementation of the Kyoto Protocol. Cost studies have been carried out using a number of international energy-economic models. For nine such studies, the range of values in impacts on the gross domestic product (GDP) of participating countries is as follows.[2] In the absence of emissions trading, estimated reductions in projected GDP in the year 2010 are between 0.2% and 2% compared with a base case with no implementation of the Protocol. With emissions trading between Annex I countries, the estimated reductions in GDP are between 0.1% and 1.1%. If emissions trading with all countries is assumed through ideal CDM (see box below) implementation, the estimated reductions in GDP are substantially less – between 0.01% and 0.7%. Although there are differences between countries, most of the large range in the results is due to differences in the models and can be considered as an expression of the large uncertainties inherent in such studies at the present stage of development.

The Kyoto Protocol is an important start to the mitigation of climate change through reductions in greenhouse gas emissions. With its complexity and its diversity of mechanisms for implementation, it also represents a considerable achievement in international negotiation and agreement. It will stem the continuing growth of emissions from many industrialised countries and achieve a reduction overall compared with 1990 from those Annex I countries that participate. The much more substantial longer-term reductions that are likely to be necessary for the decades that follow the first commitment period will be discussed later in the chapter.

Forests

We now turn to the situation of the world's forests and the contribution that they can make to the mitigation of global warming. Action here can easily be taken now and is commendable for many other reasons.

Over the past few centuries many countries, especially those at mid latitudes, have removed much of their forest cover to make room for agriculture. Many of the largest and most critical remaining forested areas are in the tropics. However, during the last few decades, the additional needs of the increasing populations of developing countries for agricultural land and for fuelwood, together with the rise in demand for tropical hardwoods by developed countries, has led to a worrying rate of loss of forest in tropical regions (see box below). In many tropical countries the development of forest areas has been the only hope of subsistence for many people. Unfortunately, because the soils and other conditions were often inappropriate, some of this forest clearance has not led to sustainable agriculture but to serious land and soil degradation.[3]

Measurements on the ground and observations from orbiting satellites have been combined to provide estimates of the area of tropical forest lost. Over the decades of the 1980s[4] and 1990s the average loss was about one per cent per year (see box below) although in some areas it was considerably higher. Such rates of loss cannot be sustained if much forest is to be left in fifty or a hundred years' time. The loss of forests is damaging, not only because of the ensuing land degradation but also because of the contribution that loss makes to carbon emissions and therefore to global warming. There is also the dramatic loss in biodiversity (it is estimated that over half the world's species live in tropical forests) and the potential damage to regional climates (loss of forests can lead to a significant regional reduction in rainfall – see box on page 173).

For every square kilometre of a typical tropical forest there is about 25 000 tonnes of biomass (total living material) above ground, containing about 12 000 tonnes of carbon.[6] It is estimated that burning or other destruction from deforestation turns about two-thirds of this carbon into carbon dioxide. Approximately the same amount of carbon is also stored below the surface in the soil. On this basis, from the destruction of about

---

The world's forests and deforestation[5]

The total area covered by forest is almost one-third of the world's land area, of which ninety-five per cent is natural forest and five per cent planted forest. About forty-seven per cent of forests worldwide are tropical, nine per cent subtropical, eleven per cent temperate and thirty-three per cent boreal.

At the global level, the net loss in forest area during the 1990s was an estimated 940 000 km^2 (2.4% of total forest area). This was the combined effect of a deforestation rate of about 150 000 km^2 per year and a rate of forest increase of about 50 000 km^2 per year. Deforestation of tropical forests averaged about one per cent per year.

The area under forest plantations grew by an average of about 3000 km^2 per year during the 1990s. Half of this increase was the result of afforestation on land previously under non-forest land use, whereas the other half resulted from conversion of natural forest.

In the 1990s, almost seventy per cent of deforested areas changed to agricultural land, predominantly under permanent rather than shifting systems.

150 000 km^2 per annum over the decades of the 1980s and 1990s (see box above) about 1.2 Gt of carbon would enter the atmosphere as carbon dioxide. Although there are substantial uncertainties in the numbers, they approximately tally with the IPCC estimate, quoted in Chapter 3 (see Table 3.1), of the carbon as carbon dioxide entering the atmosphere each year from land-use change (mostly deforestation) of 1.7 ± 0.8 Gt per year – a significant fraction of the current total emissions of carbon dioxide into the atmosphere from human activities.

Reducing deforestation can therefore make a substantial contribution to slowing the increase of greenhouse gases in the atmosphere, as well as the provision of other benefits such as guarding biodiversity and avoiding soil degradation. These benefits are being increasingly recognised and developing countries in tropical regions where there are large areas of natural forest are beginning to concentrate seriously on the management of their forests, on limiting the extent of deforestation or planning for substantial afforestation. Other large areas of forest lie at higher latitudes where developed countries are also taking action to increase forest area so as to contribute to the mitigation of global warming.

Let us look at the possibilities for afforestation. For every square kilometre, a growing forest fixes between about 100 and 600 tonnes of carbon per year for a tropical forest and between about 100 and 250 tonnes for a boreal forest.[7] To illustrate the effect of afforestation on atmospheric carbon dioxide, suppose that an area of 100 000 km^2, a little more than the area of the island of Ireland, were planted each year for forty years – starting now. By the year 2045, 4 000 000 km^2 would have been planted; that is roughly half the area of Australia. During that forty years, the forests would continue to grow and uptake carbon for twenty to fifty years or more after planting (the actual period depending on the type of forest and site conditions) – and, assuming a mixture of tropical, temperate and boreal forest, between about 20 and 50 Gt of carbon from the atmosphere would have been sequestered. This accumulation of carbon in the forests is equivalent to between about five and ten per cent of the likely emissions due to fossil fuel burning up to 2045.

But is such a tree planting programme feasible and is land on the scale required available? The answer is almost certainly, yes. Studies have been carried out that have identified land which is not presently being used for croplands or settlements, much of which has supported forests in the past, of an area totalling about 3 500 000 km^2.[8] About 2 200 000 km^2 of this total is land that is technically suitable at mid and high latitudes – all of this is deemed to be available. In tropical regions, of the 22 000 000 km^2 actually deemed suitable, only six per cent or 1 300 000 km^2 is considered to be actually available because of additional cultural, social and economic constraints. These studies have also considered in detail how much carbon could be sequestered between the years 1995 and 2050 by a programme of afforestation on this land. It is estimated to be between 50 and 70 Gt of carbon, to which a further 10–20 Gt can be added if the rate of tropical deforestation were to be slowed. Estimates of the cost of carrying out the programme have also emerged from the studies; they are considerably lower than those estimated earlier in the 1990s. When expressed per tonne of carbon sequestered they typically fall between $US 1 and 10 (the lower values in developing countries) not including land and transaction costs, but also not including the value of local benefits (for instance, watershed protection, maintenance of biodiversity, education, tourism and recreation) which might be derived from the programme and which, in some circumstances, might offset most of the programme's cost. Compare this figure with the estimate given in Chapter 9 of between $US 50 and 100 for the cost per tonne of carbon of the likely damage due to global warming. The programme therefore appears as a potentially attractive one for alleviating the rate of change of climate due to increasing greenhouse gases in the relatively short term.

Let me insert here a note of caution. As with many environmental projects the situation, however, may not be as simple as it seems at first. One complicating factor is that introducing forest can change the albedo[9] of the Earth's surface. Dark green forests absorb more of the incoming solar radiation than arable cropland or grassland and so tend to warm the surface. This is particularly noticeable in winter months when unforested areas may possess highly reflecting snow cover. Calculations show that, particularly at high latitudes, the warming due to this 'albedo effect' can offset a significant fraction of the cooling that arises from the additional carbon sink provided by the forest.[10]

A possible afforestation programme has been presented in order to illustrate the potential for carbon sequestration. Once the trees are fully grown, of course, the sequestration ceases. What happens then depends on the use that may be made of them. They may be 'protection' forests, for instance for the control of erosion or for the maintenance of biodiversity; or they may be production forests, used for biofuels or for industrial timber. If they are used as fuel for energy generation (see Chapter 11), they add to the atmospheric carbon dioxide but, unlike fossil fuels, they are a renewable resource. As with the rest of the biosphere where natural recycling takes place on a wide variety of timescales, carbon from wood fuel can be continuously recycled through the biosphere and the atmosphere.

However, although there is a useful potential contribution from afforestation to the mitigation of climate change, it can only provide a small part of what is required. An approximate upper bound for the reduction that could be achieved in the twenty-first century through the enhancement of carbon uptake by land-use change has been estimated at 40–70 ppm[11] (equivalent to a storage of 85–150 Gt). Compare this with the range in carbon dioxide concentration in 2100 of about 400 ppm that results from different SRES emission scenarios (Figure 6.1) or with the increase of up to 300 ppm by 2100 because of the possible effect of climate-carbon cycle feedbacks (see Figure 3.5).

Reduction in the sources of methane

Methane is a less important greenhouse gas than carbon dioxide, contributing perhaps fifteen per cent to the present level of global warming. The stabilisation of its atmospheric concentration would contribute a small but significant amount to the overall problem. Because of its much shorter lifetime in the atmosphere (about twelve years compared with 100–200 years for carbon dioxide), only a relatively small reduction in the anthropogenic emissions of this gas, about eight per cent, would be required to stabilise its concentration at the current level.

In Figures 6.1 and 6.2 are shown the emissions and the atmospheric concentrations of methane estimated for the various SRES scenarios, assuming no special action to reduce them. Of the various sources of methane listed in Table 3.3, there are three sources arising from human activities that could rather easily be reduced at small cost.[12] Firstly, methane emission from biomass burning would be cut by, say, one-third if deforestation were drastically curtailed.

Secondly, methane production from landfill sites could be cut by at least a third if more waste were recycled or used for energy generation by incineration or if arrangements were made on landfill sites for the collection of methane gas (it could then be used for energy production or if the quantity were insufficient it could be flared, turning the methane into carbon dioxide which molecule-for-molecule is less effective than methane as a greenhouse gas). Waste management policies in many countries already include the encouragement of such measures.

Thirdly, the leakage from natural gas pipelines from mining and other parts of the petrochemical industry could at little cost (probably even at a saving in cost) also be reduced by, say, one-third. An illustration of the scale of the leakage is provided by the suggestion that the closing down of some Siberian pipelines, because of the major recession in Russia, has been the cause of the fall in the growth of methane concentration in the atmosphere from 1992 to 1993. Improved management of such installations could markedly reduce leakage to the atmosphere, perhaps by as much as one-quarter overall.

Fourthly, with better management, options exist for reducing methane emissions from sources associated with agriculture.[13]

Reductions from these four sources could reduce anthropogenic methane emissions by over 60 000 000 tonnes per annum which would be more than adequate to stabilise the concentration of methane in the atmosphere at about or below the current level. Put another way, the reduction in methane emissions from these sources would be equivalent to a reduction in annual carbon dioxide emissions producing about one-third of a gigatonne of carbon[14] or a little less than five per cent of total greenhouse gas emissions – a useful contribution towards the solution of the global warming problem.

Because the lifetime of methane in the atmosphere is relatively short, a small reduction in methane emissions will quickly lead to its stabilisation as required by the Climate Convention objective. The same, however, is not true of the stabilisation of carbon dioxide concentration with its much longer and rather complicated lifetime. It is to that we shall now turn.

Stabilisation of carbon dioxide concentrations

Carbon dioxide, as we have seen, is the most important of the greenhouse gases that result from human activities. Under all the SRES scenarios, the concentration of carbon dioxide rises continuously throughout the twenty-first century and apart from scenario B1 none come anywhere near to stabilisation of concentration by 2100.

What sort of emissions scenario would stabilise the carbon dioxide concentration? Suppose for instance that it were possible to keep global emissions for the whole of the twenty-first century at the same level as in the year 2000, would that be enough? Stabilising concentrations is, however, very different from stabilising emissions. With constant emissions after the year 2000, the concentration in the atmosphere would continue to rise and would approach 500 ppm by the year 2100. After that carbon cycle models predict that, because of the long time constants involved, the carbon dioxide concentration would still continue to increase, although more slowly, for many centuries.

Examples of scenarios that would lead to stabilisation of atmospheric carbon dioxide concentration at different levels are shown in Figure 10.1. Note that stabilisation at any level shown in the figure, even at an extremely high level, requires that anthropogenic carbon dioxide emissions eventually fall to a small fraction of current emissions. This highlights the fact that to maintain a constant future carbon dioxide concentration, emissions must be no greater than the level of persistent natural sinks. The main known such sink is due to the dissolution of calcium carbonate from the oceans into ocean sediments that, for high levels of carbon dioxide concentration, is probably less than 0.1 Gt per year.[15]

In the work presented in Figure 10.1, many different pathways to stabilisation could have been chosen. The particular emission profiles illustrated in Figure 10.1 begin by following the current average rate of increase of emissions and then provide a smooth transition to the time of stabilisation. To a first approximation, the stabilised concentration level depends more on the accumulated amount of carbon emitted up to the time of stabilisation than on the exact concentration path followed en route to stabilisation. This means that alternative pathways that might assume higher emissions in earlier years would require steeper reductions in later years. Table 10.3 lists the accumulated emissions for the period 2001–2100 for the different stabilisation profiles and also those for the SRES scenarios. It shows that if the atmospheric concentration of carbon dioxide is to remain below about 500 ppm, the future global annual emissions averaged over the twenty-first century cannot exceed the current level of global annual emissions. Figure 10.1(c) shows the projected global mean surface temperature response to the carbon dioxide concentration profiles shown in Figure 10.1(a)

The main results shown in Table 10.3 do not include the effect of climate feedbacks on the carbon cycle (see box in Chapter 3 on page 40). Two of the feedbacks are important in the context of the consideration of stabilisation scenarios; namely, increased respiration from the soil as the temperature rises and die-back especially from forests as the climate changes. As we saw in Chapter 3, the effect of these feedbacks could lead to the biosphere becoming a substantial source of carbon dioxide during the twenty-first century. The size of that source will depend on the amount of climate change. To take it into account, the accumulated amount of that source has to be subtracted from the figures in Table 10.3 to arrive at the emissions from fossil fuel burning that would lead to different stabilisation levels. Some of the estimates of what would need to be subtracted are large – for instance, for the 450 ppm and 550 ppm stabilisation scenarios, they are as large as 200 Gt and 300 Gt respectively during the twenty-first century.[16] In which case, if these estimates are confirmed, emissions scenarios that are aiming at 450 ppm stabilisation (but not allowing for the feedbacks), when the feedbacks are included, would in fact achieve around 550 ppm, and aiming at 550 ppm would in fact achieve around 750 ppm.

It is instructive also to look at annual emissions of carbon dioxide expressed per capita. Averaged over the world in 2000 they were just over one tonne (t) (as carbon) per capita but they varied very much from

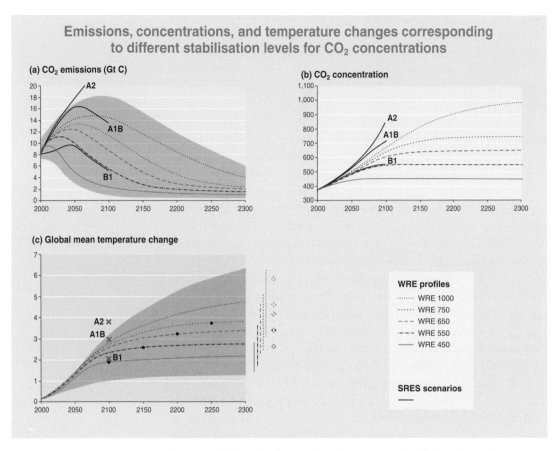

Figure 10.1 (a) Emission profiles of carbon dioxide that would lead to stabilisation of carbon dioxide concentration in the atmosphere at levels of 450, 550, 650, 750 and 1000 ppm according to the concentration profiles shown in (b), estimated from carbon cycle models, without the effects of climate carbon cycle feedbacks included. The shaded area illustrates the range of uncertainty in the estimates that includes the effects of climate carbon cycle feedbacks (e.g. the low boundary of the shading is the profile of the 450 ppm stabilisation curve with the feedbacks included). Also shown are three of the SRES emissions scenarios (A1B, A2 and B1) and the concentrations that would result from them. (c) Global mean temperature changes for the stabilisation profiles in (a) estimated in the same way as for Figure 6.4. The black spots indicate the year in which stabilisation of carbon dioxide concentration is achieved. It is assumed that emissions of gases other than carbon dioxide follow the SRES A1B scenario until the year 2100 and are constant thereafter. The shaded area indicates the effect of a range of climate sensitivity across five stabilisation cases (see caption to Figure 6.4) and the bars on the right-hand side show the range at the year 2300 for the different profiles. The diamonds show the equilibrium (very long-term) warming for each stabilisation level using average climate model results. Also shown for comparison are temperature increases in the year 2100 estimated for the three SRES scenarios.

Table 10.3 *Total anthropogenic carbon dioxide emissions in Gt carbon accumulated from 2001–2100 inclusive for SRES scenarios and for stabilisation scenarios (calculated using the Bern carbon cycle model*[a] *with no carbon cycle feedbacks)*

Case	Accumulated CO_2 emissions (GtC) 2001 to 2100
SRES Scenarios	
A1B	1415
A1T	985
A1FI	2105
A2	1780
B1	900
B2	1080
Stabilisation scenarios	
450 ppm	600
550 ppm	900
650 ppm	1100
750 ppm	1200
1000 ppm	1300

[a] Adapted from Table 5 of Technical summary. In Houghton, J. T., Meira Filho, L. G., Callander, B. A., Harris, N., Kattenberg, A., Maskell, K. (eds.) 1996. *Climate Change 1995*: *The Science of Climate Change*. Cambridge: Cambridge University Press

country to country (Figure 10.2). For developed countries and transitional economy countries in 2000 they averaged 2.8 t (ranging downwards from about 5.5 t for the USA) while for developing countries they averaged about 0.5 t. Looking ahead to the years 2050 and 2100, even if the world population rises to only about seven billion (as with SRES scenarios A1 and B1) under the profiles of carbon dioxide emissions leading to stabilisation at concentrations of 450 ppm and 550 ppm (Figure 10.1) the per capita annual emissions averaged over the world would be about 0.6 t and 1.1 t respectively for 2050 and 0.3 and 0.7 t respectively for 2100[17] – much less than the current value of about 1 t.

The choice of stabilisation level

The last few sections have addressed the main greenhouse gases and how their concentrations might be stabilised. To decide how the appropriate stabilisation levels should be chosen as targets for the future we look to the guidance provided by the Climate Convention Objective (see box

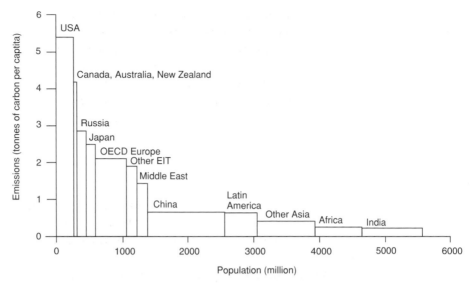

Figure 10.2 Carbon dioxide emissions in 2000 from different countries or groups of countries expressed per capita and population.

on page 243), which states that the levels and the timescales for their achievement should be such that dangerous interference with the climate system must be prevented, that ecosystems should be able to adapt naturally, that food production must not be threatened and that economic development can proceed in a sustainable manner. We do not yet know enough to pick precisely the levels or the timescales under the criteria the Climate Convention is prescribing, but perhaps already some limits can be set.

Firstly, considering the most important greenhouse gas, carbon dioxide, as we have already noted its long life in the atmosphere provides severe constraints on the future emission profiles that lead to stabilisation at any level. It will be clear, for instance, from Figure 10.1 that stabilisation below about 400 ppm would require an almost immediate drastic reduction in emissions. Such reduction could only be achieved at a large cost and with some curtailment of energy availability and would almost certainly breach the criterion which requires 'that economic development can proceed in a sustainable manner'.

What about the upper end of the choice of level? Here we refer to the likely impacts of climate change under a situation in which the atmospheric concentration of carbon dioxide has doubled from its pre-industrial value of 280 ppm to about 560 ppm. Many of the impacts described in Chapter 7 with their associated costs apply to this situation. We also noted there that in estimating these costs there were components of the damage that could not be quantified in money terms. But even if only the costs that can be estimated in terms of money are considered, in Chapter 9 it was pointed out that estimates of the cost of the likely damage of the impacts at that level of climate change were larger than the costs of stabilising carbon dioxide concentration at levels above about 500 ppm (Figure 9.4). We also noted that, beyond the doubled carbon dioxide situation, the damage due to greenhouse gas climate change is likely to rise substantially more rapidly as the amount of carbon dioxide in the atmosphere increases. A further factor is the rate of climate change (see Figure 10.1(c)) which, with all the profiles except possibly the two lowest, is likely to be such that some important ecosystems may not be able to adapt to it (see Chapter 7). Studies[18] show that stabilisation below 550 ppm should avoid some of the worst impacts; for instance, some of the large-scale die-back of forests and the transition of the biosphere from a source to a sink for carbon dioxide (see box in Chapter 3 on page 40) which would otherwise occur around the middle of the twenty-first century. Considering carbon dioxide alone, these considerations suggest that the range between about 400 ppm and 550 ppm is where further careful consideration of the choice of the target stabilisation level should be made.

Although carbon dioxide is the most important greenhouse gas, other gases also make a contribution to climate change. The combined effect of the increases to 1990 of the gases methane, nitrous oxide and the CFCs[19] is to add a forcing equivalent to that from an additional 60 ppm or so of carbon dioxide (see Chapter 6, page 124). The effect of these other gases also needs to be taken into account in our overall discussion of the Climate Convention Objective of stabilisation. Even if there were no further increase in these minor gases, the 1990 forcing would still require to be added to future projections of change. The effect of this, if turned into equivalent amounts of carbon dioxide, would be that the 450 ppm carbon-dioxide-only level would become about 520 ppm and the 550 ppm level would become about 640 ppm of equivalent carbon dioxide.[20] This means that, if it is considered that the climate effects of doubled pre-industrial carbon dioxide concentration should be an upper limit, when the increases in other gases are allowed for, the stabilisation limit for carbon dioxide only is about 490 ppm.

How realistic is it to assume that the concentration of the other gases will not change? We saw earlier that the Montreal Protocol should ensure that the CFCs are stabilised in concentration over the next decade or two. We also saw, for methane, that means are available that are not costly and that, if taken, could stabilise methane concentrations at about today's levels. There is more uncertainty about nitrous oxide as its sources and sinks are not well known. However, it is only a small contributor to the forcing to date (equivalent to about 10 ppm of carbon dioxide); any increase in the future is not likely to have a large effect.

In this simple argument regarding the influence of other gases on the choice of a concentration level for carbon dioxide stabilisation that might be acceptable under the terms of the Climate Convention Objective, the concept of equivalent carbon dioxide concentration has proved a useful tool. But it must not be used blindly. For any detailed consideration of the choice of level, there are other scientific factors to be included. Firstly, there are the other contributors to radiative forcing and climate change; for instance, tropospheric ozone and aerosols that are very inhomogeneous in their distribution. Their likely effect, although small compared with that of carbon dioxide, also needs to be taken into account. Secondly, there are the different regional climate responses and different timescales of responses that result from the different greenhouse gases or from aerosols. Thirdly, there are the effects of particular feedbacks (e.g. carbon dioxide fertilisation) or impacts (e.g. acid rain from aerosols).

The choice of a target stabilisation level for greenhouse gases according to the criteria listed in the Objective of the Climate Convention involves scientific, economic, social and political factors. In Chapter 9 (see box on page 237) the concept of Integrated Assessment and Evaluation was introduced that involves employment of the whole range of disciplines in the natural and social sciences. Taking all factors into consideration will involve different kinds of analysis, cost benefit analysis (which was considered briefly in Chapter 9), multicriteria analysis (which takes into account factors that cannot be expressed in monetary terms) and sustainability analysis (which considers avoidance of particular thresholds of stress or of damage). Further, because much uncertainty is associated both with many of the factors that have to be included and with the methods of analysis, the process of choice is bound to be an evolving one subject to continuous review – a process often described as sequential decision making.

Taking account of considerations such as those above, let me mention statements that have come from two very different bodies regarding their view of where the choice of a stabilisation level at the present time could or should be made. Firstly, the European Union has proposed setting a limit for the rise in global average temperature of 2 °C.[21] Since the best estimate of global average temperature rise for doubled pre-industrial carbon dioxide (560 ppm) is 2.5 °C, a rise of 2 °C would occur with a carbon dioxide concentration of about 430 ppm allowing for the effect of other gases at their 1990 levels. The second statement comes from Lord John Browne, the Group Chief Executive of British Petroleum, one of the world's largest oil companies. He recognises the dangers of global warming and the challenge it presents and has stated that, for carbon dioxide, 'stabilisation in the range 500–550 ppm is possible, and with care could be achieved without disrupting economic growth.'[22]

Realising the Climate Convention Objective

Having decided on a choice of stabilisation level, a large question remains: how can the nations of the world work together to realise it in practice?

The Objective of the Climate Convention is largely concerned with factors associated with the requirement for sustainable development. In Chapter 9, four principles were enunciated that should be at the basis of negotiations concerned with future emissions reductions to mitigate climate change. One of these was the Principle of Sustainable Development. The others were the Precautionary Principle, the Polluter-Pays Principle and the Principle of Equity. This latter Principle includes *intergenerational* equity, or weighing the needs of the present generation against those of future generations, and *international* equity, or weighing the balance of need between industrial and developed nations and the developing world. Striking this latter balance is going to be particularly difficult because of the great disparity in current carbon dioxide emissions between the world's richest nations and the poorest nations (Figure 10.2), the continuing demand for fossil fuel use in the developed world and the understandable desire of the poorer nations to escape from poverty through development and industrialisation. This latter is particularly recognised in the Framework Convention on Climate Change (see box at the beginning of the chapter) where the growing energy needs of developing nations as they achieve industrial development are clearly stated.

An example of how the approach to stabilisation for carbon dioxide might be achieved is illustrated in Figure 10.3. It is based on a proposal called 'Contraction and Convergence' that originates with the Global Commons Institute (GCI),[23] a non-governmental organisation based in the UK. The envelope of carbon dioxide emissions is one that leads to stabilisation at 450 ppm (without climate feedbacks included), although the rest of the proposal does not depend on that actual choice of level. Note that, under this envelope, global fossil-fuel emissions rise by about fifteen per cent to about 2025; they then fall to less than half the current level by 2100. The figure illustrates the division of emissions between major countries or groups of countries as it has been up to the present. Then the simplest possible solution is taken to the sharing of emissions between countries and proposes that, from some suitable date (in the

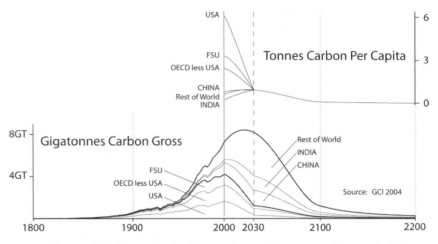

Figure 10.3 Illustrating the 'Contraction and Convergence' proposal of the Global Commons Institute for achieving stabilisation of carbon dioxide concentration. The envelope of carbon dioxide emissions illustrated is one that leads to stabilisation at 450 ppm (but the effect of climate carbon cycle feedbacks is not included). For major countries or groups of countries, up to the year 2000, historic emissions are shown. After 2030 allocations of emissions are made on the basis of equal shares per capita on the basis of population projections for that date. From now until 2030, smooth 'convergence' from the present situation to that of equal shares is assumed to occur. In the upper part of the diagram the per capita contributions that apply to different countries or groups of countries are shown. For OECD and FSU see Glossary.

figure, 2030 is chosen), emissions are allocated on the basis of equal shares per capita. From now until 2030 the division is allowed to converge from the present situation to that of equal per capita shares. Hence the 'contraction and convergence'. The further proposal is that arrangements to trade the carbon dioxide allocations are made.

The 'Contraction and Convergence' proposal addresses all of the four principles mentioned above. In particular, through its equal per capita sharing arrangements it addresses head-on the question of international equity – and the proposed trading arrangements ensure that the greatest 'polluters' pay. Its simple and appealing logic means that it is a strong candidate for providing a long-term solution. What has yet to be worked is how the 'convergence' part of the proposal can be implemented, but then any proposal for a solution will have to address the problem of 'convergence'.

Another example of a pathway to stabilisation during the twenty-first century of carbon dioxide concentration is set out in a study sponsored by the World Energy Council and published in 1993.[24] An 'ecologically driven scenario' – Scenario C – of global carbon dioxide emissions is described that leads to stabilisation at about 450 ppm (without carbon feedbacks included) – see Figure 11.4. Under that scenario, global carbon dioxide emissions grow by about ten per cent (from 1990 levels) by the year 2050; they then fall by sixty per cent by 2100 (Table 11.2). For the first two decades of the twenty-first century, the World Energy Council provide detailed projections for Scenario C that recognise the requirement for international equity. Up to the year 2020, emissions from fossil fuels in the developing world are allowed to approximately double, while those from developed countries fall by about thirty per cent (Figure 11.5). In 2020, global emissions from developing countries would be sixty per cent of the total for the world compared with about one-third in 1990. After 2020 reductions in emissions in all countries would be required.

As the World Energy Council point out in their report, achievement of such a scenario will be far from easy. It requires three essential ingredients. The first is an aggressive emphasis on energy saving and conservation. Much here can be achieved at zero net cost or even at a cost saving. Though much energy conservation can be shown to be economically advantageous, it is unlikely to be undertaken without significant incentives. However, it is clearly good in its own right, it can be started in earnest now and it can make a significant contribution to the reduction of emissions and the slowing of global warming. The second ingredient is an emphasis on the development of appropriate non-fossil fuel energy sources leading to very rapid growth in their implementation. The third is the transfer of technologies to developing countries that will enable them to apply the most appropriate and the most efficient technologies to their industrial development, especially in the energy sector.

Summary of the action required

This chapter has suggested some actions that can be taken to slow climate change and ultimately to stabilise it as required by the internationally agreed Climate Convention.

Some actions have already been taken that have an effect on global emissions of greenhouse gases, namely:

- the reduction by some countries of carbon dioxide emissions in the year 2000 to 1990 levels, and
- the provisions of the Montreal Protocol regarding the emissions of CFCs and CFC substitutes.

Other actions that can be taken now to slow climate change, that can be done at little or no net cost and that are good to do for other reasons are the following:

- a reduction of deforestation,
- a substantial increase in afforestation,
- some relatively easy-to-do reductions in methane emissions,
- an aggressive increase in energy saving and conservation measures,
- increased implementation of renewable sources of energy supply.

For the longer term, as well as increased emphasis on these actions, the world needs to begin to follow an energy scenario that will lead to the stabilisation of carbon dioxide concentration in the atmosphere. The choice of a target stabilisation level following the guidance of the Climate Convention involves the consideration of many factors and, because of the uncertainties, will necessarily be subject to continuous review. We have presented arguments suggesting that, at the current state of knowledge, the range 400–500 ppm in carbon dioxide concentration is where further detailed consideration of costs and impacts should be concentrated. A proposal called 'Contraction and Convergence' meets the requirement for international equity through eventual agreement for equal allocations per capita coupled with arrangements for allocations' trading. A study by the World Energy Council has detailed an energy scenario that would lead to the stabilisation of carbon dioxide concentration by about 2100. Its realisation will require very rapid growth in the implementation of appropriate non-fossil fuel energy sources; it will also require that means be provided to enable developing countries to apply appropriate and efficient technologies to their industrial development, especially in the energy sector – matters that will be addressed in detail in the next chapter.

Questions

1 From Figure 10.1, what are the rates of change of global average temperature for the profiles shown that lead to stabilisation of carbon dioxide concentration at different levels? From information in Chapter 7 or from elsewhere, can you suggest a criterion involving rate of change that might assist in the choice of a stabilisation level for carbon dioxide concentration as required by the Objective of the Climate Convention?

2 From the formula in Note 20 and the information in Figure 3.8 and Table 6.1, calculate the contributions from the various components of radiative forcing (including aerosol) to the equivalent carbon dioxide concentration in 1990. How valid do you think is it to speak of equivalent carbon dioxide for components such as aerosol and tropospheric ozone?

3 From the information in Table 6.1 and the formula in Note 20, calculate the equivalent carbon dioxide concentration including the well mixed greenhouse gases for SRES scenarios A1B and A2 in 2050 and 2100.

4 Associated with the choice of stabilisation level under the criteria of the Objective of the Climate Convention, different kinds of analysis were mentioned; cost–benefit analysis, multicriteria analysis and sustainability analysis. Discuss which analysis is most applicable to each of the criteria in the Objective. Suggest how the analyses might be presented together so as to assist in the overall choice.

5 From the information available in previous chapters and using the criteria laid out in the Climate Convention Objective, what stabilisation levels of greenhouse gas concentrations do you think should be chosen?

6 The arguments concerning the choice of stabilisation level and the action to be taken have concentrated on the likely costs and impacts of climate change before the year 2100. Do you think that information about continuing climate change or sea level rise (see Chapter 7) after 2100 should be included and taken into account by decision makers, or is that too far ahead to be of importance?

7 The international response to global warming is likely to lead to decisions being taken sequentially over a number of years as knowledge regarding the science, the likely impacts and the possible responses becomes more certain. Describe how you think the international response might progress over the next twenty years. What decisions might be taken at what time?

8 Explain how the 'Contraction and Convergence' proposal meets the four principles listed in Chapter 9 and elaborated in Chapter 10. Suggest the political or economic arguments that might be used to argue against the proposal. Can you suggest other ways of sharing emissions between countries that might achieve agreement more easily?

9 Find out the details of any plans for afforestation in your country. What actions or incentives could make it more effective?

10 Assume a snow-covered area at latitude 60° with an albedo of fifty per cent is replaced by partially snow-covered forest with an albedo of twenty per cent. Make an approximate comparison between the 'cooling' effect of the carbon sink provided by the forest and the 'warming' effect of the added solar radiation absorbed, averaged over the year.

Notes for Chapter 10

1 More details of the Kyoto Protocol and of the detailed arrangements for the inclusion of carbon sinks can be found in Watson, R. T., Noble, I. R., Bolin, B., Ravindranath, N. H., Verardo, D. J., Dokken, D. J. (eds.) 2000. *Land Use, Land-Use Change and Forestry. A Special Report of the IPCC*. Cambridge: Cambridge University Press and on the FCCC web site www.unfccc.int/resource/convkp.html.

2   More details in Watson, R. *et al.* (eds.) 2001. *Climate Change 2001: Synthesis Report. Contribution of Working Groups I, II and III to the Third Assessment Report of the Intergovernmental Panel on Climate Change.* Cambridge: Cambridge University Press, question 7. Also, Hourcade, J.-C., Shukla, P. *et al.* Global, regional and national costs and ancillary benefits of mitigation. In Metz, B., Davidson, O., Swart, R., Pan, J. (eds.) 2001. *Climate Change 2001: Mitigation. Contribution of Working Group III to the Third Assessment Report of the Intergovernmental Panel on Climate Change.* Cambridge: Cambridge University Press, Chapter 8.

3   More detail in *Global Environmental Outlook 2000, UNEP.* London: Earthscan. Also eds. Tolba, M. K., El-Kholy, O. A. (eds.) 1992. *The World Environment 1972–1992.* London: Chapman and Hall, pp. 157–82.

4   See Tolba, *The World Environment 1972–1992*, p. 169.

5   From *Global Environmental Outlook 3 (UNEP Report).* 2002. London: Earthscan. pp. 91–2.

6   Bolin, B., Sukumar, R. *et al.* 2000. Global perspective. In Watson, *Land Use*, Chapter 1. Also Salati *et al.* 1991. In Jager, J., Ferguson, H. L. (eds.) *Climate Change: Science, Impacts and Policy; Proceedings of the Second World Climate Conference.* Cambridge: Cambridge University Press, pp. 391–5; also Leggett, J. *et al.* 1992. Emission scenarios for the IPCC: an update. In Houghton, J. T., Callender, B. A., Varney, S. K. (eds.) *Climate Change 1992: the Supplementary Report to the IPCC Assessments.* Cambridge: Cambridge University Press, p. 89; also Houghton, R. A. 1991. *Climate Change*, **19**, pp. 99–118.

7   Bolin, B., Sukumar, R. *et al.* 2000. Global perspective. In Watson, *Land Use*, Chapter 1, p 26.

8   Brown, S. *et al.* 1996. Management of forests for mitigation of greenhouse gas emissions. In Watson, R. T., Zinyowera, M. C., Moss, R. H. (eds.) 1996. *Climate Change 1995: Impacts, Adaptations and Mitigation of Climate Change: Scientific-Technical Analyses. Contribution of Working Group II to the Second Assessment Report of the Intergovernmental Panel on Climate Change.* Cambridge: Cambridge University Press, Chapter 24. Similar areas have been quoted in Watson, *Land Use*, Policymakers Summary, and also in Kauppi, P., Sedjo, R. *et al.* Technical and economic potential of options to enhance, maintain and manage biological carbon reservoirs and geo-engineering. In Metz, B., Davidson, O., Swart, R., Pan, J. (eds.) 2001. *Climate Change 2001: Mitigation. Contribution of Working Group III to the Third Assessment Report of the Intergovernmental Panel on Climate Change.* Cambridge: Cambridge University Press, Chapter 4.

9   For definition *see* Glossary.

10   Betts, R. A. 2000. Offset of the potential carbon sink from boreal forestation by decreases in surface albedo. *Nature*, **408**, 187–90.

11   From Policymakers summary. In Houghton, J. T., Ding, Y., Griggs, D. J., Noguer, M., van der Linden, P. J., Dai, X., Maskell, K., Johnson, C. A. (eds.) *Climate Change 2001: The Scientific Basis. Contribution of Working Group I to the Third Assessment Report of the Intergovernmental Panel on Climate Change.* Cambridge: Cambridge University Press.

12 Hourcade, J. C. *et al*. 1996. A review of mitigation cost studies. In Bruce, J., Hoesung Lee, Haites, E. (eds.) 1996. *Climate Change 1995: Economic and Social Dimensions of Climate Change*. Cambridge: Cambridge University Press, Chapter 9.

13 Cole, V. *et al*. 1996. Agricultural options for mitigation of greenhouse gas emissions. In Watson, *Climate Change 1995: Impacts*.

14 This figure is calculated by multiplying the sixty million tonnes by the global warming potential for methane which, for a time horizon of 100 years, is about 23 (Table 10.2), then by 12/44 to put it into tonnes of carbon.

15 Prentice, I. C. *et al*. 2001. The carbon cycle and atmospheric carbon dioxide. In Houghton, J. T., Ding, Y., Griggs, D. J., Noguer, M., van der Linden, P. J., Dai, X., Maskell, K., Johnson, C. A. (eds.) *Climate Change 2001: The Scientific Basis. Contribution of Working Group I to the Third Assessment Report of the Intergovernmental Panel on Climate Change*. Cambridge: Cambridge University Press.

16 Cox, P. M. *et al*. 2000. Acceleration of global warming due to carbon cycle feedbacks in a coupled climate model. *Nature*, **408**, pp. 184–7; Jones, C. D. *et al*. 2003. *Tellus*, **55B**, pp. 642–58.

17 Additional climate-carbon cycle feedbacks have been ignored in this calculation.

18 Arnell, N. W. *et al*. 2002. The consequences of CO_2 stabilisation for the impacts of climate change. *Climatic Change*, **53**, pp. 413–46.

19 Allowing, for the CFCs, a reduction in their forcing because of stratospheric ozone destruction. Further, only the well-mixed greenhouse gases have been considered here. Tropospheric ozone and sulphate aerosols are not well mixed but have significant radiative forcing effects (see Figure 3.8). Their effects are of opposite sign and when globally averaged are of similar magnitude so to some degree might be considered to compensate for each other.

20 Note that, although the amount of forcing from the minor gases is the same, when turned into equivalent carbon dioxide, the amounts added increase with the carbon dioxide concentration to which the amount is added. This is because the relationship between radiative forcing (R in W m^{-2}) and concentration (C in ppm) is non-linear. The relationship is $R = 5.3 \ln (C/C_0)$ where C_0 is the pre-industrial CO_2 concentration.

21 European Commission Communication on a Community Strategy on Climate Change; Council of Ministers Conclusion, 25–26 June 1996.

22 From speech to Institutional Investors Group, London, 26 November 2003.

23 Further details on the GCI web site, www.gci.org.uk.

24 World Energy Council Commission Report, *Energy for Tomorrow's World*. London: World Energy Council, 1993. The World Energy Council is an international body that links together the World's Energy Industries.

Chapter 11
Energy and transport for the future

We flick a switch and energy flows. Energy is provided so easily for the developed world that thought is rarely given to where it comes from, whether it will ever run out or whether it is harming the environment. Energy is also cheap enough that little serious attention is given to conserving it. However, most of the world's energy comes from the burning of fossil fuels, which generates a large proportion of the greenhouse gas emissions into the atmosphere. If these emissions are to be reduced, a large proportion of the reduction will have to occur in the energy sector. There is a need, therefore, to concentrate the minds of policymakers and indeed of everyone on our energy requirements and usage. This chapter looks at how future energy might be provided in a sustainable manner.[1] It also addresses how basic energy services might be made available to the more than two billion in the world who as yet have no such provision.

World energy demand and supply

Most of the energy we use can be traced back to the Sun. In the case of fossil fuels (coal, oil and gas) it has been stored away over millions of years in the past. If wood (or other biomass including animal and vegetable oils), hydro-power, wind or solar energy itself is used, the energy has either been converted from sunlight almost immediately or has been stored for at most a few years. These latter sources of energy are renewable; they will be considered in more detail later in the chapter. The only common form of energy that does not originate with the Sun is nuclear energy; this comes from radioactive elements that were present in the Earth when it was formed.

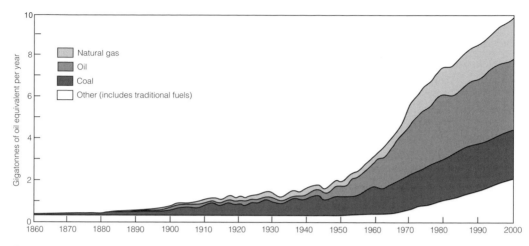

Figure 11.1 Growth in the rate of energy use and in the sources of energy since 1860 in thousand millions of tonnes of oil equivalent (Gtoe) per year. In terms of primary energy units, 1 Gtoe = 41.87 exajoules (1 exajoule (FJ) = 10^{18} J). Of the 'other' in 2000, approximately 0.9 Gtoe is attributed to traditional fuels, 0.7 Gtoe to nuclear energy and 0.6 Gtoe to hydro and other renewables (source: Report of G8 Renewable Energy task Force, July 2001).

Until the Industrial Revolution, energy for human society was provided from 'traditional' sources – wood and other biomass and animal power. Since 1860, as industry has developed, the rate of energy use has multiplied by about a factor of thirty (Figure 11. 1), at first mostly through the use of coal followed, since about 1950, by rapidly increasing use of oil and then more recently by the use of natural gas. In 2000 the world consumption of energy was about 10 000 million tonnes of oil equivalent (toe). This can be converted into physical energy units to give an average rate of energy use of about thirteen million million watts (or 13 terawatts = 13×10^{12} W).[2]

Great disparities exist in the amount of energy used per person in various parts of the world. The two billion poorest people in the world (less than $US 1000 annual income per capita) each use an average of only 0.2 toe of energy annually while the billion richest in the world (more than $US 22 000 annual income per capita) use nearly twenty-five times that amount at 5 toe per capita annually.[3] The average annual energy use per capita in the world is about 1.7 toe, an average consumption of energy of about 2.2 kilowatts (kW). The highest rates of energy consumption are in North America where the average citizen consumes an average of about 11 kW. Over one-third of the world's population rely wholly on traditional fuels (wood, dung, rice husks, other forms of 'biofuels') and do not currently have access to commercial energy in any of its forms.

It is interesting to see how the energy we consume is used.[4] Taking the world average for commercial energy (i.e. omitting 'traditional energy'), about twenty-two per cent of primary energy is used in transportation, about forty-one per cent by industry, about thirty-four per cent in buildings (two-thirds in residential buildings and one-third in commercial buildings) and about three per cent in agriculture. It is also perhaps interesting to know how much energy is used in the form of electricity. Rather more than one-third of primary energy goes to make electricity at an average efficiency of conversion of about one-third. Of this electrical power about half, on average, is utilised by industry and the other half in commercial activities and in homes.

How much is spent on energy? Taking the world as a whole, the amount spent per year by the average person for the 1.7 toe of energy used, is about five per cent of annual income. Despite the very large disparity in incomes, the proportion spent on primary energy is much the same in developed countries and developing ones.

How about energy for the future? If we continue to generate most of our energy from coal, oil and gas, do we have enough to keep us going? Current knowledge of proven recoverable reserves (Figure 11.2) indicates that known reserves of fossil fuel will meet demand for the period up to 2020 and substantially beyond. Before mid century, if demand continues to expand, oil and gas production will come under increasing pressure. Further exploration will be stimulated, which will lead to the exploitation of more sources, although increased difficulty of extraction can be expected to lead to a rise in price. So far as coal is concerned, there are operating mines with resources for production for well over a hundred years.

Estimates have also been made of the ultimately recoverable fossil fuel reserves, defined as those potentially recoverable assuming high but not prohibitive prices and no significant bans on exploitation. Although these are bound to be somewhat speculative,[5] they show that, at current rates of use, reserves of oil and gas are likely to be available for 100 years and of coal for more than 1000 years. In addition to fossil fuel reserves considered now to be potentially recoverable there are reserves not included in Figure 11.2, such as the methane hydrates, which are probably very large in quantity but from which extraction would be much more difficult.

Likely reserves of uranium for nuclear power stations should also be included in this list. When converted to the same units (assuming their use in 'fast' reactors) they are believed to be at least 3000 and possibly as high as 12 000 Gtoe, substantially greater than likely fossil fuel reserves.

For at least the twenty-first century, sufficient fossil fuels in total are available to meet likely energy demand. It is considerations other than

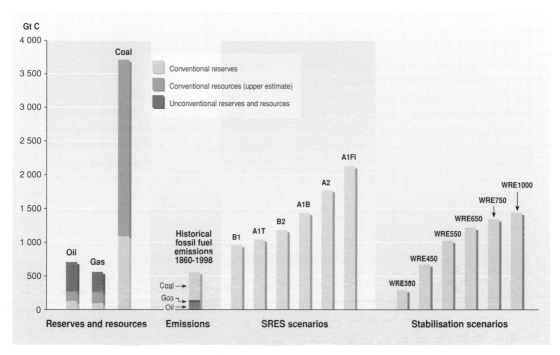

Figure 11.2 Carbon in oil, gas and coal reserves and resources compared with historic fossil fuel carbon emissions over the period 1860–1998 and with cumulative carbon emissions up to the year 2100 from a range of SRES scenarios and scenarios leading to stabilisation of carbon dioxide concentrations. Data for current estimates of reserves and resources are shown in the left-hand columns. Unconventional oil and gas includes tar sands, shale oil, other heavy oil, coal bed methane, deep geopressured gas, gas in aquifers, etc. Gas hydrates (clathrates) that amount to an estimated 12 000 GtC are not shown. Note that if by the year 2100 cumulative emissions associated with SRES scenarios are equal to or smaller than those for stabilisation scenarios, this does not imply that these scenarios equally lead to stabilisation.

availability, in particular environmental considerations, that will provide limitations on fossil fuel use.

Future energy projections

In Chapter 6 were described the SRES scenarios sponsored by the IPCC that detail, for the twenty-first century, a range of possibilities regarding future energy demand (based on a range of assumptions concerning population, economic growth and social and political development), how that demand might be met and what carbon dioxide emissions might result. In that chapter were also described the implications for those scenarios regarding climate change. Chapter 10 presented the imperative set out by the Framework Convention on Climate Change (FCCC) in its Objective that carbon dioxide concentrations in the atmosphere must be stabilised so that continued anthropogenic climate change can be

Energy intensity and carbon intensity

An index that provides an indication of a country's energy efficiency is the ratio of annual energy consumption to gross domestic product (GDP) known as the *energy intensity*. Figure 11.3 shows that from 1971 to 1996 for Organisation for Economic Co-operation and Development (OECD) countries GDP increased by a factor of two while energy consumption increased by about fifty per cent, the result being a decrease in energy intensity of about twenty per cent or an average of about one per cent per year. Within the OECD there are substantial differences between countries. Denmark, Italy and Japan have the lowest energy intensities and Canada and the USA the highest, with more than a factor of two between the lowest and the highest.

Of importance too in the context of this chapter is the *carbon intensity*, which is a measure of how much carbon is emitted for a given amount of energy. This can vary with different fuels. For instance, the carbon intensity of natural gas is twenty-five per cent less than that of oil and forty per cent less than that of coal. For renewable sources the carbon intensity is small and depends largely on that which originates during manufacture of the equipment making up the renewable source (e.g. during manufacture of solar cells).

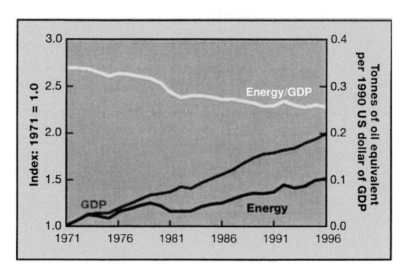

Figure 11.3 Energy intensity averaged over OECD countries 1971–96.

avoided. Scenarios of carbon dioxide emissions that would be consistent with various stabilisation levels were presented there and arguments put forward for a target level for atmospheric concentration of carbon dioxide in the range of 450 to 500 ppm. How the world's energy producers and consumers can meet the challenge of this target is addressed by this chapter.

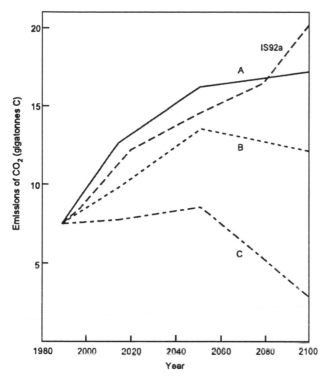

Figure 11.4 Scenarios of carbon dioxide emissions during the twenty-first century from fossil fuel burning for WEC scenarios A, B and C (details in Tables 11.1 and 11.2). It is only scenario C that leads to stabilisation of carbon dioxide concentrations as required by the Objective of the Climate Convention.

One of the bodies that has considered how such a target can be met is the World Energy Council (WEC), who have constructed four detailed energy scenarios (Figure 11.4) for the period to 2020, extended with less detail to 2100.[6] Three of the scenarios (more details in box on pages 276–7) make assumptions (more details in the box below) that fall within the range of those made by the SRES scenarios. The fourth scenario C, which is described as 'ecologically driven', assumes that environmental pressures have a large influence on energy demand and growth. For all these scenarios, except WEC scenario C, atmospheric concentrations of carbon dioxide continue to rise throughout the twenty-first century. It is WEC scenario C, if achieved, that is consistent with carbon dioxide stabilisation with a target in the range 450–500 ppm.

Details of the scenarios to the year 2020 are shown in Figures 11.5 and 11.6. As can be seen from Figure 11.5, it is only in the developed world that there is potential for containing future energy demand.

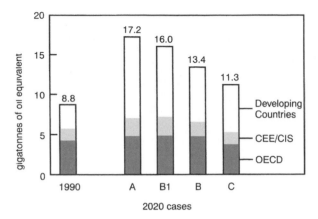

Figure 11.5 Primary energy demand by economic groupings of countries for WEC scenarios.

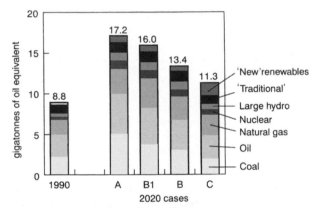

Figure 11.6 Primary energy supply mix for WEC scenarios.

Population growth and the need for economic development in developing countries make it inevitable that they will, for many decades, consume increased amounts of energy. For all the scenarios to 2020, fossil fuels continue to dominate the energy mix (Figure 11.6). The contribution from nuclear power is assumed to grow in all the cases. New renewable energy sources play an increasing role, although apart from Case C their contribution is small.

For the 'ecologically driven' WEC scenario C the energy demand in 2020 is about thirty per cent more than that in 1990 and thirty per cent less than that for scenario A. Scenario C assumes that there will be large increases in efficiency (or a large decrease in energy intensity) leading to a reduced energy demand and also, following the results of

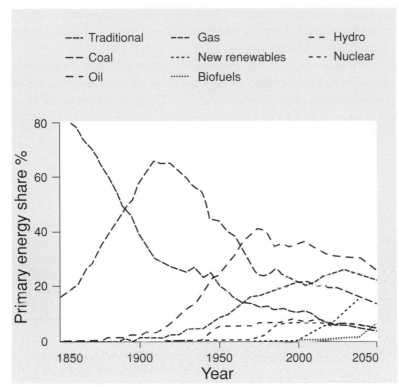

Figure 11.7 Energy transitions over two centuries. Under 'dynamics as usual' energy supplies continue to evolve from high to low carbon fuels and towards electricity as the dominant energy carrier – from increasingly distributed sources – driven by demands for security, cleanliness and sustainability.

a WEC study on renewable energy,[7] a substantial growth in the share of primary energy supply coming from new renewable energy sources ('modern' biomass, solar, wind and so on). A growth in world energy supply from these new renewable sources from two per cent in 1990 to twelve per cent in 2020 (by when 1.4 Gtoe per year would be coming from these sources) is considered feasible if their development is given sufficient support. By the year 2050 under scenario C, twenty per cent of energy supply is assumed to come from new renewable sources and by 2100, fifty per cent. The WEC report points out that 'cost effective research, development and installation involving financing which only governments can supply will be needed if these sources of energy are to be implemented on the large scale shown in the Ecologically Driven Case C'. Renewable energy sources will be discussed further later in the chapter.

During the last few years many organisations have developed mitigation energy scenarios for the twenty-first century under a wide variety of assumptions regarding the growth of total energy, renewable energy and energy efficiency.[8] As one example of these, Figure 11.7 shows a scenario developed by the Shell Oil Company which is called 'Dynamics

as Usual'. The figure illustrates the large transitions that have occurred
in energy sources in the past and those that will have to take place in the
future.

The following sections address how increased energy conservation
and efficiency can be achieved and what developments can be realised in
new renewable energy sources; these are the technical means through
which the necessary reductions in carbon dioxide emissions will be
achieved in the energy sector.

World Energy Council scenarios

The World Energy Council (WEC) is an international non-governmental
organisation with representation from all parts of the energy industry and
from over ninety countries. The Council has developed four energy scenarios for the period to 2020, each representing different assumptions in
terms of economic development, energy efficiencies, technology transfer
and the financing of development round the world. The WEC emphasises
that they have been developed to illustrate future possibilities and they
should not be considered as predictions.

All four cases assume that there will be significant environmental and
economic pressures to achieve major improvements in energy efficiency
compared to historic performance, although to different degrees within
the various economic groupings of countries. One of them, scenario
C, assumes very strong pressure to reduce the emissions of greenhouse
gases in order to combat global warming. Table 11.1 presents the detailed
assumptions underlying the four scenarios.

Table 11.1 refers to the 'energy intensity' (see box on page 272),
which is a measure of energy efficiency. When averaged over the world,
over the past fifty years it has been falling by about one per cent per year. A
more demanding rate of reduction in energy intensity than this is assumed
for all the scenarios; for case C, the ecologically driven scenario, the rate
assumed is considered very demanding indeed. The main difference
between the modified reference case B1 and the reference case B is that,
in B1, the rate of reduction of energy intensity assumed for the economies
in transition is less than in case B and for the developing countries is
only half that in case B.

With somewhat less detail, scenarios A, B and C have been extended
to the year 2100 (Figure 11.4). Global energy demand can be expected to
continue to increase, but by that time the availability of fossil fuels will
be more limited and new renewables will contribute substantially to the
energy mix for all scenarios. Some of the characteristics of the scenarios
out to 2100 are listed in Table 11.2 – similar details of the scenarios for
2020 are in Figure 11.6.

Table 11.1 *Assumptions underlying the four WEC energy scenarios. See glossary for explanation of abbreviations*

Assumptions	A (High growth)	B1 (Modified reference	B (Reference)	C (Ecologically driven
			Case (name)	
Economic growth % p.a.	High	Moderate	Moderate	Moderate
OECD	2.4	2.4	2.4	2.4
CEE/CIS	2.4	2.4	2.4	2.4
DCs	5.6	4.6	4.6	4.6
World	3.8	3.3	3.3	3.3
Energy intensity				
reduction % p.a.	High	Moderate	High	Very high
OECD	−1.8	−1.9	−1.9	−2.8
CEE/CIS	−1.7	−1.2	−2.1	−2.1
DCs	−1.3	−0.8	−1.7	−2.4
World	−1.6	−1.3	−1.9	−2.4
Technology transfer	High	Moderate	High	Very high
Institutional				
Improvements (world)	High	Moderate	High	Very high
Possible total demand	Very high	High	Moderate	Low
(Gtoe)	17.2	16.0	13.4	11.3

From *Energy for Tomorrow's World: the Realities, the Real Options and the Agenda for Achievement.* WEC Commission Report. New York: World Energy Council, p. 27.

Table 11.2 *Some characteristics of the WEC scenarios out to the year 2100*

		A		B		C	
	1990	2050	2100	2050	2100	2050	2100
				Case			
Global energy demand (Gtoe)	8.8	27	42	23	33	15	20
Fossil fuels (% of primary energy)	77	58	40	57	33	58	15
Nuclear (% of primary energy)	5	14	29	15	28	8	11
New renewables							
(% of primary energy)	2	15	24	14	26	20	50
Annual CO_2 emissions							
from fossil fuels (Gt carbon)	6.0	14.9	16.6	12.2	11.7	7.3	2.5
Annual CO_2 emissions from							
fossil fuels (% change on 1990)		152	181	107	98	24	−59

From *Energy for Tomorrow's World: the Realities, the Real Options and the Agenda for Achievement.* WEC Commission Report. New York: World Energy Council, p. 304.

Energy conservation and efficiency in buildings

If we turn lights off in our homes when we do not need them, if we turn down the thermostat by a degree or two so that we are less warm or if we add more insulation to our home, we are conserving or indeed saving energy. But are such actions significant in overall energy terms? Is it realistic to plan for really worthwhile savings in our use of energy?

To illustrate what might be possible, let us consider the efficiency with which energy is currently used. The energy available in the coal, oil, gas, uranium, hydraulic or wind power is *primary energy*. It is either used directly, for instance as heat, or it is transformed into motor power or electricity that in turn provides for many uses. The process of energy conversion, transmission and transformation into its final useful form involves a proportion of the primary energy being wasted. For example, to provide one unit of electrical power at the point of use typically requires about three units of primary energy. An incandescent light bulb is about three per cent efficient in converting primary energy into light energy;

Thermodynamic efficiencies

When considering the efficiency of energy use, it can be important to distinguish between efficiency as defined by the First Law of Thermo-dynamics and efficiency as defined by the Second Law. The second particularly applies when energy is used for heating.

A furnace used to heat a building may deliver to heating the building say eighty per cent of the energy released by full combustion of the fuel, the rest being lost through the pipes, flue, etc. That eighty per cent is a First Law efficiency. An ideal thermodynamic device delivering 100 units of energy as heat to the inside of a building at a temperature of 20 °C from the outside at a temperature of 0 °C would only require just under seven units of energy. So the Second Law efficiency of the furnace is less than six per cent.

Heat pumps (refrigerators or air conditioners working in reverse) are devices that make use of the Second Law and deliver more energy as heat than the electrical energy they use.[12] Although typically their Second Law efficiencies are only about thirty per cent, they are still able to deliver more heat energy than the primary energy required to generate the electricity they use. Because of their comparatively high capital and maintenance costs, however, heat pumps have not been widely used. An example of their substantial use is their contribution to district heating in the city of Uppsala in Sweden where 4 MW of electricity is employed to extract heat from the river and deliver 14 MW of heat energy.

unnecessary use of lighting reduces the overall efficiency to perhaps no more than one per cent.[9] Assessments have been carried out across all energy uses comparing actual energy use with that which would be consumed by ideal devices providing the same services. Although there is some difficulty in defining precisely the performance of such 'ideal' devices (see box below for a discussion of thermodynamic efficiencies), assessments of this kind come up with world average end-use energy efficiencies of the order of three per cent. That sort of figure suggests that there is a large amount of room for improvement in energy efficiency, perhaps by at least threefold.[10] In this section we look at the possibilities for energy saving in buildings; in later sections we shall consider possible savings in transport and in industry.[11]

To be comfortable in buildings we heat them in winter and cool them in summer. In the United States, for instance, about thirty-six per cent

---

Efficiency of appliances

There is large potential for reducing the electricity consumption from appliances used in domestic or commercial buildings. If, in replacing appliances, everyone bought the most efficient available, their total electricity consumption could easily drop by more than half.

Take lighting for instance. One-fifth of all electricity used in the USA goes directly into lighting. This can easily be reduced by the wider use of compact fluorescent light bulbs that are as bright as ordinary light bulbs, but use a quarter of the electricity and last eight times as long before they have to be replaced – with significant economic savings to the user. For instance, a 20-W compact fluorescent bulb (equivalent to a 100-W ordinary incandescent light bulb) costing £5 or less will use about £20 worth of electricity over its lifetime of twelve years. To cover the same period eight ordinary bulbs would be needed costing about £4 but using £100 worth of electricity. The net saving is therefore about £80. A further large increase in the efficiency of lighting will occur when light emitting diodes (LEDs) giving out white light become commonplace.[14] The latest such device that is about one square centimetre in size and consuming only 3-W, produces the same light as a 60-W incandescent bulb.

The average daily electricity use from the appliances in a home (cooker, washing machine, dishwasher, refrigerator, freezer, TVs, lighting) for typical appliances bought in the early 1990s amounts to about 10 kWh per day. If these were replaced by the most efficient available now, electricity use would fall by about two-thirds. The extra cost of the purchase of efficient appliances would soon be recovered in the savings in running cost. Similar calculations can be carried out for other appliances.

Insulation of buildings

About one and a half thousand million people live in cold climates where some heating in buildings is required. In most countries the energy demand of space heating in buildings is far greater than it need be if the buildings were better insulated.

Table 11.3 provides as an example details of two houses, showing that the provision of insulation in the roof, the walls and the windows can easily lead to the energy requirement for space heating being more than halved (from 5.8 kW to 2.65 kW). The cost of the insulation is small and is quickly recovered through the lower energy cost.

If a system for circulating air through the house is also installed, the number of necessary air changes with outside air is less and the total heating requirement further reduced. In this case it is worthwhile to add more insulation to reduce the heating requirement still further.

Table 11.3 *Two assumptions (one poorly insulated, and one moderately well insulated) regarding construction of a detached, two-storey house with ground floor of size 8 m × 8 m, and the accompanying heat losses (U-values express the heat conduction of different components in watts per square metre per °C)*

	Poorly insulated	Moderately well insulated
Walls (150 m² total area)	Brick + cavity + block U-value 0.7	Brick + cavity + block with insulation in cavity of 75 mm thickness: U-value 0.3
Roof (85 m² area)	Uninsulated U-value 2.0	Covered with insulation of thickness 150 mm: U-value 0.2
Floor (64 m²)	Uninsulated U-value 1.0	Includes insulation of thickness 50 mm: U-value 0.3
Windows (12 m² total area)	Single glazing U-value 5.7	Double glazing with low emissivity coating: U-value 2.0
Heat losses (in kW) with 10 °C temperature difference from inside to outside	roof 1.7, walls 1.1, windows 0.7, floor 0.7	roof 0.2, walls 0.45, windows 0.2, floor 0.2
Total heat loss (kilowatts)	4.2	1.05
Add heat (in kilowatts) needed for air changes (1.5 per hour)	1.60	1.60
Total heating required (kilowatts)	5.8	2.65

of the total use of energy is in buildings (about two-thirds of this in electricity), including about twenty per cent for their heating (including water heating) and about three per cent for cooling.[13] Energy demand in the buildings sector grew by about three per cent per year averaged worldwide from 1970 to 1990 and, apart from countries with economies in transition, has been growing during the last decade by about 2.5% per year. How can these trends be reversed?

Substantial energy savings can be made in buildings by improving their insulation (see box above) and by improving the efficiency of appliances (see box on page 279). Many countries, including the UK and the USA, still have relatively poor standards of building insulation compared, for instance, with Scandanavian countries. Improvements in building design to make better use of energy from sunlight can also help (see box on page 301). There are also large possibilities for the improvement of the efficiency of appliances at relatively small cost.

The results of a study in the USA have identified some of the large savings that could be made in the electricity used in buildings. The cost of such action would be less than the cost of the energy savings; overall therefore there would be a substantial net saving (Figure 11.8). The twelve options in Figure 11.8 together cover about forty-five per cent of the amount of electricity used in residential buildings in the USA, which in 1990 was about 1700 TWh or about ten per cent of the USA's total energy use. The four options that provide the largest savings (together adding up to sixty per cent of the savings) are in the areas of commercial lighting, commercial air conditioning, residential appliances and residential space heating. Electricity companies in some parts of the USA are contracting to implement some of these energy-saving measures as an alternative to the installation of new capacity – at significant profit both to the companies and its customers. Similar savings would be possible in other developed countries. Major savings at least as large in percentage terms could also be made in countries with economies in transition and in developing countries if existing plant and equipment were used more efficiently.

Further large savings can be realised when buildings are being planned and designed by the employment of *integrated building design*. When buildings are designed, the designs of the systems for heating, air conditioning and ventilation are commonly carried out separately from the main design. The value of integrated building design is that energy-saving opportunities can be taken up associated with the synergies between many aspects of the overall design and the design (including the sizing) of the systems where much of the energy use occurs. Many examples exist of low energy buildings, where integrated building design has been employed, that consume less than half the energy and are often

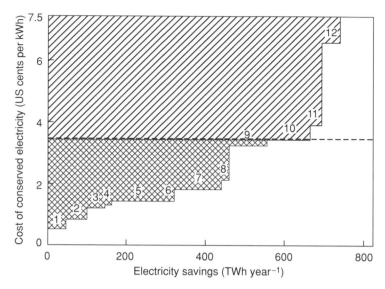

Figure 11.8 Cost of various options (at 1989 prices) for saving electricity in buildings. If the cost of conservation is less than the cost of the electricity saved over the lifetime of the application, a net saving results. The various options are: (1) use of white surfaces to reduce the need for air conditioning; (2) residential lighting; (3) residential water heating; (4) commercial water heating; (5) commercial lighting; (6) commercial cooking; (7) commercial cooling; (8) commercial refrigeration; (9) residential appliances; (10) residential space heating; (11) commercial and industrial space heating; (12) commercial ventilation. The shaded areas are all below 7.5 cents per kWh (the all-sector average electricity price) and 3.5 cents per kWh (typical operating cost of US electricity generation). The total savings in the figure add up to about forty-five per cent of the electricity use.

more acceptable and user-friendly than those that have been designed in more traditional ways.[15] Some recent examples demonstrate the possibility of more radical building designs that aim at Zero Emission (fossil-fuel) Developments (ZED). The box illustrates a recent development in the UK along these lines.

Recent studies[17] suggest that with aggressive implementation of energy-efficient policies and measures, carbon dioxide emissions from buildings in both developed and developing countries could be reduced by about twenty-five per cent in 2010 and about fifty per cent in 2050. If however growth in energy demand in the buildings sector continues to increase at the current rate, these savings in emissions due to increased efficiency will mostly go to compensate for the growth in demand. Further increases in efficiency, however, could be achieved by new technologies that are in prospect such as the use of LEDs for lighting (see box above on appliances). What is clearly necessary also is a switch to non-fossil-fuel energy sources to which we shall be turning in later sections.

Example of a ZED (Zero Emission (fossil-fuel) Development)

BedZED is a mixed development urban village constructed on a brown-field wasteland in the London Borough of Sutton, providing eighty-two dwellings in a mixture of apartments, maisonettes and town houses together with some work/office space and community facilities.[16] The combination of super-insulation, a wind-driven ventilation system incorporating heat recovery and passive solar gain stored within each unit in thermally massive floors and walls reduces the energy needs so that a 135-kW wood-fuelled combined heat and power (CHP) plant is sufficient to meet the village's energy requirements. A 109-kW peak photovoltaic installation provides enough solar electricity to power forty electric cars, some pool, some taxi, some privately owned. The community has the capacity to lead a carbon neutral lifestyle – with all energy for buildings and local transport being supplied from renewable sources.

Energy savings in transport

Transport is responsible for nearly one-quarter of greenhouse gas emissions worldwide. It is also the sector where emissions are growing most rapidly. Road transport accounts for the largest proportion of this, over eighty per cent in industrialised countries; air transport is next at thirteen per cent. Since 1970, the number of motor vehicles in the United States has grown at an average rate of 2.5% per year, in the rest of the world the growth has been almost twice as rapid at nearly five per cent per year (Figure 11.9). The latter trend will continue or increase as there remain very large differences in the degree of car ownership between different countries – for instance about 1.5 persons per car in the USA and a little over 100 persons per car in India and China. The advantages conferred by the motor car, the convenience, freedom and flexibility that it brings, mean that growth in its use is bound to continue. Increased prosperity also brings with it increased movement of freight. In the transport sector the achievement of reductions in carbon dioxide emissions will be particularly challenging.

There are three types of action that can be taken to curb the energy use of motor transport.[18] The first is to increase the efficiency of fuel use. We cannot expect the average car to compete with the vehicle which, in 1992, set a record by covering over 12 000 km on one gallon of petrol – a journey which serves to illustrate how inefficiently we use energy for transport! However, it is estimated that the average fuel consumption of the current fleet of motor cars could be halved through the use of existing technology

Figure 11.9 Growth of world motor vehicle population, 1946–96.

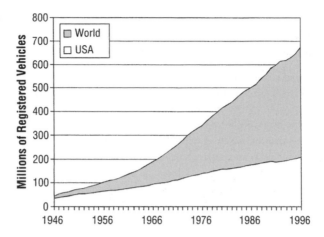

(see box below) – more efficient engines, lightweight construction and low-air-resistance design – while maintaining an adequate performance. The second action is to plan cities and other developments so as to lessen the need for transport and to make personalised transport less necessary – work, leisure and shopping should all be easily accessible by public transport, or by walking or cycling. Such planning needs also to be linked with a recognition of the importance of ensuring that public transport is reliable, convenient, affordable and safe. The third action is to increase the energy efficiency of freight transport by making maximum use of the most energy-efficient forms of freight transport, e.g. rail or water rather than road or air and by eliminating unnecessary journeys.

Air transport is growing even faster than motor transport. Global passenger air travel, as measured in passenger-km, is projected to grow at about five per cent per year over the next decade or more and total aviation fuel use – including passenger, freight and military – is projected to increase by about three per cent per year, the difference being due largely to increased fuel efficiency.[19] Further increases in fuel efficiency are expected but they are unlikely to keep up with the increase in the volume of air transport. A further problem with air transport, as mentioned in Chapter 3 page 52, is that its carbon dioxide emissions are not the only contributor to global warming; increased high cloudiness due to other emissions produce an effect of similar or even greater magnitude. Further research directed at understanding the climatic effects of aircraft and how they may be reduced is urgently required.

Energy savings in industry

Substantial opportunities exist for efficiency savings in industry. The installation of relatively simple control technology often provides large

Technologies for reducing carbon dioxide emissions from motor cars[20]

An important recent development is that of the hybrid electric motor car that combines an internal combustion engine with an electric drive train and battery. The gains in efficiency and therefore fuel economy achieved by hybrid vehicles are typically around fifty per cent. They mainly arise from: (1) use of regenerative braking (with the motor used as a generator and captured electricity stored in the battery), (2) running on the battery and electric traction only when in slow moving or congested traffic, (3) avoiding low efficiency modes of the internal combustion engine and (4) downsizing the internal combustion engine through the use of the motor/battery as a power booster. Both Toyota and Honda have introduced commercially available hybrid models and other manufacturers are not far behind.

Other significant efficiency improvements have come from the use of lower weight structural materials, improvements in low-air-resistance design and the availability of direct injection diesel engines, long used in heavy trucks, for automobiles and light trucks.

Biofuels generated from crops can be employed to fuel motor vehicles thereby avoiding fossil fuel use. For instance, ethanol has been extensively produced from sugarcane in Brazil and from maize in the USA. Biodiesel is also becoming more widely available. However, such fuels can as yet only compete with fossil fuels if they are strongly subsidised.

During the next few years we will begin to see the introduction of vehicles driven by fuel cells (see Figure 11.15) based on hydrogen fuel that can potentially be produced from renewable sources (see page 310). This new technology has the potential to have a large influence on the transport sector.

potential for energy reduction at a substantial net saving in cost. The co-generation of heat and power, which already enables electricity generators to make better use of heat which would otherwise be wasted, is particularly applicable to some industrial plants where large amounts of both heat and power can be required. To take an example: British Sugar in 1992 with an annual turnover of £700 million spent £21 million p.a. on energy. Through low-grade heat recovery, co-generation schemes and better control of heating and lighting, in 1992 the spend on energy per tonne of sugar had been reduced by forty-one per cent from that in 1980.[21] Other potential decreases in carbon dioxide emissions can occur through the recycling of materials, the use of waste as an energy source

Table 11.4 *Estimates of potential global greenhouse gas emission reductions in 2010 and in 2020*

Sector		Historic emission in 1990 (MtC_{eq} yr^{-1})	Historic C_{eq} annual growth rate in 1990–1995(%)	Potential emission reductions in 2010 (MtC_{eq} yr^{-1})	Potential emission reductions in 2020 (MtC_{eq} yr^{-1})	Net direct costs per tonne of carbon avoided
Buildings[a]	CO_2 only	1650	1.0	700–750	1000–1100	Most reductions are available at negative direct costs.
Transport	CO_2 only	1080	2.4	100–300	300–700	Most studies indicate net direct costs less than $US 25/tC but two suggest net direct costs will exceed $US 50/tC.
Industry Energy efficiency Material efficiency	CO_2 only	2300	0.4	300–500 ~200	700–900 ~600	More than half available at net negative direct costs. Costs are uncertain.
Industry	Non-CO_2 gases	170		~100	~100	N_2O emissions reduction costs are $US 0–10/$tC_{eq}$.
Agriculture[b]	CO_2 only Non-CO_2 gases	210 1250–2800	n.a	150–300	350–750	Most reductions will cost between $US 0 and 100/$tC_{eq.}$ with limited opportunities for negative net direct cost options.
Waste[b]	CH_4 only	240	1.0	~200	~200	About 75% of the savings as methane recovery from landfills at net negative direct cost; 25% at a cost of $US 20/$tC_{eq.}$

Montreal Protocol replacement applications	Non-CO_2 gases	0	n.a.	~ 100	n.a.	About half of reductions due to difference in study baseline and SRES baseline values. Remaining half of the reductions available at net direct costs below $US 200/t$C_{eq}$.
Energy supply and conversion[c]	CO_2 only	(1620)	1.5	50–150	350–700	Limited net negative direct cost options exist; many options are available for less than $US 100/t$C_{eq}$.
Total		6900–8400[d]		1900–2600[e]	3600–5050[e]	

[a] Buildings include appliances, buildings, and the building shell.

[b] The range for agriculture is mainly caused by large uncertainties about CH_4, N_2O and soil related emissions of CO_2. Waste is dominated by landfill methane and the other sectors could be estimated with more precision as they are dominated by foss l CO_2.

[c] Included in sector values above. Reductions include electricity generation options only (fuel switching to gas/nuclear, CO_2 capture and storage, improved power station efficiencies, and renewables).

[d] Total includes all sectors for all six gases. It excludes non-energy related sources of CO_2 (cement production, 160 MtC; gas flaring, 60 MtC; and land use change, 600–1400 MtC) and energy used for conversion of fuels in the end-use sector totals (630MtC). Note that forestry emissions and their carbon sink mitigation options are not included.

[e] The baseline SRES scenarios (for six gases included in the Kyoto Protocol) project a range of emissions of 11 500–14 000 MtC_{eq} for 2010 and of 12 000–16 000 MtC_{eq} for 2020. The emissions reduction estimates are most compatible with baseline emissions trends in the SRES-B2 scenario. The potential reductions take into account regular turn-over of capital stock. They are not limited to cost-effective options, but exclude options with costs above $US 100/t$C_{eq}$ (except for Montreal Protocol gases) or options that will not be adopted through the use of generally accepted policies.

Source: Table SPM-1 from Metz, *Climate Change 2001: Mitigation,* Chapter 3. Further information in Moomaw and Moreira *et al.,* Chapter 3 in Metz *et al. Climate Change 2001: Mitigation.*

and through switching to less carbon intensive fuels. Many studies in industrialised countries indicate that savings of thirty per cent or more could be made in the industrial sector at a net saving in overall economic terms.[22]

Given appropriate incentives, substantial savings of carbon dioxide emissions can also be realised in the petrochemical industry that can result in significant savings in cost. For instance, British Petroleum has set up a carbon emissions trading system within the company that encourages the elimination of waste and leaks from their operations and the application of technology to eliminate the venting of methane. In its first three years of operation, 600 million US dollars were saved and carbon emissions reduced to ten per cent below 1990 levels.[23]

There is also room for increased efficiency in large power stations or other installations burning fossil fuels. The efficiency of coal-fired power stations, for instance, has improved from about thirty-two per cent, a typical value of twenty years ago, to about forty-two per cent for a pressurised, fluidised bed combustion plant of today. Gas turbine technology has also improved providing efficiency improvements such that efficiencies approaching sixty per cent are reached by large modern gas-turbine-combined cycle plants. Such improvements are very significant in environmental terms and it is important that means be provided for the latest, most efficient technology to be available and attractive to rapidly industrialising countries such as China and India. Substantial further gains in overall efficiency can be realised by making sure that the large quantities of low-grade heat generated by power stations is not wasted but utilised, for instance in combined heat and power (CHP) schemes. For such co-generation, the efficiencies attainable in the use of the energy from combustion of the fuel are typically around eighty per cent.

Table 11.4 provides summary estimates of the contributions different sectors or industries could make to greenhouse gas reductions by 2010 and 2020 respectively. In total they amount to about half of the emissions from these industries or sectors in 1990. Many of the contributions summarised in the table and most of the proposals described in this section fall into the category of 'no regrets' proposals – mentioned in Chapter 9. In other words, not only do they lead to substantial reductions in greenhouse gas emissions but they are good to do for other reasons – they lead to increased efficiency, cost savings or improvements in performance or comfort. It remains the case, however, that basic energy is generally so cheap, that without both encouragement and incentives, progress with the implementation of many of the proposals will be limited. Some of the policy instruments mentioned later can address this issue.

Capture and storage of carbon dioxide

An alternative to moving away from fossil fuel sources of energy is to prevent the carbon dioxide from fossil fuel burning from entering the atmosphere. This can be done either by removing it from the flue gases in a power station, or the fossil fuel feedstock could, in a gasification plant, be converted through the use of steam,[24] to carbon dioxide and hydrogen. The carbon dioxide is then relatively easy to remove and the hydrogen can be used as a versatile fuel. The latter option will become more attractive when the technical and logistic problems of the large-scale use of hydrogen in fuel cells to generate electricity have been overcome – we mention this again later in the chapter.

Various options are possible for the disposal (or sequestration) of the very large amounts of carbon dioxide which result. For instance, the carbon dioxide can be pumped into spent oil or gas wells, into deep saline reservoirs or into unminable coal seams.[25] Other suggestions have also been made such as pumping it into the deep ocean, but these are more speculative and need careful research and assessment before they can be realistically put forward. In the most favourable circumstances (for instance when power stations are close to oil or gas fields and when the extraction cost is relatively small), the cost of removal, although significant, is only a small fraction of the total energy cost. For instance, in Norway where there is carbon tax of $US 15 per tonne of carbon, a company is finding it economic to pump over one million tonnes per year of carbon dioxide removed from a natural gas stream into storage under the North Sea. In other circumstances estimates of the cost are larger (perhaps up to 100% on top of the energy cost) – the cost of extraction being generally larger than the cost of storage.

The technology of carbon capture and storage could enable continuing use of fossil fuels without the deleterious effects of carbon dioxide emissions. The global potential for underground carbon dioxide storage is large. For instance, it has been estimated that over 200 Gt of carbon as carbon dioxide could be stored in geological reservoirs in north west Europe alone. How much it is used will depend more on the cost than on the availability of suitable storage sites.

Renewable energy

To put our energy use in context it is interesting to realise that the energy incident on the Earth from the Sun amounts to about 180 thousand million million watts (or 180 000 terawatts, 1 TW $= 10^{12}$ W). This is about 14 000 times the world's average energy use of about 13 million million watts (13 TW). As much energy arrives at the Earth from the Sun in forty minutes as we use in a whole year. So, providing we can harness it satisfactorily and economically, there is plenty of renewable energy coming in from the Sun to provide for all the demands human society can conceivably make.

There are many ways in which solar energy is converted into forms that we can use; it is interesting to look at the efficiencies of these conversions. If the solar energy is concentrated, by mirrors for instance, almost all of it can be made available as heat energy. Between one and two per cent of solar energy is converted through atmospheric circulation into wind energy, which although concentrated in windy places is still distributed through the whole atmosphere. About twenty per cent of solar energy is used in evaporating water from the Earth's surface which eventually falls as precipitation, giving the possibility of hydropower. Living material turns sunlight into energy through photosynthesis with an efficiency of around one per cent for the best crops. Finally, photovoltaic (PV) cells convert sunlight into electricity with an efficiency that for the best modern cells can be over twenty per cent.

Around the year 1900, very early in the production of commercial electricity, water power was an obvious source and from the beginning made an important contribution. Hydroelectric schemes now supply about six per cent of the world's commercial energy. Other renewable sources of commercial energy, however, have been dependent on recent technology for their implementation. In 1990, only about two per cent of the world's commercial energy came from renewable sources other than large hydro[26] (these are often collectively known as 'new renewables'). Of this two per cent (Table 11.5), about three-quarters was from 'modern' biomass (called 'modern' when it contributes to commercial energy to distinguish it from traditional biomass), the other 0.5% being shared between solar, wind energy, geothermal and small hydro sources.

Returning to commercial energy generation, in order to put renewable sources into context, it is useful to inspect the detailed projection of the WEC (Table 11.5) for the contributions from different 'new renewable' sources which make up the twelve per cent of total energy supply in the year 2020 assumed for the WEC scenario C. The main growth expected is in energy from 'modern' biomass and from solar and wind energy sources. Table 11.6 provides detailed summary information about the status and cost of different renewable energy sources.

In the following paragraphs, the main renewable sources are described in turn and their possibilities for growth considered.[27] Most of them are employed for the production of electricity through mechanical means (for hydro and wind power), through heat engines (for biomass and solar thermal) and through direct conversion from sunlight (solar PV). In the case of biomass, liquid or gaseous fuels can also be produced.

Table 11.5 *Contributions to world energy supply (in millions of tonnes of oil equivalent) from renewable sources in 1990 and as assumed under the WEC scenario C in 2020*

	1990		2020	
	Mtoe	% of world energy	Mtoe	% of world energy
'Modern' biomass	121	1.4	561	5.0
Solar	12	0.1	355	3.1
Wind	1	0.0	215	1.9
Geothermal	12	0.1	91	0.8
'Small' hydro	18	0.2	69	0.6
Tides, waves and tidal streams	0	0.0	54	0.5
Total (new renewable sources)	164	1.8	1345	11.9
'Large' hydro	465	5.3	661	5.8
'Traditional' biomass	930	10.6	1060	9.3
Total (all renewables)	1559	17.7	3066	27.0

From *Energy for Tomorrow's World: the Realities, the Real Options and The Agenda for Achievement*. WEC Commission Report. 1993. London: World Energy Council, p. 94.

Hydro-power

Hydro-power, the oldest form of renewable energy, is well established and is competitive economically with electricity generated by other means. Some hydroelectric schemes are extremely large. The world's largest, the Three Gorges project on the Yangtze river in China, when completed will generate about 20 000 MW of electricity. Two other large schemes, each of over 10 000 MW capacity, are in South America at Guri in Venezuela and at Itaipu on the borders of Brazil and Paraguay. It is estimated[28] that there is potential for further exploitation of hydroelectric capacity to three or four times the amount that has currently been developed, much of this undeveloped potential being in the former Soviet Union and in developing countries. Large schemes, however, can have significant social impact (such as the movement of population from the reservoir site), environmental consequences (for example, loss of land, of species and of sedimentation to the lower reaches of the river), and problems of their own such as silting up, which have to be thoroughly addressed before they can be undertaken.

Table 11.6 *Current status and potential future costs of renewable energy technologies. The costs can be compared with typical current costs of fossil fuel supplied energy of 3 to 6 ¢US kWh^{-1}*

Technology	Increase in installed capacity in last five years (% p.a.)	Operating capacity – end year 1998 GW^a	Capacity factor (%)	Energy production, 1998 TWh^a	Turnkey investment costs ($US kW^{-1})	Energy cost in year 1999 (¢US kWh^{-1})	Potential future energy cost (¢US kWh^{-1})
Hydro electricity							
Large	~ 2	640 (e)	35–60	2510 (e)	1000–3500	3–8	3–8
Small	~ 3	23 (e)	20–70	90 (e)	1200–3000	5–10	4–10
Biomass energy							
Electricity	~ 3	40 (e)	25–80	160 (e)	900–3000	5–15	4–10
Heat^b	~ 3	>200 (th)	25–80	>700 (th)	250–750	1–5	1–5
Ethanol	~ 3	18 × 10^9 litres		120 (= 420 PJ)		3–9	2–4
Wind electricity	~ 30	10 (e)	20–30	18 (e)	1100–1700	5–13	3–10
Solar PV electricity	~ 30	500 (e)	8–20	0.5 (e)	5000–10 000	25–125	5 or 6–25
Solar thermal electricity	~ 5	400 (e)	20–35	1.0 (e)	3000–4000	12–18	4–10
Low temperature solar heat	~ 8	18 (th)	8–20	14 (th)	500–1700	3–20	2 or 3–10
Geothermal energy							
Electricity	~ 4	8 (e)	45–90	46 (e)	800–3000	2–10	1 or 2–8
Heat	~ 6	11 (th)	20–70	40 (th)	200–2000	0.5–5	0.5–5
Marine							
Tidal	0	300 (e)	20–30	0.6 (e)	1700–2500	8–15	8–15
Current^c			25–35		2000–3000	8–15	5–7
Wave^c			20–35		1500–3000	8–20	

^{a} (e) refers to electrical energy and (th) to thermal energy.

^{b} Heat embodied in steam (or hot water in district heating) often produced by CHP using various forms of biomass.

^{c} Still in experimental phase.

Source: Goldemberg, J. (ed.) *World Energy Assessment: Energy and the Challange of Sustainability.* Table 4 from Overview.

But hydroelectric schemes do not have to be large; Table 11.5 distinguishes between large and small hydroelectric sources. Many units exist generating a few kilowatts only that may supply one farm or a small village. The attractiveness of small schemes is that they provide a locally based supply at modest cost. Substantial growth in 'small hydro' has occurred during the last decade or so, from around 20 000 MW in 1990 to about 40 000 MW in 2000.[29] Installations in China account for about half of this latter figure where the growth has been about twice as rapid as in the rest of the world. Many more possibilities still exist for the exploitation of the potential of small rivers and streams in many parts of the world.

An important facility provided by some hydro schemes is that of pumped storage. Using surplus electricity available in off-peak hours, water can be pumped from a lower reservoir to a higher one. Then, at other times, by reversing the process, electricity can be generated to meet periods of peak demand. The efficiency of conversion can be as high as eighty per cent and the response time a few seconds, so reducing the need to keep other generating capacity in reserve. In 1990 about 75 000 MW of pumped storage capacity was available worldwide with a further 25 000 MW under construction.[30]

Biomass as fuel[31]

Second in current importance as a renewable energy source is the use of biomass as a fuel. The annual global primary production of biomass of all kinds expressed in energy units is about 4500 EJ ($=$ 107 Gtoe). About one per cent of this is currently turned into energy mostly in developing countries – we have labelled it 'traditional biomass'. It has been estimated that about six per cent of the total could become available for energy crops taking into account the economics of production and the availability of suitable land.[32] The energy so generated would represent about seventy-five per cent of current world energy consumption – a substantial contribution to global energy needs. It is a genuinely renewable resource in that the carbon dioxide which is emitted when the biomass is burnt is turned back into carbon, through the process of photosynthesis, in the renewed biomass when it is grown again. The word biomass not only covers crops of all kinds but also domestic, industrial and agricultural dry waste material and wet waste material, all of which can be used as fuel for heating and to power electricity generators; some are also appropriate to use for the manufacture of liquid or gaseous fuels. Since biomass is widely distributed, it is particularly appropriate as a distributed energy source suitable for rural areas.

In much of the developing world, most of the population live in areas where there is no access to modern or on-grid energy. They rely on 'traditional biomass' (fuelwood, dung, rice husks and other forms of biofuels) to satisfy their needs for cooking and heating. Ten per cent or so of world energy originates from these sources supplying over one-third of the world's population. Although these sources are renewable, it is still important that they are employed efficiently, and a great deal of room for increased efficiency exists. For instance, a large proportion of each day is often spent in collecting firewood especially by the women, increasingly far afield from their homes.

The burning of biomass in homes causes serious health problems and has been identified by the World Health Organization as one of the most serious causes of illness and mortality especially amongst children.[33] For instance, much cooking is still carried out on open fires with their associated indoor pollution and where only about five per cent of the heat reaches the inside of the cooking pot. The introduction of a simple stove can increase this to twenty per cent or with a little elaboration to fifty per cent.[34] An urgent need exists for the large-scale provision of stoves using simple technology that is sustainable – although there is often considerable consumer resistance to their introduction. Other means of reducing fuelwood demand are to encourage alternatives such as the use of fuel from crop wastes, of methane from sewage or other waste material or of solar cookers (mentioned again later on). From the existing consumption of 'traditional biomass' there is the potential to produce sustainable 'modern' energy services with much greater efficiency and much less pollution for the two billion or so people who currently rely on this basic energy source. A particular challenge is to set up appropriate management and infrastructure for the provision of these services in rural areas in developing countries (see box below).

Firstly, consider the use of waste.[37] There is considerable public awareness of the vast amount of waste produced in modern society. The UK, for example, produces each year somewhat over thirty million tonnes of domestic solid waste, or about half a tonne for every citizen; this is a typical value for a country in the developed world. Even with major programmes for recycling some of it, large quantities would still remain. If it were all incinerated for power generation (modern technology enables this to be done with negligible air pollution) about 1.7 GW could be generated, about five per cent of the UK's electricity requirement.[38] Uppsala in Sweden is an example of a city with a comprehensive district heating system, for which, before 1980, over ninety per cent of the energy was provided from oil. A decision was then made to move to renewable energy and by 1993 energy from waste incineration and from

Biomass projects in rural areas in the developing world

In much of the developing world, most of the population live in areas where there is little or no access to electricity or modern energy services. There is large potential for creating local biomass projects to provide such services. The use of sugar cane as biomass provides one example of what is beginning to be provided; three other examples are given of pilot projects in different countries,[35] all of which could be replicated many times.

Sugar cane as biomass

A sugar cane factory produces many different byproducts that can be efficiently employed as sources of energy – either for biofuels or for electricity production.

Sugar cane production yields two kinds of biomass fuel suitable for gasification (Figure 11.10), known as bagasse and barbojo. Bagasse is the residue from crushing the cane and is thus available during the milling season; barbojo consists of the tops and leaves of the cane plant, which could be stored for use after the milling season. It has been estimated that, using these sugar cane sources, within thirty years or so, the eighty sugar-cane producing countries in the developing world could generate two-thirds of their current electricity needs at a price competitive with fossil fuel energy sources.[36]

Rural power production, India

Decentralised Energy Systems India Private Limited are piloting the first independent power projects of around 100 kW capacity in rural India owned and operated by village community co-operatives. An example is a small co-operative in Baharwari, Bihar State, where a biomass gasification power plant is used as a source of electricity for local enterprises, for instance for pumping water in the dry season. Local income is thereby generated that enables villagers to expand their micro-industries and create more jobs – all of which in turn increases the ability of people to pay for improved energy services. A 'mutuality of interest' is created between biomass fuel suppliers, electricity users and plant operators.

Integrated biogas systems, Yunnan, China

The South-North Institute for Sustainable Development has introduced a novel integrated biogas system in the Baima Snow Mountains Nature Reserve, Yunnan Province. The system links a biogas digester, pigsty, toilet and greenhouse. The biogas generated is used for cooking and replaces the burning of natural firewood, the 'greenhouse' pigsty increases the efficiency of pig-raising, the toilet improves rural environmental hygiene, and vegetables and fruits planted in the greenhouse increase the income of local inhabitants. Manure and other organic waste from the pigsty and toilet are used as the raw material for biogas generation that delivers about 10 kWh per day of useful energy. The operation of fifty such systems has considerably reduced local firewood consumption.

Biomass power generation and coconut oil pressing, the Philippines

The Community Power Corporation (CPC) has developed a modular biopower unit that can run on waste residue or biomass crops and can enable village-level production of coconut oil. CPC and local partners are using the modular biopower unit fuelled by the waste coconut shells to provide electricity to a low-cost mini-coconut-oil-mill (developed by the Philippines Coconut Authority and the University of Philippines), sixteen of which are now operating in various Philippine villages. Furthermore, the biopower unit generates waste heat which is essential to drying the coconuts prior to pressing.

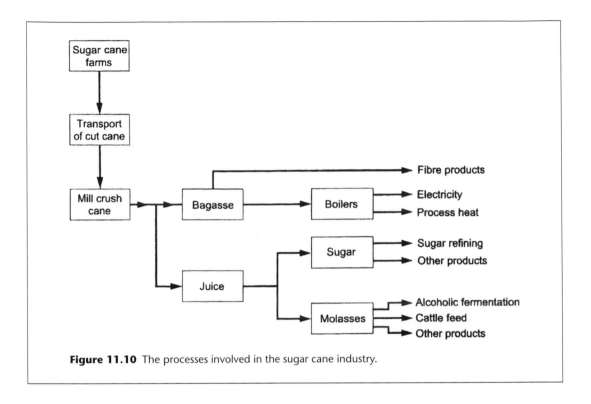

Figure 11.10 The processes involved in the sugar cane industry.

other biomass fuel sources provided nearly eighty per cent of what is required for the city's heating.

But what about the greenhouse gas generation from waste incineration? Carbon dioxide is of course produced from it, which contributes to the greenhouse effect. However, the alternative method of disposal is landfill (most of the waste in the UK currently is disposed of that way). Decay of the waste over time produces carbon dioxide and methane in roughly equal quantities. Some of the methane can be collected and used as a fuel for power generation. However, only a fraction of it can be captured; the rest leaks away. Because methane is a much more effective greenhouse gas, molecule for molecule, than carbon dioxide, the leaked methane makes a substantial contribution to the greenhouse effect. Detailed calculations show that if all UK domestic waste were incinerated for power generation rather than landfilled, the net saving per year in greenhouse gas emissions would be equivalent to about ten million tonnes of carbon as carbon dioxide.[39] Since this is about five per cent of the total UK greenhouse gas emissions, we can infer that power generation from waste could be a significant contribution to the reduction in overall emissions.

Other wastes resulting from human or agricultural activity are wet wastes such as sewage sludge and farm slurries and manures. Bacterial fermentation in the absence of oxygen (anaerobic digestion) of these wastes produces biogas, which is mostly methane and which can be used as a fuel to produce energy. There is room for an increasing contribution from these sources. If the potential for power generation from agricultural and industrial waste was taken into account, the savings in emissions arising from domestic waste already mentioned could be approximately doubled.

Turning now to the use of crops as a fuel, the potential is large. Many different crops can be employed as biomass for energy production. In Brazil, for instance, since the 1970s large plantations of sugar cane have produced alcohol for use as a fuel mainly in transport, generating, incidentally, much less local pollution than petrol or diesel fuel from fossil sources. A lot of potential has been recognised for the sugar cane industry to produce both sugar and energy together with other byproducts as well (see box above). Biomass from wood plantations on agricultural land no longer needed for food crops features as an important future source in Sweden's energy plans[10]; the most efficient use of the biomass is first to turn it into biogas and then burn it in a gas turbine to produce electricity. For the UK, trials indicate that the most promising option is willow and poplar grown in coppices.[41]

Because of the low efficiency of conversion of solar energy to biomass, the amount of land required for significant energy production by this means is large – and it is important that land is not taken over that is required for food production. However, there is in principle no shortage of land for this purpose. Plenty of suitable crops are available which could be grown on land only marginally useful for agriculture. In many developing countries biomass plantations can provide suitable fuel for local electricity generation more competitively than other means of generation.

The growth of the use of biomass for energy in industrialised countries (Table 11.6) is substantial but is limited by the cost differential that exists between energy from biomass and that generated from fossil fuels (Table 11.6). This problem is addressed later in the chapter.

Wind energy

Energy from the wind is not new. Two hundred years ago windmills were a common feature of the European landscape; for example, in 1800 there were over 10 000 working windmills in Britain. During the past few years they have again become familiar on the skyline especially in countries in western Europe (for instance, Denmark, Great Britain and Spain) and in western North America. Slim, tall, sleek objects silhouetted against the sky, they do not have the rustic elegance of the old windmills, but they are much more efficient. A typical wind energy generator will have a two- or three-bladed propeller about 50 m in diameter and a rate of power generation in a wind speed of 12 ms^{-1} (43 kmh^{-1}, 27 mph or Beaufort Force 6), of about 700 kW. On a site with an average wind speed of about 7.5 ms^{-1} (an average value for exposed places in many western regions of Europe) it will generate an average power of about 250 kW. The generators are often sited close to each other in wind farms that may include several dozen such devices.

From the point of view of the electricity generating companies the difficulty with the generation of electrical power from wind is that it is intermittent. There are substantial periods with no generation at all. The generating companies can cope with this in the context of a national electricity grid that pools electrical power from different sources providing that the proportion from intermittent sources is not too large.[42] Some public concern about wind farms arises because of loss of visual amenity. The use of more off-shore sites may therefore be more generally acceptable than too much concentration in windy sites on-shore.

Rapid growth has occurred in many countries in the installation of wind generators over the past decade – a growth that continues unabated. But most of the growth has been for electricity generation. Over 30 GW peak operating capacity has now (2002) been built worldwide. With this large growth, economies of scale have brought down the cost of the electricity generated so that it is approaching the cost of electricity generated from fossil fuels (Table 11.6). Because the power generated from the wind depends on the cube of the wind speed (a wind speed of 12.5 ms^{-1} is twice as effective as one of 10 ms^{-1}) it makes sense to build wind farms on the windiest sites available. Some of the windiest sites available are to be found in western Europe where some of the most rapid growth in wind generation has occurred. In Denmark for instance, twenty per cent of electricity is now generated by wind – increasingly being built offshore.[43] Eventually, it is envisaged the proportion could rise to forty to fifty per cent. Similar estimates of the eventual resource are being made for the UK where rapid expansion, again especially of off-shore wind energy generation, is envisaged.[44] Developing countries are also making increased use of wind energy. For instance, it has been estimated that India could be generating up to 10 GW of electrical power (about a quarter of current needs) from wind by 2030.[45] With the growth that is occurring, the proportion of global energy needs supplied from wind energy could be substantially greater than that envisaged by the projection in Table 11.5.

Wind energy is also particularly suitable for the generation of electricity at isolated sites to which the transmission costs of electricity from other sources would be unacceptable. Because of the wind's intermittency, some storage of electricity or some back-up means of generation

Wind power on Fair Isle

A good example[46] of a site where wind power has been put to good effect is Fair Isle, an isolated island in the North Sea north of the Scottish mainland. Until recently, the population of seventy depended on coal and oil for heat, petrol for vehicles and diesel for electricity generation. A 50-kW wind generator was installed in 1982 to generate electricity from the persistent strong winds of average speed over 8 ms^{-1} (29 kmh^{-1} or 18 mph). The electricity is available for a wide variety of purposes; at a relatively high price for lighting and electronic devices and at a lower price controlled amounts are available (wind permitting) for comfort heat and water heating. At the frequent periods of excessive wind further heat is available for heating glasshouses and a small swimming pool. Electronic control coupled with rapid switching enables loads to be matched to the available supply. An electric vehicle has been charged from the system to illustrate a further use for the energy.

With the installation of the wind generator, which now supplies over ninety per cent of the island's electricity, electricity consumption has risen about fourfold and the average electricity costs have fallen from 13 pkWh^{-1} to 4 pkWh^{-1}. A second wind turbine of 100 kW capacity was installed in 1996/7 to meet increasing demand and to improve wind capture.

has to be provided as well. The installation on Fair Isle (see box) is a good example of an efficient and versatile system. Small wind turbines also provide an ideal means for charging batteries in isolated locations; for instance, about 100 000 are in use by Mongolian herdsmen. Wind energy is often also an ideal source for water pumps – one million small wind turbines are used for this purpose worldwide.[47]

In the longer term it can be envisaged that wind generation could expand into areas remote from direct electrical connection providing an effective means for energy storage (for instance, using hydrogen; more of that possibility later in the chapter) is developed.

Energy from the Sun

The simplest way of making use of energy from the Sun is to turn it into heat. A black surface directly facing full sunlight can absorb about 1 kW for each square metre of surface. In countries with a high incidence of sunshine it is an effective and cheap means of providing domestic hot water, which is extensively employed in countries such as Australia, Israel, Japan and the southern states of the USA (see box above).

Solar water heating

The essential components of a solar water heater (Figure 11.11) are a set of tubes in which the water flows embedded in a black plate insulated from behind and covered with a glass plate on the side facing the Sun. A storage tank for the hot water is also required. A more efficient (though more expensive) design is to surround the black tubes with a vacuum to provide more complete insulation. Ten million households worldwide have solar hot water systems.[48]

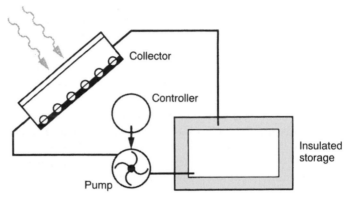

Figure 11.11 Design of a solar water heater: a solar collector connected to a storage tank through a circulating pump. Alternatively, if the storage is above the collector, the hot water will collect through gravity flow.

In tropical countries, a solar cooking stove can provide an efficient alternative to stoves burning wood and other traditional fuels. Thermal energy from the Sun can also be employed effectively in buildings (it is called passive solar design), in order to provide a modest boost towards heating the building in winter and, more importantly, to provide for a greater degree of comfort and a more pleasant environment (see box).

Solar heat can also be employed to provide heating to produce steam for the generation of electricity. To produce significant quantities of steam, the solar energy has to be concentrated by using mirrors. One arrangement employs trough-shaped mirrors aligned east-west which focus the Sun on to an insulated black absorbing tube running the length of the mirror. A number of such installations have been built, particularly in the USA, where solar thermal installations provide over 350 MW of commercial electricity. The high capital cost of such installations,

Solar energy in building design

All buildings benefit from unplanned gains of solar energy through windows and, to a lesser extent,
through the warming of walls and roofs. This is
called 'passive solar gain'; for a typical house in
the UK it will contribute about fifteen per cent of
the annual space heating requirements. With 'passive solar design' this can relatively easily and in-
expensively be increased to around thirty per cent,
while increasing the overall degree of comfort and
amenity. The main features of such design are to
place, so far as is possible, the principal living
rooms with their large windows on the south side of
the house, with the cooler areas such as corridors,
stairs, cupboards and garages with the minimum of
window area arranged to provide a buffer on the
north side. Conservatories can also be strategically
placed to trap some solar heat in the winter.

The wall of a building can be designed specifically to act as a passive solar collector, in which
case it is known as 'solar wall' (Figure 11.12).[49] Its
construction enables sunlight, after passing through
an insulating layer, to heat the surface of a wall of
heavy building blocks that retain the heat and slowly
conduct it into the building. The insulating layer, although allowing sunlight to pass through, prevents
thermal radiation from passing out. A retractable
reflective blind can be placed in front of the insulation at night or during the summer when heating of
the building is not required. A set of student residences for 376 students at Strathclyde University in
Glasgow in southwest Scotland has been built with
a 'solar wall' on its south-facing side. Even under
the comparatively unfavourable conditions during
winter in Glasgow (the average duration of bright
sunshine in January is only just over one hour per
day) there is a significant net gain of heat through
the wall to the building

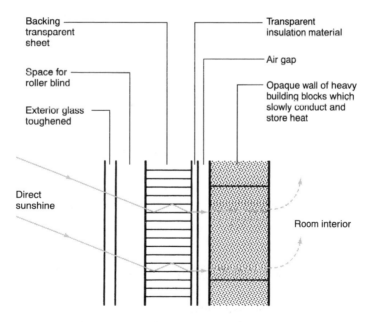

Figure 11.12 Construction of a 'solar wall'. The insulation material is about 100 mm thick and consists of open honeycomb channels of transparent polycarbonate material.

The photovoltaic solar cell

The silicon photovoltaic (PV) solar cell consists of a thin slice of silicon into which appropriate impurities have been introduced to create what is known as a p–n junction. The most efficient cells are sophisticated constructions using crystalline silicon as the basic material; they possess efficiencies for the conversion of solar energy into electricity typically of fifteen to twenty per cent; experimental cells have been produced with efficiencies well over twenty per cent.[51] Single crystal silicon is less convenient for mass production than amorphous silicon (for which the conversion efficiency is around ten per cent), which can be deposited in a continuous process on to thin films.[52] Other alloys (such as cadmium telluride and copper indium selenide) with similar photovoltaic properties can also be de-

posited in this way and, because they have higher efficiencies than amorphous silicon, are likely to compete with silicon for the thin film market.[53] However, since typically about half the cost of a solar PV installation is installation cost, the high efficiency of single crystal silicon, which means a smaller size, remains an important factor.

Cost is of critical importance if PV solar cells are going to make a significant contribution to energy supply. This has been coming down rapidly. More efficient methods and larger scale production are bringing the cost of solar electricity down to levels where it can compete with other sources. Projections up to the year 2020 of the likely cost of generating electricity from PV sources are shown in Figure 11.13.

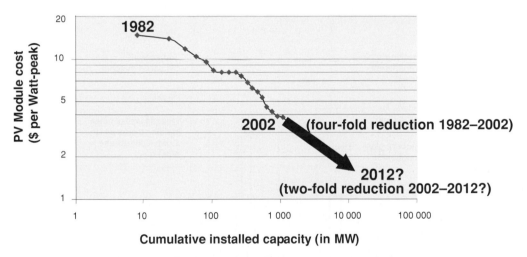

Figure 11.13 The falling cost of PV modules over the last twenty years and as projected for the next twenty years. Note the twenty per cent reduction in cost for every doubling of installed capacity.

however, assuming a reasonable pay-back period, translates into an electricity cost which, at the moment, is at least three times that from most conventional sources. Generating plants which incorporate integrated solar and fossil fuel heat sources in combined cycle operation are currently under development that promise significantly lower costs.[50]

Sunlight can be converted directly into electricity by means of photo-voltaic (PV) solar cells (see box on page 302). Solar panels on spacecraft have provided electrical power for spacecraft from the earliest days of space research nearly fifty years ago. They now appear in a variety of ways in everyday life; for instance, as power sources for small calcu-lators or watches. Their efficiency for conversion of solar energy into electrical energy is now generally between just under ten per cent and twenty per cent. A panel of cells of area one square metre facing full sunlight will therefore deliver between 100 and 200 W of electrical power. A cost-effective way of mounting PV modules is on the surface of manufactured items or built structures rather than as free-standing arrays. In the fast-growing building-integrated-PV (BIPV) sector, the PV façade replaces and avoids the cost of conventional cladding. In-stalled on rooftops in cities, they provide a way for city dwellers to contribute renewably to their energy needs. Japan has done the most to encourage rooftop solar installations and by 2000 had installed 320 MW capacity. The USA and Germany follow as countries with large rooftop programmes, the USA with a target by 2010 of one million roofs and Germany with programme for 100 000. The cost of energy from solar cells has reduced dramatically over the past twenty years (see box below); so much so that they can now be employed for a wide range of appli-cations and can also begin to contribute to the large-scale generation of electricity.

Small PV installations are also suitable, especially for developing countries, to provide local sources of electricity in rural areas. About a third of the world's population have no access to electricity from a central source. Their predominant need is for small amounts of power for light-ing, for radio and television, for refrigerators (for example, for vaccines at a health clinic) and for pumping water. The cost of PV installations for these purposes is now competitive with other means of generation (such as diesel units). Over the twenty years to the year 2000, approximately 1.1 million 'Solar Home Systems'and 'Solar Lanterns' had been in-stalled in Asia, Africa and South American countries.[54] Solar Home Systems provide typically 15–75 W from a solar array (Figure 11.14) and cost in the range of $US 200–1200. Smaller 'solar lanterns' (typi-cally 10–20 W) provide lighting only. Larger installations are required for public buildings, although they need not be that much larger. Many small hospitals can benefit from an electrical power source as small as 1–2 kW. For instance, by 1995, seventy small hospitals in Sri Lanka, through assistance from the Australian government, had installed 1.3-kW solar arrays, backed up by 2200 amp-hour batteries, to provide for light-ing, refrigeration for vaccines, autoclave sterilisation, pumping for hot water (produced through a solar-thermal system) and radio. Over

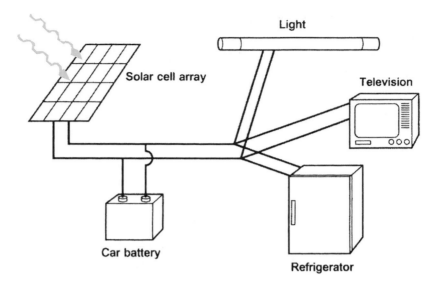

Figure 11.14 A simple 'solar home system' now being marketed in many countries in Africa, Asia and South America for a cost of a few hundred US$. An array of thirty-six solar cells, covering an area of 60 cm × 60 cm, provides around 40 W of peak power. This is sufficient to charge a car battery that can power up to three 9-W fluorescent lights and three hours of radio and one hour of television per day. With more restricted use of these devices or with a larger solar array, a small refrigerator can be added to the system.

20 000 water pumps are now powered by solar PV and thousands of communities receive drinking water from solar-PV-powered purifiers/pumps. The potential for further growth and development of solar systems is clearly very large. For instance, mini-electrical grids powered by a combination of solar PV, wind, biomass and diesel are beginning to emerge especially in the remoter parts of China and India.[55]

The total installed world capacity of PV grew from about 500 MW peak in 1998 to about 1500 MW peak in 2002, an increase of about thirty per cent per year. With that rate of continued growth it should be possible to more than meet the projected contribution by PV solar cells to world energy supply in WEC scenario C of at least 150 GW[56] by the year 2020 (Table 11.5). In the short term, increased development of local installations is likely to have priority; later, with the expectation of a significant cost reduction (Figure 11.13), penetration into large-scale electricity generation will become more possible. Eventually, because of its simplicity, convenience and cleanliness, it is expected that electricity from solar PV sources will become one of the largest – if not the largest – of the world's energy sources.

Other renewable energies

We have so far covered the renewable energy sources for which there is potential for growth on a scale that can make a substantial contribution to overall world energy demand. We should also mention briefly other renewable energy technologies which contribute to global energy production and which are of particular importance in certain regions, namely geothermal energy from deep in the ground and energies from the tides, currents or waves in the ocean.

The presence of geothermal energy from deep down in the Earth's crust makes itself apparent in volcanic eruptions and less dramatically in geysers and hot springs. The energy available in favourable locations may be employed directly for heating purposes or for generating electrical power. Although very important in particular places, for instance in Iceland, it is currently only a small contributor (about 0.3%) to total world energy; its contribution could rise to the order of one per cent during the next few decades (Table 11.5).

Large amounts of energy are in principle available in movements of the ocean; but in general they are not easy to exploit. Tidal energy is the only one currently contributing significantly to commercial energy production. The largest tidal energy installation is a barrage across the estuary at La Rance in France; the flow from the barrage is directed through turbines as the tide ebbs so generating electricity with a capacity of up to 240 MW. Several estuaries in the world have been extensively studied as potential sites for tidal energy installations. The Severn Estuary in the UK, for instance, possesses one of the largest tidal ranges in the world and has the potential to generate a peak power of over 8000 MW or about six per cent of the total UK electricity demand. Although the long-term cost of the electricity generated from the largest schemes could be competitive, the main deterrents to such schemes are the high capital up-front cost and the significant environmental impacts that can be associated with them. More recent proposals, rather than making use of estuaries, have been based on the construction of tidal 'lagoons' in suitable shallow regions off shore where there is a large tidal range.[57] Turbines in the lagoon walls generate electricity as water flows in and out of the lagoons. Some of the environmental and economic problems of barrages built in estuaries are therefore avoided.

The energy in currents (including tidal streams) in the ocean can be exploited in much the same way as wind energy from the atmosphere is harnessed. Although the speeds in water are lower than that of the wind, the greater density of sea water results in higher energy densities and requires smaller turbine diameters. Substantial energy is also present in ocean waves and a number of ingenious devices have been designed to turn this into electrical energy.[58] The waters around the British Isles contain some of the best opportunities to exploit all forms of tidal and wave energy. Because of the hostile ocean environment, early exploitation of some of these may be comparatively costly, but the size of the potential resource is large. What is urgently needed is the necessary research and development.

The support and financing of renewable energy

Renewable energy on the scale required to meet any stabilisation scenario for carbon dioxide (for instance WEC scenario C – see Table 11.5) will only be realised if it is competitive in cost with energy from other sources. Table 11.6 provides a summary of the status and costs, present and future, of the main sources of renewable energy. Under some circumstances renewable energy sources are already competitive in cost, for instance in providing local sources of energy where the cost of transporting electricity or other fuel would be significant; some examples of this (such as Fair Isle in Scotland – see box above) have been given. However, when there is direct competition with fossil fuel energy from oil and gas, as Table 11.6 shows, many renewable energies at the present compete only marginally. In due course, as easily recoverable oil and gas reserves begin to run out, those fuels will become more expensive enabling renewable sources to compete more easily.[59] That is some decades away and, since estimates of recoverable fossil fuel reserves have always tended to be low, it may be well into the second half of the twenty-first century before any substantial limitation in oil and gas resources occurs. Before then in order that renewables begin to displace fossil fuels to the extent required, appropriate financial incentives must be introduced to bring about the change.

As we saw in Chapter 9, the basis of such incentives would be the principle that the polluter should pay by the allocation of an environmental cost to carbon dioxide emissions. There are three main ways in which this can be done. Firstly, through a direct subsidy being provided by governments to renewable energy. Secondly, through the imposition of a carbon tax. Suppose, for instance, that through taxes or levies an additional cost of between $US 50 and 100 per tonne of carbon (figures mentioned in the context of environmental costs towards the end of Chapter 9) were to be associated with carbon dioxide emissions, between 0.5 and 2.5 cents per kWh would be added to the price of electricity from fossil fuel sources (Table 11.6) – which could bring some renewables (for instance, biomass and wind energy) into competition with them.[60] It is interesting to note that in many countries substantial subsidies are attached to energy – worldwide they amount on average to the equivalent of about $US 40 per tonne of carbon.[61] A start with incentives would therefore be made if subsidies were removed from energy generated from fossil fuel sources.

A third way of introducing an environmental cost to fossil fuel energy is through tradeable permits in carbon dioxide emissions, as are being introduced under arrangements for the management of the Kyoto Protocol (Chapter 10, page 248). These control the total amount of carbon dioxide that a country or region may emit while providing the means for industries to trade permits for their allowable emissions within the overall total.

These fiscal measures are relatively easy to apply in the electricity sector. Electricity, however, only accounts for about one-third of the world's primary energy use. They also need to be applied to solid, liquid or gaseous fuels that are used for heating, industry and transport. It has already been mentioned that, currently, liquid fuels such as ethanol derived from biomass are about twice as expensive as those derived from oil. Although there is an expectation that the processing of biomass will become more efficient[62] – the rapid development of technologies in bioengineering will help – it is unlikely that in the short term the substitution of biomass-derived fuels will occur on a significant scale without the application of appropriate financial incentives.

There is a further crucial area where incentives are also required if renewable energy sources are going to come on stream sufficiently rapidly to meet the need. That is the area of research and development (R & D) – the latter is especially vital. Government R & D, averaged worldwide, currently runs at about ten billion US dollars per year or about one per cent of worldwide capital investment in the energy industry of about one trillion (million million) dollars per year (about three per cent of GWP). On average, in developed countries it has fallen by about a factor of two since the mid 1980s. In some countries the fall has been much greater. This is particularly true in the UK where government-sponsored energy R & D fell by about a factor of ten from the mid 1980s to 1998 when, in proportion to GDP, it was only one-fifth of that in the USA and one-seventeenth of that in Japan.[63] It is surprising – and concerning – that such falls in R & D have occurred at a time when the need to bring new renewable energy sources on line is greater than it has ever been. Energy R & D should be substantially increased to well over one per cent of the energy's total investment so as to enable promising renewable technologies to be introduced more quickly.

More appreciation of the value of renewable energy sources and their potential together with signals such as increased R & D in renewable technologies will provide necessary encouragement for the rapid increase in investment in these sources that is required. We have already mentioned that sustained growth of thirty per cent or more per year is needed in wind and solar energy in order to meet the targets for 2020 set by carbon dioxide stabilisation scenarios (e.g. WEC scenario C). Increased growth in biomass sources is also necessary. This means that an increasing and substantial fraction of capital investment in the energy industry will have to go into new renewable sources. In the box below are listed some of the policy instruments that need to be applied for this revolution in the way we generate our energy to really get under way.

In a recent speech, Lord John Browne, the Group Chief Executive of British Petroleum, has emphasised the importance of actively planning for the long term. After explaining the steps to be taken to combat change in the energy sector and the major investments that will be required he goes on to say[64]:

> If such steps are to be taken, it is important to demonstrate the real value of taking a long-term approach which transcends the gap in time between the costs of investment and the delivery of the benefits. Political decisions are often taken on a very short-term basis and the challenge is to demonstrate the benefits of the actions which need to be taken for the long term...
> The role of business is to transform the possibilities into reality. And that means being severely practical – undertaking very focused research and then experimenting with the different possibilities. The advantage of the fact that the energy business is now global is that international companies can both access the knowledge around the world and can then apply it very quickly throughout their operations.

Nuclear energy

An energy source mentioned rather little so far is nuclear energy. It is not strictly a renewable source, but it has considerable attractiveness from the point of view of sustainable development because it does not produce greenhouse gas emissions (apart from a small amount, which is used in making the materials employed in the construction of nuclear power stations) and because the rate at which it uses up resources of radioactive material is small compared to the total resource available. It is only efficiently generated in large units, so is suitable for supplying power to national grids or to large urban connurbations, but not for small, more localised supplies. An advantage of nuclear energy installations is that the technology is known; they can be built now and therefore contribute to the reduction of carbon dioxide emissions in the short term. The cost of nuclear energy compared with energy from fossil fuel sources is often a subject of debate; exactly where it falls in relation to the others depends on the return expected on the up-front

Policy instruments

Action in the energy sector on the scale required to mitigate the effects of climate change through reduction in the emissions of greenhouse gases will require significant policy initiatives by governments in co-operation with industry. Some of these initiatives are the following[65]:

- putting in place appropriate institutional and structural frameworks;
- energy pricing strategies (carbon or energy taxes and reduced energy subsidies);
- reducing or removing other subsidies (e.g. agricultural and transport subsidies) that tend to increase greenhouse gas emissions;
- tradeable emissions permits[66];
- voluntary programmes and negotiated agreements with industry;
- utility demand-side management programmes;
- regulatory programmes, including minimum energy efficiency standards (e.g. for appliances and fuel economy);
- stimulating research and development to make new technologies available;
- market pull and demonstration programmes that stimulate the development and application of advanced technologies;
- renewable energy incentives during market build-up;
- incentives such as provisions for accelerated depreciation or reduced costs for consumers;
- information dissemination for consumers especially directed towards necessary behavioural changes;
- education and training programmes;
- technological transfer to developing countries;
- provision for capacity building in developing countries;
- options that also support other economic and environmental goals.

capital cost and on the cost of decommissioning spent power stations (including the cost of nuclear waste disposal), which represent a significant element of the total. Recent estimates are that the cost of nuclear electricity is similar to the cost of electricity from natural gas when the additional cost of capture and sequestration of carbon dioxide is added.[67]

The continued importance of nuclear energy is recognised in the WEC energy scenarios, which all assume growth in this energy source in the twenty-first century. How much growth will be realised will depend to a large degree on how well the nuclear industry is able to satisfy the general public of the safety of its operations; in particular that the risk of accidents from new installations is negligible, that nuclear waste can be safely disposed of, that the distribution of dangerous nuclear material can be effectively controlled and that it can be prevented from getting into the wrong hands.

A further nuclear energy source with great potential depends on fusion rather than fission (see box below in next section, page 312, 'Power from Nuclear Fusion').

Technology for the longer term

This chapter has concentrated mostly on what can be achieved with available and proven technology during the next few decades. It is also interesting to speculate about the more distant future and what relatively new technologies may become dominant during the twenty-first century. In doing so, of course, we are almost certainly going to paint a more conservative picture than will actually occur. Imagine how well we would have done if asked in 1900 to speculate about technology change by 2000! Technology will certainly surprise us with possibilities not thought of at the moment. But that need not deter us from being speculative!

There is general agreement that a central component of a sustainable energy future is the fuel cell that with high efficiency converts hydrogen and oxygen directly into electricity (see box below). In the fuel cell the electrolytic process of generating hydrogen and oxygen from water is reversed – the energy released by recombination of the hydrogen and oxygen is turned back into electrical energy. Fuel cells can have high efficiency of fifty to eighty per cent and they are pollution free; their only output other than electricity is water. They offer the prospect of high efficiency, small-scale power generation. They can be made in a large range of sizes suitable for use in transport vehicles or to act as local sources of electrical power for homes, for commercial premises or for many applications in industry. Much research and development has been put into fuel cells in recent years that has confirmed their potential as an important future technology. There seems little doubt that they will come into widespread use within the next decade.

Hydrogen for fuel cells can be generated from a wide variety of renewable sources (see box above). Of these, from many points of view, the one that is most attractive is through the hydrolysis of water using electricity from photovoltaic (PV) cells exposed to sunlight (Figure 11.16) – a very efficient process; over ninety per cent of the electrical energy can be stored in the hydrogen. There are many regions of the world where sunshine is plentiful and where suitable land not useful for other purposes would be readily available. It is a very clean, non-polluting technology, easily adaptable to mass production. The cost

Fuel cell technology

A fuel cell converts the chemical energy of a fuel directly into electricity without first burning it to produce heat. It is similar to a battery in its construction. Two electrodes (Figure 11.15) are separated by an electrolyte which transmits ions but not electrons. A fuel cell has a theoretical efficiency of one hundred per cent. Fuel cells have been constructed with efficiencies in the range of forty to eighty per cent.

Hydrogen for fuel cells may be supplied from a wide variety of sources, from coal or other biomass (see Note 24), from natural gas,[68] or from the hydrolysis of water using electricity generated from renewable sources such as wind power or solar photovoltaic (PV) cells (see box on page 302).

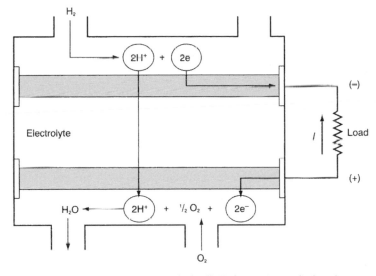

Figure 11.15 Schematic of a hydrogen-oxygen fuel cell. Hydrogen is supplied to the porous anode (negative electrode) where it dissociates into hydrogen ions (H$^+$) and electrons. The H$^+$ ions migrate through the electrolyte (typically an acid) to the cathode (positive electrode) where they combine with electrons (supplied through the external electrical circuit) and oxygen to form water.

of PV electricity has been coming down rapidly (Figure 11.13) – a trend that will continue with technological advances and with increased scale of production.

Hydrogen is also important for other reasons. It provides a medium for energy storage and it can easily be transported by pipeline or bulk transport. The main technical problem to be overcome is to find efficient and compact ways of storing hydrogen. Present technology (primarily in cylinders at high pressure) is bulky and heavy, especially for use in transport vehicles. A number of other possibilities are being explored.

Most of the technology necessary for a solar-hydrogen energy economy is available now, although the cost of energy supplied this way

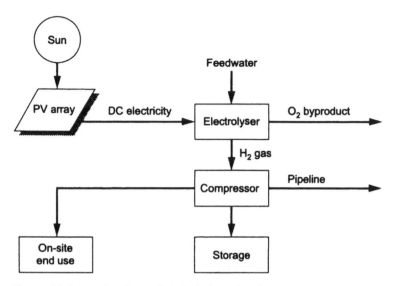

Figure 11.16 A solar photovoltaic (PV) electrolytic hydrogen system.

would at the moment be several times that from fossil fuel sources.[69] As the technology for its further development progresses and as larger-scale production becomes possible, the cost will undoubtedly reduce substantially. If its attractiveness from an environmental point of view

Power from nuclear fusion

When at extremely high temperatures the nuclei of hydrogen (or one of its isotopes, deuterium or tritium) are fused to form helium, a large amount of energy is released. This is the energy source that powers the Sun. To make it work on Earth, deuterium and tritium are used; from 1 kg, 1 GW can be generated for one day. The supply of material is essentially limitless and no unacceptable pollution is produced. A temperature of one hundred million degrees celsius is required for the reaction to occur. To keep the hot plasma away from the walls of the reaction vessel, it is confined by strong magnetic fields in a 'magnetic bottle' called a Tokamak. The challenges are to create effective confinement and a robust vessel.

Fusion power[70] has been produced on Earth at levels up to 16 MW. This has generated the confidence in a consortium of countries to build a new power-station-scale device called ITER capable of 500 MW with the object of demonstrating commercial viability. If this is successful, it is estimated that the first commercial plant could be in operation within thirty years.

Energy policy in the UK

Three important reports concerned with energy policy have been published in the UK in the last few years.

The first of these is *Energy in a Changing Climate* published in 2000 by the Royal Commission on Environmental Pollution (RCEP)[71] – an expert body that provides advice to government. It supported the concept of 'contraction and convergence' (Figure 10.3) as the best basis for future international action to reduce greenhouse gas emissions and pointed out that application of this concept would imply a goal of sixty per cent reduction in UK emissions of greenhouse gases by 2050. National quotas calculated on this basis should be combined with international trading in emission permits. To achieve such a goal more effective measures are needed to increase energy efficiency (especially in buildings) and to encourage the growth of renewable energy sources for instance by greatly increased research and development.

The second report is an *Energy Review* by the Policy and Innovation Unit (PIU) of the UK Cabinet Office[72] published in 2002. This review provided an important input into the third report which is a statement of energy policy published by the UK government in 2003[73] known as the Energy White Paper and entitled *Our Energy Future: Creating a Low Carbon Economy*. The White Paper accepts the need for a UK strategy to meet the goal set by the RCEP of a sixty per cent reduction in emissions by 2050. The main pillars of the strategy, that must be implemented immediately, will be the aggressive promotion of energy efficiency (targets in the domestic sector of twenty per cent improvement in 2010 and a further twenty per cent by 2020) and expanding the role of renewables (target of twenty per cent electricity generated from renewable sources by 2020). In addition, the options of new investment in nuclear power and in clean coal (through carbon sequestration) need to be kept open and further explored.

An estimate is provided in the review of the cost to the UK economy of realising the RCEP goal. The cost estimate is not large; it is expressed as a possible slowing in the growth of the UK economy of six months over the fifty-year period.

The picture of the future that these reports present is one where even by 2020 large changes will have occurred. There will be moves to more local energy supplies (much of it from renewable sources), to widespread use of vehicles driven by hybrid engines or by fuel cells and to the beginning of the development of an energy infrastructure based on hydrogen rather than on coal, oil or gas.

were recognised as a dominant reason for its rapid development, a solar-hydrogen economy could take off more rapidly than most energy analysts are currently predicting.

Iceland is a country that is in the forefront of the development of a hydrogen economy and aspires to be largely free of the use of fossil fuels by 2030–2040. Much of its electricity already comes from hydroelectric or geothermal sources. The first hydrogen fuel station in Iceland was opened in April 2003 and several buses powered by fuel cells are its first customers.

Finally, in this section looking at the longer term, there is the possibility of power from nuclear fusion, the energy that powers the Sun (see box). If this can be harnessed, virtually limitless supplies of energy could be provided. The result of the next phase in this programme of work will be watched with great interest.

Summary

This chapter has outlined the ways in which energy for human life and industry is currently provided. Growth in conventional energy sources at the rate required to meet the world's future energy needs will generate greatly increased emissions of greenhouse gases that will lead to unacceptable climate change. Such would not be consistent with the agreements reached at the United Nations Conference on Environment and Development at Rio de Janeiro in June 1992 when the countries of the world committed themselves to the action necessary to address the problems of energy and the environment. During the twenty-first century, emissions of carbon dioxide must be substantially reduced (e.g. as in WEC scenario C) as required by the Objective of the Climate Convention so that the concentration in the atmosphere of carbon dioxide is stabilised by the end of the century. In order to achieve the large changes required, four areas of action are essential:

- Many studies have shown that in most developed countries improvements in energy efficiency of thirty per cent or more can be achieved at little or no net cost or even at some overall saving. But industry and individuals will require not just encouragement, but modest incentives if the savings are to be realised.
- Much of the necessary technology is available for renewable energy sources (especially 'modern' biomass, wind and solar energy) that can go a long way towards replacing energy from fossil fuels to be developed and implemented. For this to be done on an adequate scale, an economic framework with appropriate incentives will need to be set up. Policy options available include the removal of subsidies, carbon or

energy taxes (which recognise the environmental cost associated with the use of fossil fuels) and tradeable permits coupled with capping of emissions.

- Arrangements are needed to ensure that technology is available for all countries (including developing countries through technology transfer) to develop their energy plans with high efficiency and to deploy renewable energy sources (for instance, local solar energy or wind generators) as widely as possible.

- With world investment in the energy industry running at around one million million US dollars per year, there is a great responsibility on both governments and industry to ensure that energy investments (including an adequate level of research and development) take long-term environmental requirements fully into account.

These actions require clear policies, commitment and resolve on the part of governments, industries and individual consumers. Because of the long life time of large energy infrastructure (e.g. power stations) and also because of the time required for the changes required to be realised, there is an urgency about the actions required. As the World Energy Council point out 'the real challenge is to communicate the reality that the switch to alternative forms of supply will take many decades, and thus the realization of the need, and commencement of the appropriate action, must be *now*' (their italics).[74]

Questions

1 Estimate how much energy you use per year in your home or your apartment. How much of this comes from fossil fuels? What does it contribute to emissions of carbon dioxide?

2 Estimate how much energy your car uses per year. What does this contribute to emissions of carbon dioxide?

3 Look up estimates made at different times over the last thirty years of the size of world reserves of coal, oil and gas. What do you deduce from the trend of these estimates?

4 Estimate the annual energy saving for your country as a result of: (1) unnecessary lights in all homes being switched off; (2) all homes changing all light bulbs to low energy ones; (3) all homes being maintained 1 °C cooler during the winter.

5 Find out for your country the fuel sources which contribute to electricity supply. Suppose a typical home heated by electricity in the winter is converted to gas heating, what would be the change in annual carbon dioxide emissions?

6 Find out about the cost of heat pumps and building insulation. For a typical building, compare the costs (capital and running costs) of reducing by seventy-five per cent the energy required to heat it by installing heat pumps or by adding to the insulation.

7 Visit a large electrical store and collate information relating to the energy consumption and the performance of domestic appliances: refrigerators, cookers, microwave ovens and washing machines. Which do you think are the most energy efficient and how do they compare with the least energy efficient? Also how well labelled were the appliances with respect to energy consumption and efficiency?

8 Consider a flat-roofed house of typical size in a warm, sunny country with a flat roof incorporating 50 mm thickness of insulation (refer to Table 11.3). Estimate the extra energy which would have to be removed by air conditioning if the roof were painted black rather than white. How much would this be reduced if the insulation were increased to a thickness of 150 mm?

9 Rework the calculations of total heating required for the building considered in Table 11.3 supposing insulation 250 mm thick (the Danish standard) were installed in the cavity walls and in the roof.

10 Look up articles about the environmental and social impact of large dams. Do you consider the benefits of the power generated by hydroelectric means are worth the environmental and social damage?

11 Suppose an area of 10 km^2 was available for use for renewable energy sources, to grow biomass, to mount PV solar cells or to mount wind generators. What criteria would determine which use would be most effective? Compare the effectiveness for each use on a typical area of your country.

12 What do you consider the most important factors which prevent the greater use of nuclear energy? How do you think their seriousness compares with the costs or damages arising from other forms of energy production?

13 In the IPCC 1995 Report chapter 19, you will find information about the LESS scenarios. In particular estimates are provided, for the different alternatives, of the amount of land that will be needed in different parts of the world for the production of energy from biomass. For your own country or region, find out how easily, on the timescale required, it is likely that this amount of land could be provided. What would be the likely consequences arising from using the land for biomass production rather than for other purposes?

14 In making arguments for a carbon tax would you attempt to relate it to the likely cost of damage from Global Warming (Chapter 9), or would you relate it to what is required to enable appropriate renewable energies to compete at an adequate level? From the information in Table 11.6 and any further data to which you have access, what level of carbon tax do you consider would be likely to enable there to be greater employment of different forms of renewable energy: (1) at the present time, (2) in 2020?

15 In discussing policy options, attention is often given to 'win-win' situations or to those with a 'double dividend', i.e. situations in which, when a particular action is taken to reduce greenhouse gas emissions, additional benefits arise as a bonus. Describe examples of such situations.

16 Of the policy options listed towards the end of the chapter, which do you think could be most effective in your country?

17 List the various environmental impacts of different renewable energy sources, biomass, wind and solar PV. How would you assess the seriousness of these impacts compared with the advantages to the environment of the contribution from these sources to the reduction of greenhouse gas emissions?

Notes for Chapter 11

1 See the relevant parts of the 1995 and 2001 IPCC Reports:
   Watson, R. T., Zinyowera, M. C., Moss, R. H. (eds.) 1996. *Climate Change 1995: Impacts, Adaptations and Mitigation of Climate Change: Scientific-Technical Analyses. Contribution of Working Group II to the Second Assessment Report of the Intergovernmental Panel on Climate Change.* Cambridge: Cambridge University Press, Summaries, Energy Primer and Chapters 19, 20, 21 and 22. Metz, B., Davidson, O., Swart, R., Pan, J. (eds.) 2001. *Climate Change 2001: Mitigation. Contribution of Working Group III to the Third Assessment Report of the Intergovernmental Panel on Climate Change.* Cambridge: Cambridge University Press, Summaries and Chapter 3.
   See also IPCC Technical Paper number 1 *Technologies, Policies and Measures for Mitigating Climate Change.* Geneva: IPCC, 1997.

2 1 toe = 11.7 MWh; 1 toe per day = 487 kW; 1 toe per year = 1.33 kW.

3 Report of G8 Renewable Energy Task Force, July 2001.

4 Moomaw, W. R., Moreira, J. R. *et al.* 2001. Technological and economic potential of greenhouse gas emissions reduction. In Metz, *Climate Change 2001: Mitigation*, Chapter 3.

5 See Moomaw, W. R., Moreira, J. R. *et al.* 2001. Technological and economic potential of greenhouse gas emissions reduction. In Metz, *Climate Change 2001: Mitigation*, Chapter 3.

6 From *Energy for Tomorrow's World: the Realities, the Real Options and the Agenda for Achievement*. WEC Commission Report. New York: World Energy Council, 1993.

7 *Renewable Energy Resources: Opportunities and Constraints 1990–2020.* Report 1993. London: World Energy Council.

8 A review of many of these can be found in Morita, T., Robinson, J. *et al.* 2001. Greenhouse gas emission mitigation scenarios and implications. In Metz, *Climate Change 2001: Mitigation*, Chapter 2.

9 From *Energy for Tomorrow's World: the Realities, the Real Options and the Agenda for Achievement*. WEC Commission Report. New York: World Energy Council, 1993, p. 122.

10 From *Energy for Tomorrow's World: the Realities, the Real Options and the Agenda for Achievement*. WEC Commission Report. New York: World Energy Council, 1993, p. 113.

11 Ways of achieving large reductions in all these sectors are described by von Weizacker, E., Lovins, A. B., Lovins, L. H. 1997. *Factor Four, Doubling Wealth: Halving Resource Use*. London: Earthscan.

12 More detail of heat pumps and their applications in Smith, P. F. 2003. *Sustainability at the Cutting Edge*. London: Architectural Press, pp. 45–50.

13 From National Academy of Sciences, *Policy Implications of Greenhouse Warming*. 1992. Washington DC: National Academy Press, Chapter 21.

14 Smith, *Sustainability*, pp. 135–7.

15 See, for instance, von Weizacker, E., Lovins, A. B., Lovins, L. H. 1997. *Factor Four, Doubling Wealth: Halving Resource Use*. London: Earthscan, pp. 28–9.

16 www.zedfactory.com/bedzed/bedzed.html.

17 See Moomaw, W. R., Moreira, J. R. *et al*. 2001. Technological and economic potential of greenhouse gas emissions reduction. In Metz, *Climate Change 2001: Mitigation*, Chapter 3, Table 3.5.

18 Royal Commission on Environmental Pollution, 18th and 20th Reports *Transport and the Environment*. London: HMSO, 1994 and 1997. Also see Moomaw, W. R., Moreira, J. R. *et al*. 2001. Technological and economic potential of greenhouse gas emissions reduction. In Metz, *Climate Change 2001: Mitigation*, Section 3.4.

19 From the Summary for Policymakers in Penner, J. *et al*. 1999. *Aviation and the Global Atmosphere. A special report of the IPCC*. Cambridge: Cambridge University Press.

20 More detail in Moomaw and Moreira, in Metz, *Climate Change 2001: Mitigation*, Section 3.4.

21 Example quoted in *Energy, Environment and Profits*. 1993. London: Energy Efficiency Office of the Department of the Environment.

22 See, for instance, *National Academy of Sciences, Policy Implications of Greenhouse Warming*. 1992. Washington DC: National Academy Press, Chapter 22; also *Energy for Tomorrow's World: the Realities, the Real Options and the Agenda for Achievement*. WEC Commission Report. New York: World Energy Council, 1993, Chapter 4; Kashiwagi, T. *et al*. 1996. Industry. In Watson, *Climate Change 1995: Impacts*, Chapter 20; Moomaw and Moreira. In Metz, *Climate Change 2001: Mitigation*, Chapter 3.

23 From speech by Lord Browne, BP Chief Executive to the Institutional Investors Group, London, 26th November 2003.

24 Carbonaceous fuel is burnt to form carbon monoxide, CO, which then reacts with steam according to the equation $CO + H_2O = CO_2 + H_2$.

25 See *Putting Carbon Back in the Ground*. Report by IEA Greenhouse Gas R & D Programme, www.ieagreen.org.uk.

26 'Large' hydro applies to schemes greater than ten megawatts in capacity; 'small' hydro to schemes smaller than ten megawatts.

27 Sources of comprehensive information about renewable energy are: Boyle, G. (ed.) 1996. *Renewable Energy Power for a Sustainable Future*. Oxford: Oxford University Press; Johansson, T. B. *et al*. 1993. *Renewable Energy*. Washington DC: Island Press; *Renewable Energy Resources*. World Energy Council Report. 1993. London: World Energy Council; Moomaw, W. R., Moreira, J. R. *et al*. 2001. Technological and economic potential of

greenhouse gas emissions reduction. In Metz, *Climate Change 2001: Mitigation*, Chapter 3.

28 Moreira, J. R., Poole, A. D. 1993. Hydropower and its constraints. In Johansson, *Renewable Energy*, Chapter 2, pp. 73–119.

29 Martinot, E. *et al*. 2002. Renewable energy markets in developing countries. *Annual Review of Energy and the Environment*, **27**, pp. 309–48. Note that the definition of 'small hydro' is usually units less than 10 MW capacity. Different numbers that may be quoted for 'small hydro' may arise from different definitions of 'small' as compared with 'large hydro'.

30 Moreira, J. R., Poole, A. D. 1993. Hydropower and its constraints. In Johansson, *Renewable Energy*, pp. 73–119.

31 See review by Loening, A. 2003. Landfill gas and related energy sources; anaerobic digesters; biomass energy systems. In *Issues in Environmental Science and Technology*, No. 19. London: Royal Society of Chemistry, pp. 69–88.

32 Moomaw and Moreira, in Metz, *Climate Change 2001: Mitigation*, Section 3.8.4.3.2.

33 From Goldemberg, J. (ed.) *World Energy Assessment: Energy and the Challenge of Sustainability*. United Nations Development programme (UNDP), United Nations Department of Economic and Social Affairs (UN-DESA) and World Energy Council (WEC), New York. Original source, *Energy Balances of OECD Countries*. Paris: International Energy Agency, 1999. www.undp.org/seed/eap/activities/wea.

34 Twidell, J., Weir, T. 1986. *Renewable Energy Resources*. London: Spon Press, p. 291.

35 These projects are supported by the Shell Foundation, a charity set up to promote sustainable energy for the Third World

36 See Mills, E., Wilson, D., Johansson, T. 1991. In Jager, J., Ferguson, H. L. 1991. *Climate Change: Science, Impacts and Policy Proceedings of the Second World Climate Conference*. Cambridge: Cambridge University Press, pp. 311–28.

37 See review by Loening, A. 2003. Landfill gas and related energy sources; anaerobic digesters; biomass energy systems. In *Issues in Environmental Science and Technology*, No. 19. London: Royal Society of Chemistry, pp. 69–88.

38 From *Report of the Renewable Energy Advisory Group, Energy Paper Number 60*. London: UK Department of Trade and Industry, November 1992.

39 See Royal Commission on Environmental Pollution, 17th Report, *Incineration of Waste*. 1993. London: HMSO, pp. 43–7.

40 Hall, D. O. *et al*. 1993. Biomass for energy: supply prospects. In Johansson, *Renewable Energy*, pp. 593–651.

41 From *Report of the Renewable Energy Advisory Group, Energy Paper Number 60*. London: UK Department of Trade and Industry, November 1992, p. A29.

42 See Infield, D., Rowley, P. 2003. Renewable energy: technology considerations and electricity integration. *Issues in Environmental*

Science and Technology, No. 19. London: Royal Society of Chemistry, pp. 49–68.

43   From Danish Wind Energy Association, www.windpower.org.

44   www.cabinet-office.gov.uk/innovation/2002/energy/report/index.htm.

45   Lal, M. 1996. Measures for reducing climate relevant gas emissions in India. Paper presented at an Indo-German seminar IIT, Dehli, 29–31 October, 1996.

46   Twidell and Weir, *Renewable Energy Resources*, p. 252.

47   Martinot, E. *et al*. 2002. Renewable energy markets in developing countries. *Annual Review of Energy and the Environment*, **27**, pp. 309–48.

48   Martinot, E. *et al*. Renewable energy markets in developing countries. *Annual Review of Energy and the Environment*, **27**, pp. 309–48.

49   Twidell, J., Johnstone, C. 1992. Glasgow gains from Strathclyde's solar residences. *Sun at Work in Europe*, **7**, No. 4 December, pp. 15–17.

50   Ishitani, H. *et al*. 1996. Energy supply mitigation options. In Watson, *Climate Change 1995: Impacts*, Chapter 19.

51   Kelly, H. 1993. Introduction to photovoltaic technology. In Johansson, T. B. *et al*. 1993. *Renewable Energy*. Washington DC: Island Press, pp. 297–336.

52   Carlson, D. E. Wagner, S. 1993. Amorphous silicon photovoltaic systems. In Johansson, *Renewable Energy*, pp. 403–36.

53   Zweibel, K., Barnett, A. M. 1993. Polycrystalline thin-film photovoltaics. In Johansson, *Renewable Energy*, pp. 437–82.

54   Martinot, E. *et al*. 2002. Renewable energy markets in developing countries. *Annual Review of Energy and the Environment*, **27**, pp. 309–48.

55   Further information on all these systems and possibilities for financing and marketing, etc. can be found with a wealth of references in Martinot *ibid*.

56   This figure assumes that about one third of total solar energy in 2020 will be PV electricity.

57   See www.tidalelectric.com.

58   See Boyle, G. 1996. *Renewable Energy Power for a Sustainable Future*. Oxford: Oxford University Press.

59   Schimel, D. *et al*. 1997. Stabilisation of atmospheric greenhouse gases: physical, biological and socio-economic implications. *IPCC Technical Paper 3*. Geneva: IPCC.

60   See Elliott, D. 2003. Sustainable energy: choices, problems and opportunities. *Issues in Environmental Science and Technology*, No. 19. London: Royal Society of Chemistry, pp. 19–47.

61   Fisher, B. S. *et al*. 1996. An economic assessment of policy instruments for combating climate change. In Bruce, J., Hoesung Lee, Haites, E. (eds.) 1996. *Climate Change 1995: Economic and Social Dimensions of Climate Change*. Cambridge: Cambridge University Press, Chapter 11.

62   Johansson, *Renewable Energy*, p. 38; also Ishitani, H. *et al*. 1996, Energy supply mitigation options. In Watson, *Climate Change 1995: Impacts*, Chapter 19.

63   *Energy the Changing Climate*. 2000. 22nd Report of the UK Royal Commission on Environmental Pollution. London: UK Stationery Office, p. 81.

64   From speech by Lord Browne, BP Chief Executive to the Institutional In-
     vestors Group, London, 26 November 2003.
65   Based on Summary for Policymakers. In Watson, *Climate Change 1995:*
     Impacts, Section 4.4.
66   See Mullins, F. 2003. Emissions trading schemes: are they a licence to
     pollute? *Issues in Environmental Science and Technology*, No. 19. London:
     Royal Society of Chemistry, pp. 89–103.
67   www.cabinet-office.gov.uk/innovation/2002/energy/report/index.htm.
68   By reacting natural gas (methane CH_4) with steam through the reaction
     $2H_2O + CH_4 = CO_2 + 4H_2$.
69   For more details of this technology see Ogden and Nitsch, 'Solar Hydrogen'
     in *Renewable Energy*, eds. T. B. Johansson *et al.*, Island Press, Washington
     DC, 1993, pp. 925–1009.
70   McCraken, G., Stott, P. *Fusion, the Energy of the Universe.* New York:
     Elsevier/Academic Press, 2004.
71   www.rcep.org.uk.
72   www.cabinet-office.gov.uk/innovation/2002/energy/report/index.htm.
73   www.dti.gov.uk/energy/whitepaper/index.shtml.
74   From *Energy for Tomorrow's World: the Realities, the Real Options and*
     the Agenda for Achievement. WEC Commission Report. New York: World
     Energy Council, 1993, p. 88.

Chapter 12
The global village

The preceding chapters have considered the various strands of the global warming story and the action that should be taken. In this last chapter I want first to present some of the challenges of global warming, especially those which arise because of its global nature. I then want to put global warming in the context of other major global problems faced by humankind.

The challenges of global warming

We have noted in the course of our discussion that global warming is not the only environmental problem. For instance, coastal regions are liable to subsidence for other reasons; water supplies in many places are already being depleted faster than they are being replenished and agricultural land is being lost through soil erosion. Many other reasons, locally and regionally, could be listed for the occurrence of environmental degradation. However, the importance of global warming is not diminished by the existence of these other environmental problems; in fact their existence will generally exacerbate its effects – as, for instance, we noted on page 150 when looking at the effect of sea level rise on Bangladesh. It is generally beneficial to tackle all related environmental problems together.

Local degradation of the environment is generally the result of particular action in the locality. To give an example, subsidence occurs because of the over-extraction of groundwater. In these cases the community where the malpractice is occurring suffers the damage that it causes and the principle that polluters should pay the cost of their pollution is relatively easy to apply.

The particular characteristic of global warming, compared with most environmental problems, is that it is *global*. However, though everybody contributes to it to a greater or lesser extent, its adverse impact will not fall uniformly. Many, especially in the developing world, will experience significant damage; some others, mostly in the developed world, may in fact gain from it. This non-uniformity of impact also applies to local pollution. But, for local pollution, the adverse effects are more apparent and immediate than is the case with global warming. It is therefore imperative that information about the effects of the burning of fossil fuels on the global climate becomes more widely available, so leading to greater awareness that an individual burning fossil fuels anywhere in the world has impact globally. And a global problem demands a global solution. That the 'polluter should pay' when the pollution is global rather than local is one of the Principles (Principle 16) enshrined in the Rio Declaration of June 1992. Chapters 9 and 10 presented some of the mechanisms that have been devised to apply this principle on a global scale.

There is already some experience in tackling an environmental problem of global scale: the depletion of stratospheric ozone because of the injection by humans of chlorofluorocarbons (CFCs) into the atmosphere possesses similar global characteristics to the global warming problem. An effective mechanism for tackling and solving the problem of ozone depletion has been established through the Montreal Protocol. All nations contributing to the damage have agreed to phase out their emissions of harmful substances. The richer nations involved have also agreed to provide finance and technology transfer to assist developing countries to comply. A way forward for addressing global environmental problems has therefore been charted.

Moving in that direction in the case of global warming will not be easy because the problem is so much larger and because it strikes so much nearer to the core of human resources and activities – such as energy and transport – upon which our quality of life depends. However, abatement of the use of fossil fuels need not destroy or even diminish our quality of life; it should actually improve it! In tackling the problem of global warming there are particular responsibilities and challenges for different communities of expertise which generally transcend national boundaries.

- For the world's *scientists* the brief is clear: to provide better information especially about the expected climate change on the regional and local level, always keeping an appropriate emphasis on the uncertainties of prediction. Not only politicians and policymakers but also ordinary people need the information provided in the clearest possible form, in all countries and at all levels of society. Information is especially required about changes that may occur in the extremes of weather and climate. Scientists also have an important role in contributing to the research necessary to underpin the technical developments, for example in the energy, transport, forestry and agriculture sectors, required by the adaptation and mitigation strategies we have described.

- In the world of *politics*, it is over twenty years since Sir Crispin Tickell drew attention to the need for international action addressing climate change.[1] Since then, a great deal of progress has been made with the signing in Rio in 1992 of the Framework Convention on Climate Change and with the setting up of the Sustainable Development Commission in the United Nations. The challenges presented to the politicians and decision makers by the Convention are, firstly, to achieve the right balance of development against environmental concern, that is to achieve sustainable development, and secondly, to find the resolve to turn the many fine words of the Convention into adequate and genuine action (including both adaptation and its mitigation) regarding climate change.

- In describing the likely impacts of global warming and the ways they can be alleviated, I have frequently stressed the role of *technology*. The necessary technology is available. The challenge of its implementation, supported by appropriate investment, needs to be picked up enthusiastically and innovatively by the world's industry. Too often environmental concerns and environmental regulation are seen by industry as a threat when, in fact, they are an opportunity. As we saw in Chapter 11, the emerging technologies associated with energy efficiency in all its aspects, renewable energy production and the efficient use and recycling of materials should bring increased employment in industry at a high level of skill and technical training. Because of increasing public awareness of the environment and of the need for its preservation, the industries that are likely to grow and flourish during the twenty-first century are those that have taken environmental considerations firmly on board.

- The responsibilities of *industry* must also be seen in the world context. It is the imagination, innovation, commitment and activity of industry that will do most to solve the problem. Industries that have a global perspective, working as appropriate with governments, need to develop a technical, financial and policy strategy to this end. An important component of this strategy is the transfer of appropriate technology between countries, especially in the energy sector. This has been specifically recognised in the Climate Convention which in Article 4, paragraph 5 states: 'The developed country Parties . . . shall take all practical steps to promote, facilitate and finance, as appropriate, the transfer of, or access to, environmentally sound technologies and know-how to other Parties, particularly developing country Parties, to enable them to implement the provisions of the Convention.'

- There are also new challenges for *economists*; for instance, that of adequately representing environmental costs (especially including those 'costs' that cannot be valued in terms of money) and the value of 'natural' capital, especially when it is of a global kind – as mentioned in Chapter 9. There is the further problem of dealing fairly with all countries. No country wants to be put at a disadvantage economically because it has taken its responsibilities with respect to global warming more seriously than others. As economic and other instruments (for instance, taxes, subsidies, capping and trading arrangements, regulations or other measures) are devised to provide the incentives for appropriate action regarding global warming by governments or by individuals, these must be seen to be both fair and effective for all nations. Economists working with politicians and decision makers need to find imaginative solutions which recognise not just environmental concerns but political realities.

- There is an important role for *communicators and educators*. Everybody in the world is involved in climate change so everybody needs to be properly informed about it. They require to understand the evidence for it, its causes, the distribution of its impacts and the action that can be taken to alleviate them. Climate change is a complex topic; the challenge to educators and the media is to inform in ways that are understandable, comprehensive, honest and balanced.

- All countries will need to adapt to the climate change that applies in their region. For many developing countries this will not be easy because of increased floods, droughts or significant rise in sea level. Reductions in risks from disasters are some of the most important adaptation strategies. A challenge for *aid agencies* therefore is to prepare for more frequent and intense disasters in vulnerable countries; the International Red Cross has already taken the lead in this.[2]

Finally, it is important to recognise that the problem is not only global but long-term – the time scales of climate change, of major infrastructure change in energy generation or transport or of major changes in programmes such as forestry are of the order of several decades. The programme of action must therefore be seen as both urgent and evolving, based on the continuing scientific, technical and economic assessments. As the IPCC 1995 Report states, 'The challenge is not to find the best policy today for the next 100 years, but to select a prudent strategy and to adjust it over time in the light of new information'.[3]

Not the only global problem

Global warming is not the only global problem. There are other issues of a global scale and we need to see global warming in their context. Four problems of particular importance impact on the global warming issue.

The first is population growth. When I was born there were about 2000 million people in the world. At the beginning of the twenty-first century there were 6000 million. During the lifetime of my grandchildren it is likely to rise to at least 8000 million. Most of the growth will be in developing countries; by 2020 they will contain over eighty per cent of the world's people. These new people will all make demands for food, energy and work to generate the means of livelihood – all with associated implications for global warming.

The second issue is that of poverty and the increasing disparity in wealth between the developed and the developing world. The gap between the rich nations and the poor nations is becoming wider. The flow of wealth in the world is from the poorer nations to the richer ones. Increasingly there are demands that more justice and equity be realised within the world's communities. The Prince of Wales has drawn attention to the strong links that exist between population growth, poverty and environmental degradation (see box below).

Poverty and population growth

The Prince of Wales, in addressing the World Commission on Environment and Development on 22 April 1992, spoke as follows[7]:

I do not want to add to the controversy over cause and effect with respect to the Third World's problems. Suffice it to say that I don't, in all logic, see how any society can improve its lot when population growth regularly exceeds economic growth. The factors which will reduce population growth are, by now, easily identified: a standard of health care that makes family planning viable, increased female literacy, reduced infant mortality and access to clean water. Achieving them, of course, is more difficult – but perhaps two simple truths need to be writ large over the portals of every international gathering about the environment: we will not slow the birth rate until we address poverty. And we will not protect the environment until we address the issue of poverty and population growth in the same breath.

The third global issue is that of the consumption of resources, which in many cases is contributing to the problem of global warming. Many of the resources now being used cannot be replaced, yet we are using them at an unsustainable rate. In other words, because of the rate at which we are depleting them, we are seriously affecting their use even at a modest level by future generations. Further, over eighty per cent of resources are consumed by twenty per cent of the world's population and to propagate modern western patterns of consumption into the developing world is just not realistic. An important component of sustainable development, therefore, is *sustainable consumption*[4] of all resources.

The fourth issue is that of global security. Our traditional understanding of security is based on the concept of the sovereign state with secure borders against the outside world. But communications, industry and commerce increasingly ignore state borders, and problems like that of global warming and the other global issues we have mentioned transcend national boundaries. Security therefore also needs to take on more of a global dimension.

The impacts of climate change may well pose a threat to security. One of the most recent wars has been fought over oil. It has been suggested that wars of the future could be fought over water.[5] The threat of conflict must be greater if nations lose scarce water supplies or the means of livelihood as a result of climate change. A dangerous level of tension could easily arise, with large numbers of environmental refugees. As has been pointed out by Admiral Sir Julian Oswald,[6] who has been deeply concerned with British defence policy, a broader strategy regarding security needs to be developed which considers *inter alia* environmental threats as a possible source of conflict. In addressing the appropriate action to combat such threats, it may be better overall and more cost-effective in security terms to allocate resources to the removal or the alleviation of the environmental threat rather than to military or other measures to deal head-on with the security problem itself.

The conception and conduct of environmental research

While completing the writing of this last chapter I attended the opening of the Zuckerman Centre for Connective Environmental Research at the University of East Anglia – a centre devoted to interdisciplinary research on the environment. An opening lecture was given by William Clark, Professor of International Science, Public Policy and Human Development at Harvard University.[8] I was particularly struck by his remarks concerning the changes that are necessary in the way research is conceived and conducted if science (both natural and social) and technology are going to provide more adequate support to environmental sustainability. He pointed out the need to address all aspects of a problem both in the conception of the research and in its conduct and particularly emphasised the following four requirements:

- An *integrative, holistic approach* that considers the interactions be-tween multiple stresses and between various possible solutions. Such an approach also seeks to integrate perspectives from both the natural and the social sciences, so as to understand better the dynamical inter-play by which environment shapes society and society in turn reshapes environment. And these various integrations must also be in a global context.
- A *goal of finding solutions not just of characterising problems.* There is a tendency amongst scientists to talk forever about problems but leave solutions to others. Applied research seeking solutions is just as challenging and worthy as so-called fundamental research identifying and describing the problems.
- *Ownership by both scientists and stakeholders.*[9] People are more pre-pared to change their behaviour or beliefs in response to knowledge that they have had a hand in researching or shaping.
- *Scientists must see themselves more as facilitators of social learning and less as sources of social guidance.* The problems faced in envi-ronmental research are such that solutions will only be reached after a long and iterative learning process in which many sectors of society as well as scientists must be included.

Two other qualities that need to govern our attitude to research that have often received emphasis in this book are those of *honesty* (especially accuracy and balance in the presentation of results) and *humility* (see, for instance, the fourth bullet in the last paragraph and the quotation from Thomas Huxley in Chapter 8, page 211). Together with the theme of *Holism* from the last paragraph, they make up *3 Hs*, an alliteration that assists in keeping them all in mind.

The goal of environmental stewardship

In the western world there are many material goals: economic growth, social welfare, better transport, more leisure and so on. But for our fulfil-ment as human beings we desperately need not just material challenges, but challenges of a moral or spiritual kind. There are strong connections, which I drew out in Chapter 8, between our basic attitudes, including religious belief, and environmental concern. I drew a picture of humans as stewards or gardeners of the Earth. Many people in the world are already deeply involved in a host of ways in matters of environmental concern. Such concern could, however, with benefit to us all, be elevated to a higher public and political level. The United Nations, so far as it is able, has laid out a course of action. In an article in *Time Magazine* at the time of the Summit on Sustainable Development in Johannesburg last August, Kofi Annan, the Secretary General of the United Nations presented 'Competing Futures' in the following terms[10]:

> Imagine a future of relentless storms and floods; islands and heavily inhabited coastal regions inundated by rising sea levels; fertile soils rendered barren by drought and the desert's advance; mass migrations of environmental refugees; and armed conflicts over water and precious natural resources.
>
> Then, think again – for one might just as easily conjure a more hopeful picture: of green technologies; liveable cities; energy-efficient homes, transport and industry; and rising standards of living for all the people not just a fortunate minority. The choice between these competing visions is ours to make.

What the individual can do

I have spelled out the responsibilities of experts of all kinds – scientists, economists, technologists, politicians, industrialists, communicators and educators. There are important contributions also to be made by ordinary individuals to help to mitigate the problem of global warming.[12] Some of these are to:

- ensure maximum energy efficiency in the home – through good insulation (see box on page 280) against cold in winter and heat in summer and by making sure that rooms are not overheated and that light is not wasted;
- as consumers, take energy use into account, e.g. by buying goods that last longer and from more local sources and buying appliances with high energy efficiency;
- support, where possible, the provision of energy from non-fossil-fuel sources; for instance, purchase 'green' electricity (i.e. electricity from renewable sources) wherever this option is available[13];
- drive a fuel-efficient car and choose means of transport that tend to minimise overall energy use; for instance, where possible, walking or cycling;
- check, when buying wood products, that they originate from a renewable source;
- contribute to projects that reduce carbon dioxide emissions – this can be a way of compensating for some of the emissions to which we contribute, e.g. from aircraft journeys[14];
- through the democratic process, encourage local and national governments to deliver policies which properly take the environment into account.

An encouraging development is the growing interest of some of the world's largest companies in tackling the problems posed by global warming. Many are aggressively pursuing (e.g. through internal trading arrangements) the reduction of carbon dioxide emissions within their operations. Also many (e.g. two of the largest oil companies, Shell and British Petroleum) are putting strong investment into renewable energies. John Browne, the chief executive of BP, has said:[11]

> No single company or country can solve the problem of climate change. It would be foolish and arrogant to pretend otherwise. But I hope we can make a difference – not least to the tone of the debate – by showing what is possible through constructive action.

The challenge is indeed for everybody, from individuals, communities, industries and governments through to multinationals, especially for those in the relatively affluent Western world, to take on board thoroughly this urgent task of the environmental stewardship of our Earth. And none of us should argue that there is nothing we can usefully do. Edmund Burke, a British parliamentarian of 200 years ago, said:

> Noone made a greater mistake than he who did nothing because he could do so little.

Questions

1 List and describe the most important environmental problems in your country. Evaluate how each might be exacerbated under the type of climate change expected with global warming.

2 It is commonly stated that my pollution or my country's pollution is so small compared with the whole, that any contribution I or my country can make towards solving the problem is negligible. What arguments can you make to counter this attitude?

3 Speak to people you know who are involved with industry and find out their attitudes to local and global environmental concerns. What are the important arguments that persuade industry to take the environment seriously?

4 Al Gore, Vice-President of the United States in 1996–2000, has proposed a plan for saving the world's environment.[15] He has called it 'A Global Marshall Plan' paralleled after the Marshall Plan through which the United States assisted western Europe to recover and rebuild after the Second World War. Resources for the plan would need to come from the world's major wealthy countries. He has proposed five strategic goals for the plan: (1) the stabilisation of world population; (2) the rapid creation and development of environmentally appropriate technologies; (3) a comprehensive and ubiquitous change in the economic 'rules of the road' by which we measure the impact of our decisions on the environment; (4) the negotiation of a new generation of international agreements, that must be sensitive to the vast differences

of capability and need between developed and developing nations; (5) the establishment of a cooperative plan for educating the world's citizens about our global environment. Consider these five goals. Are they sufficiently comprehensive? Are there important goals that he has omitted?

5   How do you think governments can best move forward towards strategic goals for the environment? How can citizens be persuaded to contribute to government action if it involves making sacrifices, for example paying more in tax?

6   Can you add to the list in the box at the end of the chapter of contributions that the individual can make?

7   The Jubilee 2000 campaign has worked towards the cancellation of Third World debt possibly in return for appropriate environmental action. Discuss whether this is a good idea and how it might be made more successful.

8   Millions of people (especially children) die in the world's poorer countries because they lack clean water. It is sometimes argued that the resources that might be used in reducing carbon dioxide emissions would be better used in making sure that everyone has access to clean water. Do you agree with this argument? If so how could the result be realised in practice?

9   It has been suggested that anthropogenic climate change should be considered as a Weapon of Mass Destruction. Discuss the validity of this comparison.

10 Consider the requirements for the conception and conduct of research that are detailed on page 327–8. Do you consider that they could be components of a check list against which research proposals might be judged? How far does the research are in which you are engaged or do the research programmes with which you are connected fulfil these requirements?

Notes for Chapter 12

1   Tickell, C. 1986. *Climatic Change and World Affairs*, second edition. Boston: Harvard University Press.

2   The International Red Cross/Red Crescent has set up a Climate Centre based in The Netherlands as a bridge between Climate Change and Disaster Preparedness. The activities of the Centre are concerned with Awareness (information and education), Action (development of climate adaptation in the context of Disaster Preparedness programmes) and Advocacy (to ensure that policy development takes into account the growing concern about the impacts of climate change and utilises existing experience with climate adaptation and Disaster Preparedness).

3   *Synthesis of Scientific-Technical Information Relevant to Interpreting Article 2 of the UN Framework Convention on Climate Change*. 1995. Geneva: IPCC, p. 17.

4   Many of the world's national academies of science led by the Royal Society in London have joined together in a report pointing this out. See Appendix B in *Towards Sustainable Consumption: a European Perspective*. 2000. London: Royal Society.

5   The former United Nations Secretary-General, Boutros Boutros-Ghali, has said that 'the next war in the Middle East will be fought over water, not politics'.

6   Oswald, J. Defence and environmental security. 1993. In Prins, G. (ed.) 1993. *Threats Without Enemies*. London: Earthscan.

7   HRH the Prince of Wales, in the First Brundtland Speech, 22 April 1992, published in Prins, *Threats Without Enemies*, pp. 3–14.

8   Clark, W. C. 2003. Sustainability Science: Challenges for the New Millennium. An address at the official opening of the Zuckerman Institute for Connective Environmental Research, University of East Anglia, Norwich, UK, 4 September 2003. http://sustainabilityscience.org/ists/docs/clark_zicer_opening030904.pdf.

9   This is illustrated by the experience of the IPCC, as described on page 221.

10 From Kofi Annan, *Time Magazine*, 26 August 2002.

11 From a speech given by Lord John Browne in Berlin, 30 September 1997.

12 Some useful websites: Sierra Club USA, www.sierraclub.org/sustainable.consumption/; Union of Concerned Scientists, www.ucsusa.org; Energy Saving Trust, www.est.org.uk; Ecocongregation, www.encams.org; Christian Ecology Link, www.christian-ecology.org.uk; John Ray Initiative, www.jri.org.uk.

13 With changes in the organisation of electricity supply companies in some countries, it is becoming possible to purchase electricity, delivered by the national grid, from a particular generating source, see for instance for the UK. www.greenelectricity.org or www.good-energy.co.uk.

14 See for instance Climate Care website, www.climatecare.org.uk.

15 Expounded in the last chapter of Gore, A. 1992. *Earth in the Balance*. New York: Houghton Mifflin Company.

Glossary

Afforestation   Planting of new forests on lands that historically have not contained forests

Agenda 21   A document accepted by the participating nations at UNCED on a wide range of environmental and development issues for the twenty-first century

Albedo   The fraction of light reflected by a surface, often expressed as a percentage. Snow-covered surfaces have a high albedo level, vegetation-covered surfaces have a low albedo, because of the light absorbed for *photosynthesis*

Anthropic principle   A principle which relates the existence of the Universe to the existence of humans who can observe it

Anthropogenic effects   Effects which result from human activities such as the burning of *fossil fuels* or *deforestation*

AOGCM   Atmosphere–Ocean Coupled General Circulation Model

Atmosphere   The envelope of gases surrounding the Earth or other planets

Atmospheric pressure   The pressure of atmospheric gases on the surface of the planet. High atmospheric pressure generally leads to stable weather conditions, whereas low atmospheric pressure leads to storms such as cyclones

Atom   The smallest unit of an *element* that can take part in a chemical reaction. Composed of a nucleus which contains *protons* and *neutrons* and is surrounded by *electrons*

Atomic mass   The sum of the numbers of *protons* and *neutrons* in the nucleus of an *atom*

Biodiversity   A measure of the number of different biological species found in a particular area

Biological pump   The process whereby carbon dioxide in the *atmosphere* is dissolved in sea water where it is used for *photosynthesis* by *phytoplankton* which are eaten by *zooplankton*. The remains of these microscopic organisms sink to the ocean bed, thus removing the carbon from the *carbon cycle* for hundreds, thousands or millions of years

Biomass   The total weight of living material in a given area

Biome   A distinctive ecological system, characterised primarily by the nature of its vegetation

Biosphere   The region on land, in the oceans and in the *atmosphere* inhabited by living organisms

Business-as-usual   The scenario for future world patterns of energy consumption and *greenhouse gas* emissions which assumes that there will be no major changes in attitudes and priorities

C3, C4 plants   Groups of plants which take up *carbon dioxide* in different ways in *photosynthesis* and are hence affected to a different extent by increased atmospheric carbon dioxide. Wheat, rice and soya bean are C3 plants; maize, sugarcane and millet are C4 plants

Carbon cycle   The exchange of carbon in various chemical forms between the *atmosphere*, the land and the oceans

Carbon dioxide fertilisation effect   The process whereby plants grow more rapidly under an atmosphere of increased carbon dioxide concentration. It affects *C3 plants* more than *C4 plants*

Carbon dioxide   One of the major greenhouse gases. Human-generated carbon dioxide is caused mainly by the burning of fossil fuels and deforestation

Celsius   Temperature scale, sometimes known as the Centigrade scale. Its fixed points are the freezing point of water ($0\,^{\circ}$C) and the boiling point of water ($100\,^{\circ}$C)

CFCs   Chlorofluorocarbons; synthetic compounds used extensively for refrigeration and aerosol sprays until it was realised that they destroy ozone (they are also very powerful greenhouse gases) and have a very long lifetime once in the *atmosphere*. The Montreal Protocol agreement of 1987 is resulting in the scaling down of CFC production and use in industrialised countries

Chaos   A mathematical theory describing systems that are very sensitive to the way they are originally set up; small discrepancies in the initial conditions will lead to completely different outcomes when the system has been in operation for a while. For example, the motion of a pendulum when its point of suspension undergoes forced oscillation will form a particular pattern as it swings. Started from a slightly different position, it can form a completely different pattern, which could not have been predicted by studying the first one. The weather is a partly chaotic system, which means that even with perfectly accurate forecasting techniques, there will always be a limit to the length of time ahead that a useful forecast can be made

CIS   Commonwealth of Independent States (former USSR)

Climate sensitivity   The global average temperature rise under doubled *carbon dioxide* concentration in the *atmosphere*

Climate   The average weather in a particular region

Compound   A substance formed from two or more elements chemically combined in fixed proportions

Condensation   The process of changing state from gas to liquid

Convection   The transfer of heat within a fluid generated by a temperature difference

Coppicing   Cropping of wood by judicious pruning so that the trees are not cut down entirely and can regrow

Cryosphere   The component of the climate system consisting of all snow, ice and permafrost on and beneath the surface of the earth and ocean

Daisyworld   A model of biological feedback mechanisms developed by James Lovelock (see also *Gaia hypothesis*)

DC   Developing country – also Third World country

Deforestation   Cutting down forests; one of the causes of the enhanced *greenhouse effect*, not only when the wood is burned or decomposes, releasing *carbon dioxide*, but also because the trees previously took carbon dioxide from the *atmosphere* in the process of photosynthesis

Deuterium   Heavy *isotope* of hydrogen

Drylands   Areas of the world where precipitation is low and where rainfall often consists of small, erratic, short, high-intensity storms

Ecosystem   A distinct system of interdependent plants and animals, together with their physical environment

El Niño   A pattern of ocean surface temperature in the Pacific off the coast of South America, which has a large influence on world *climate*

Electron   Negatively charged component of the *atom*

Element   Any substance that cannot be separated by chemical means into two or more simpler substances

Environmental refugees   People forced to leave their homes because of environmental factors such as drought, floods, sea level rise

EU   European Union

Evaporation   The process of changing state from liquid to gas

FAO   The United Nations Food and Agriculture Organization

Feedbacks   Factors which tend to increase the rate of a process (positive feedbacks) or decrease it (negative feedbacks), and are themselves affected in such a way as to continue the feedback process. One example of a positive feedback is snow falling on the Earth's surface, which gives a high *albedo* level. The high level of reflected rather than absorbed solar radiation will make the Earth's surface colder than it would otherwise have been. This will encourage more snow to fall, and so the process continues

Fossil fuels   Fuels such as coal, oil and gas made by decomposition of ancient animal and plant remains which give off *carbon dioxide* when burned

FSU   Countries of the former Soviet Union

Gaia hypothesis   The idea, developed by James Lovelock, that the *biosphere* is an entity capable of keeping the planet healthy by controlling the physical and chemical environment

Geoengineering   Artificial modification of the environment to counteract *global warming*

Geothermal energy   Energy obtained by the transfer of heat to the surface of the Earth from layers deep down in the Earth's crust

Global warming   The idea that increased *greenhouse gases* cause the Earth's temperature to rise globally (see *greenhouse effect*)

Green Revolution   Development of new strains of many crops in the 1960s which increased food production dramatically

Greenhouse effect   The cause of *global warming*. Incoming *solar radiation* is transmitted by the *atmosphere* to the Earth's surface, which it warms. The energy is retransmitted as *thermal radiation*, but some of it is absorbed by *molecules of greenhouse gases* instead of being retransmitted out to space, thus warming the atmosphere. The name comes from the ability of greenhouse glass to transmit incoming solar radiation but retain some of the outgoing thermal radiation to warm the interior of the greenhouse. The 'natural' greenhouse effect is due to the greenhouse gases present for natural reasons, and is also observed for the neighbouring planets in the solar system. The 'enhanced' greenhouse effect is the added effect caused by the greenhouse gases present in the atmosphere due to human activities, such as the burning of *fossil fuels* and *deforestation*

Greenhouse gas emissions   The release of *greenhouse gases* into the atmosphere, causing *global warming*

Greenhouse gases   *Molecules* in the Earth's *atmosphere* such as *carbon dioxide* (CO_2), methane (CH_4) and CFCs which warm the atmosphere because they absorb some of the *thermal radiation* emitted from the Earth's surface (see the *greenhouse effect*)

GtC   Gigatonnes of carbon (C) (1 gigatonne $= 10^9$ tonnes). 1 GtC $= 3.7$ Gt carbon dioxide

GWP   Global warming potential: the ratio of the enhanced *greenhouse effect* of any gas compared with that of *carbon dioxide*

Heat capacity   The amount of heat input required to change the temperature of a substance by $1\,°C$. Water has a high heat capacity so it takes a large amount of heat input to give it a small rise in temperature

Hectopascal (hPa)   Unit of atmospheric pressure equal to *millibar*. Typical pressure at the surface is 1000 hPa

Hydrological (water) cycle   The exchange of water between the atmosphere, the land and the oceans

Hydro-power   The use of water-power to generate electricity

IPCC   Intergovernmental Panel on Climate Change – the world scientific body assessing *global warming*

Isotopes   Different forms of an *element* with different atomic masses; an element is defined by the number of *protons* its nucleus contains, but the number of *neutrons* may vary, giving different isotopes. For example, the nucleus of a carbon atom contains six protons. The most common isotope of carbon is ^{12}C, with six neutrons making up an atomic mass of 12. One of the other isotopes is ^{14}C, with eight neutrons, giving an atomic mass of 14. Carbon-containing compounds such as *carbon dioxide* will contain a mixture of ^{12}C and ^{14}C isotopes. See also *deuterium, tritium*

Latent heat   The heat absorbed when a substance changes from liquid to gas (*evaporation*), for example when water evaporates from the sea surface using the Sun's energy. It is given out when a substance changes from gas to liquid (*condensation*), for example when clouds are formed in the *atmosphere*

Milankovitch forcing   The imposition of regularity on climate change triggered by regular changes in distribution of *solar radiation* (see *Milankovitch theory*)

Milankovitch theory   The idea that major ice ages of the past may be linked with regular variations in the Earth's orbit around the sun, leading to varying distribution of incoming *solar radiation*

Millibar (mb)   Unit of atmospheric pressure equal to *hectopascal*. Typical pressure at the surface is 1000 mb

MINK   Region of the United States comprising the states of Missouri, Iowa, Nebraska and Kansas, used for a detailed climate study by the US Department of Energy

Mole Fraction   (or mixing ratio) The ratio of the number of moles of a constituent in a given volume to the total number of moles of all constituents in that volume. It differs from volume mixing ratio (expressed for instance in ppmv etc.), by the corrections for non ideality of gases – that is significant relative to measurement precision for many greenhouse gases

Molecule   Two or more *atoms* of one or more *elements* chemically combined in fixed proportions. For example, atoms of the elements carbon (C) and oxygen (O) are chemically bonded in the proportion one to two to make *molecules* of the compound *carbon dioxide* (CO_2). Molecules can also be formed of a single element, for example ozone (O_3)

Monsoon   Particular seasonal weather patterns in sub-tropical regions which are connected with particular periods of heavy rainfall

Neutron   A component of most atomic nuclei without electric charge, of approximately the same mass as the *proton*

OECD   Organization for Economic Cooperation and Development; a consortium of thirty countries (including the members of the European Union, Australia, Canada, Japan and the USA) that share commitment to democratic government and market economy

Ozone hole   A region of the *atmosphere* over Antarctica where, during spring in the southern hemisphere, about half the atmospheric ozone disappears

Paleoclimatology   The reconstruction of ancient *climates* by such means as ice-core measurements. These use the ratios of different *isotopes* of oxygen in different samples taken from a deep ice 'core' to determine the temperature in the *atmosphere* when the sample *condensed* as snow in the clouds. The deeper the origin of the sample, the longer ago the snow became ice (compressed under the weight of more snowfall)

Parameterisation   In climate models, this term refers to the technique of representing processes in terms of an algorithm (a process of step by step calculation) and appropriate quantities (or parameters)

Passive solar design   The design of buildings to maximize use of *solar radiation*. A wall designed as a passive solar energy collector is called a solar wall

Photosynthesis   The series of chemical reactions by which plants take in the sun's energy, *carbon dioxide* and water vapour to form materials for growth, and give out oxygen. Anaerobic photosynthesis takes place in the absence of oxygen

Phytoplankton   Minute forms of plant life in the oceans

ppb   parts per billion (thousand million) – measurement of mixing ratio (see *mole fraction*) or concentration

ppm   parts per million – measurement of mixing ratio (see *mole fraction*) or concentration

Precautionary Principle   The principle of prevention being better than cure, applied to potential environmental degradation

Primary energy   Energy sources, such as *fossil fuels*, nuclear or wind power, which are not used directly for energy but transformed into light, useful heat, motor power and so on. For example, a coal-fired power station which generates electricity uses coal as its primary energy

Proton   A positively charged component of the atomic nucleus

PV   Photovoltaic: a solar cell often made of silicon which converts *solar radiation* into electricity

Radiation budget   The breakdown of the radiation which enters and leaves the Earth's *atmosphere*. The quantity of *solar radiation* entering the atmosphere from space on average is balanced by the *thermal radiation* leaving the Earth's surface and the atmosphere

Radiative forcing   The change in average net radiation at the top of the *troposphere* (the lower *atmosphere*) which occurs because of a change in the concentration of a *greenhouse gas* or because of some other change in the overall climate system. Cloud radiative forcing is the change in the net radiation at the top of the troposphere due to the presence of the cloud

Reforestation   Planting of forests on lands that have previously contained forests but that have been converted to some other use

Renewable energy   Energy sources which are not depleted by use, for example *hydro-power*, PV solar cells, wind power and *coppicing*

Respiration   The series of chemical reactions by which plants and animals break down stored foods with the use of oxygen to give energy, *carbon dioxide* and water vapour

Sequestration   Removal and storage, for example, *carbon dioxide* taken from the *atmosphere* into plants via *photosynthesis*, or the storage of carbon dioxide in old oil or gas wells

Sink   Any process, activity or mechanism that removes a greenhouse gas, aerosol or precursor of a greenhouse gas or aerosol from the atmosphere

Solar radiation   Energy from the Sun

Sonde   A device sent into the *atmosphere* for instance by balloon to obtain information such as temperature and *atmospheric pressure*, and which sends back information by radio

Stewardship   The attitude that human beings should see the Earth as a garden to be cultivated rather than a treasury to be raided. (See also *sustainable development*)

Stratosphere   The region of the region of the atmosphere between about 10 and 50 km altitude where the temperature increases with height and where the ozone layer is situated

Sustainable development   Development which meets the needs of the present without compromising the ability of future generations to meet their own needs

Thermal radiation   Radiation emitted by all bodies, in amounts depending on their temperature. Hot bodies emit more radiation than cold ones

Thermodynamics   The First Law of thermodynamics expresses that in any physical or chemical process energy is conserved (i.e. it is neither created or destroyed). The Second Law of thermodynamics states that it is not possible to construct a device which only takes heat energy from a reservoir and turns it into other forms of energy or which only delivers the heat energy to another reservoir at a different temperature. The Law further provides a formula for the maximum efficiency of a heat engine which takes heat from a cooler body and delivers it to a hotter one

Thermohaline Circulation (THC)   Large-scale density driven circulation in the ocean caused by differences in temperature and salinity

Transpiration   The transfer of water from plants to the *atmosphere*

Tritium   Radioactive *isotope* of hydrogen, used to trace the spread of radioactivity in the ocean after atomic bomb tests, and hence to map ocean currents

Tropical cyclone   A storm or wind system rotating around a central area of low *atmospheric pressure* and occurring in tropical regions. They can be of great strength and are also called hurricanes and typhoons. Tornadoes are much smaller storms of similar violence

Troposphere   The region of the lower *atmosphere* up to a height of about 10 km where the temperature falls with height and where convection is the dominant process for transfer of heat in the vertical

UNCED   United Nations Conference on Environment and Development, held at Rio de Janeiro in June 1992, after which the United Nations Framework Convention on Climate Change was signed by 160 participating countries

UNEP   United Nations Environmental Programme – one of the bodies that set up the *IPCC*

UV   Ultra-violet radiation

Watt   Unit of power

WEC   World Energy Council – an international body with a broad membership of both energy users and the energy industry

Wind farm   Grouping of wind turbines for generating electric power

WMO   World Meteorological Organization – one of the bodies that set up the *IPCC*

Younger Dryas event   Cold climatic event that occurred for a period of about 1500 years, interrupting the warming of the Earth after the last ice age (so called because it was marked by the spread of an Arctic flower, *Dryas octopetala*). It was discovered by a study of paleoclimatic data

Zooplankton   Minute forms of animal life in the ocean

Index